Bad Buying

'Filled with examples of painful, unbelievable, funny and downright stupid buying by organizations, and the people in them. A brilliantly unique and insightful read from one of the most experienced individuals in this space'
Jonathan O'Brien, author and CEO Positive Purchasing Ltd

'Factually rich, funny and full of practical hard-earned wisdom, this book is a revelation . . . actually it's full of them! All costly catastrophes. If you don't read this book, you may find yourself in the next one!' Dr Richard Russil, author and coach

'*Bad Buying* tells story after story of bad buying for both novices and experts alike – and shows us a roadmap to doing it better'
Jason Busch, Managing Director, Azul Partners
and founder, co-author of *Spend Matters*

'A great opportunity to learn from other people's mistakes, not your own. Through exceptionally well researched examples, Peter teaches both new and established professionals how to avoid the same mistakes. I thoroughly recommend this book to anybody involved in, or responsible for spending money for their organization'
Garry Mansell, former GM, Source-to-Contract, Coupa Software

'An unusual business book in that it is both useful to readers and also genuinely entertaining with fascinating stories of failure and fraud from around the world and every sector' Shirley Cooper, Commercial Director, Tapestry Compliance and NED, Ministry of Justice

'A hilarious, enlightening and brilliant look at how organizations – public and private – have been guilty of horrendous buying failures. Not content with just lifting the lid on some of the most egregious excesses in history, Peter Smith provides insightful and practical advice to avoid repeating such disasters. This book will make you think twice about buying anything – but do buy this'
Antonio Weiss, bestselling author of *101 Business Ideas that Will Change the Way You Work*, Director, The PSC

'A fascinating account of the biggest buying blunders by private and public sectors alike. Written in his trademark dashing and fluent style, it doesn't just expose hilarious dodgy stories with Schadenfreude but offers insightful and practical advice on how to avoid career-limiting mistakes' Bernhard Raschke, Partner and Head of EMEA Supply Chain Centre of Excellence, Korn Ferry

'A great effort to dig the challenges out of the back office of procurement and bring to the fore the opportunities for improvement . . . If you want to improve your chances of not being done by bad buying, there's no better place than this book to see what the risks are and begin to understand how to avoid them' Charles Findlay, Director, State of Flux

'Covid-19 restrictions have put the spotlight on global supply chains and the difficulties caused when everybody wants to buy the same thing at the same time. Peter's book highlights where buying can go catastrophically wrong and how to avoid these pitfalls. Had this been published pre-Covid, some of the recent c*ck-ups and waste might have been avoided. It's a must-read for the public and private sector alike' Lt-Gen. Sir Andrew Gregory, Controller, SSAFA

'In turns informative, shocking and amusing, *Bad Buying* explores the career-limiting catastrophes to avoid and sets out a vision for better buying, not only stewarding finances responsibly but also supporting the firm's purpose and place in the community' Stuart Brocklehurst, CEO, Applegate Marketplace Ltd

'*Bad Buying* has a rich set of examples of both corrupt practices and unintentional but costly and wasteful mistakes made by business professionals. With trillions spent by organizations buying goods and services, every executive who is involved in or oversees those processes needs to make this a must-read' Raj Sharma, founder and CEO, Public Spend Forum

'A must-read for boards, CEOs and governments. The case studies remind us all that there is no such thing as a free lunch and "if it seems too good to be true it usually is". Buyer beware, and never underestimate your stakeholder or your suppliers. Instead, build relationships and trust' Lucy Harding, Partner and Global Head of Practice, Odgers Berndtson

'Purchasing plays such an important role in business success, but is also one of the least understood activities. Using case studies from around the globe, *Bad Buying* has illuminated how and why organizations can get it wrong when it comes to spending money with suppliers. This is a timely, informative and highly entertaining read!' Nandini Basuthakur, CEO, Procurement Leaders

ABOUT THE AUTHOR

Peter Smith has over 35 years' experience in procurement and supply chain as a manager, procurement director, consultant, analyst and writer. Peter is the Managing Director of Procurement Excellence Ltd, a leading specialist consulting firm. He is recognized as one of the UK's leading experts in public and private sector procurement performance improvement. Peter has an MA in Mathematics from Cambridge University where he is now a Fellow. His first book, *Buying Professional Services*, was published by Profile in 2010.

Bad Buying

How organizations waste billions through failures, frauds and f*ck-ups

PETER SMITH

BUSINESS

PENGUIN BUSINESS

UK | USA | Canada | Ireland | Australia
India | New Zealand | South Africa

Penguin Business is part of the Penguin Random House group of companies whose addresses can be found at global.penguinrandomhouse.com.

First published 2020

001

Copyright © Peter Smith, 2020

The moral right of the author has been asserted

Set in 12/14.75pt Dante MT Std
Typeset by Jouve (UK), Milton Keynes
Printed and bound in Great Britain by Clays Ltd, Elcograf S.p.A.

A CIP catalogue record for this book is available from the British Library

ISBN: 978–0–241–43459–8

Follow us on LinkedIn: https://www.linkedin.com/company/penguin-connect

www.greenpenguin.co.uk

 Penguin Random House is committed to a sustainable future for our business, our readers and our planet. This book is made from Forest Stewardship Council® certified paper.

To Jane, Ginny and everyone involved in buying
(procurement, supply chain, logistics) in the healthcare sector,
and all those they support

Contents

Acknowledgements   xi
Introduction   xiii

PART I: FAILURE

1. **Getting the Specification Right**: Irish printers, Easter-egg rockets and Aussie train toilets   3
2. **Understanding the Market**: Cocoa beans, fake followers and rehabilitating offenders   16
3. **Choosing Suppliers**: Dodgy T-shirts, working with mates and banking fiascos   29
4. **Don't Get Too Dependent**: Seat covers, hospital rip-offs and bigger isn't always better   42

    Bad Buying Award – Schlitz Beer   56

5. **How to Negotiate**: Charlie Hurley, missile interceptors and consultants' lunches   59
6. **Understanding Incentives**: Cultivating coca, Dutch traffic jams and Birmingham call centres   73
7. **How *Not* to Be Stupid (particularly if you're a politician . . .)**: Misplaced airports, imaginary ferries and Indian offsets   86

    Bad Buying Award – NHS National IT Programme (NPfIT)   100

8. **Trust No One (at least not suppliers)**: Lulu the dog, French concert halls and US Navy ships   104
9. **Coping with Change**: Technology disasters, fried-chicken shortages and Crossrail delays   118

Contents

10. **What's the Risk?**: The wrong fish, Japanese earthquakes and running bears — 130

    Bad Buying Award – Berlin Brandenburg (not yet an) Airport — 144

11. **The Joys of Contract Management**: Bollards, Xmas parties and big IT overspends — 148

PART 2: FRAUD & CORRUPTION

12. **The Fundamentals of Fraud**: Power stations, hotel bills and the greyhound-racing mafia — 163

13. **Who Am I Really Buying From?**: Marine hoses, Indian brewers and the 'Ndrangheta — 171

14. **Fixing the Supplier Selection**: Canadian politics, working with Mum and painting the NHS — 183

    Bad Buying Award – Fat Leonard and the US Navy — 196

15. **What Am I Really Buying?**: Bomb detection, Frenchified kiwi fruit and spaceship resilience — 203

16. **Spending Someone Else's Money**: Florida dogs, owl jars and sex lairs — 215

17. **What Am I Paying For?**: Pricey potatoes, horse semen and shops in Wolverhampton — 224

18. **Politics and Fraud**: Ski-jumping, bribing dictators and Austrian promises — 236

    Bad Buying Award – Petrobas and Odebrecht — 246

19. **Preventing Fraud**: Collusion, checking and commitments — 250

PART 3: HOW TO AVOID THE F*CK-UPS

20. **Ten Principles for Good Buying** — 263

    Bibliography — 281
    Notes — 283
    Index — 301

Acknowledgements

After many years in which buying – bad and good – has been at the heart of my career, I'm grateful to everyone I've met who has helped me learn and understand more about the subject, including colleagues, academics and other writers. It would take far too long to name them all but two bosses, Mike Allaway and Doug Greene, taught me great lessons during my relatively early days in business; and Dr Dick Russill was an inspiration as a writer and educator, helping me realize that communication skills were just as important as the technical side of management!

In terms of *Bad Buying* inspiration, it would seem perverse to thank everyone who has contributed to this book through their incompetence, ignorance or criminality. But genuine appreciation goes to the UK's National Audit Office, which provided excellent analysis of some very 'interesting' case studies that have found their way into the book, and to Transparency International, whose focus on fighting fraud and corruption globally is valuable to every citizen around the world.

More thanks to the team at Penguin Books, including Martina, Celia, Natalie and Mandy, who have been a pleasure to work with at all stages. And, finally, thanks to my daughter and son-in-law, and particularly to my wife Jane, who all put up with me seeing the chance for a story about *Bad Buying* in pretty much everything we saw, did or talked about!

Introduction

An American space scientist was boasting to his Russian counterpart at a conference that the US had spent years, and many millions of dollars, inventing a pen that would enable astronauts to write notes in the zero-gravity environment of a spacecraft, with a clever mechanism for pumping ink to the nib when gravity couldn't help, as it does on Earth.

'Ah, yes,' said the Russian thoughtfully. 'That is certainly a problem, and your solution is very clever. But we use a pencil.'

This is a simple but powerful example of how an organization can waste time, money and resources buying something it doesn't really need, or that costs far more than it should. In organizations, government departments and businesses of all shapes and sizes, all round the world, money is being lost, wasted, spent inappropriately, defrauded or stolen. What is the cause of this epidemic? Let's just call it bad buying, because, at its simplest, that's what it is.

These issues are truly global, and the case studies in this book reflect that. No industry or country is immune from bad buying; it exists in every nation in the world, and in almost every organization. When Kentucky Fried Chicken runs out of chicken, to the horror of its customers, or a firm such as Skanska pays large sums of money to fraudsters through invoice misdirection, we can see the result of bad-buying practices or processes. Even in Singapore,[1] widely considered to be one of the least corrupt and most efficient governments in the world, you can find problems in this field. Singapore's Auditor General's report for 2018/19 identified government money wasted through multiple issues, including serious tendering process irregularities, poor bid-evaluation techniques, a lack of control around contract changes, and so on.[2] Almost every government around the world is guilty of similar failures.

Introduction

How much is spent on buying by organizations worldwide? That's not a number that is easy to find or calculate, but there was an estimate in 2012 that all the businesses in the world had a combined revenue of $64 trillion.[3] Increase that by 20 per cent to allow, conservatively, for growth and inflation since then, which gets us to $77 trillion. Let's say, conservatively, that 50 per cent of that is used to buy from other organizations. That gives some $38 trillion of 'buying spend'. It would take an economist to determine exactly what the effect would be if that expenditure could be executed more effectively and efficiently. But clearly even a small 'saving' of a couple of per cent on that huge number would bring major benefits to organizations and would improve the overall efficiency of the global economy.

A recent estimate puts the total global public-sector procurement spend at $13 trillion.[4] A 5 per cent improvement in the value obtained from this money would release another $650 billion every year. That is money that could be used by governments to alleviate hunger, cure diseases or improve education in the developing world. And 5 per cent is not unrealistic, given the scale of fraud and corruption in many countries, as well as the opportunities from improving conventional buying performance. In the UK that percentage saving would mean the potential to double the defence budget, or increase total spending on education by 25 per cent.

In the private sector, no less than 69 per cent of an average commercial business's revenue is spent on external suppliers.[5] How this money is used has fascinating implications for the profitability of the company. Assume that the company has a revenue of $1 billion and makes a profit of $100 million – a 10 per cent margin – and that 60 per cent (below the average) of its revenue goes to suppliers. If the firm can improve the efficiency of that spend by just 5 per cent, the $600 million external spend could drop to $570 million, and profit would grow by $30 million. That immediate increase in profit of 30 per cent, from $100 to $130 million, is very significant indeed.

Given how much money is being spent with suppliers, it is perhaps not surprising that it goes wrong occasionally. But sometimes

Introduction

it goes *very* wrong. In fact at times it can bankrupt the company or, in the case of government, lead to political turmoil, front-page news or even revolutions and resignations.

In this book we will discover why organizations, both public and private, waste billions through bad buying. In Part 1 we will look at the many cases where it is down to simple incompetence – laziness, a lack of knowledge, understanding or information – but the end result will have a cost to the organization, sometimes a significant one.

Occasionally, as we will discover in Part 2, bad buying has darker, criminal motivations. Ranging from the clerk who creates a fictitious company as a 'supplier' and channels a few thousand into his own bank account, to huge multinational scandals that have led to prime ministers, admirals or CEOs languishing in prison, we will uncover the fraud and corruption at the heart of many buying scandals.

In Part 3 I propose ten principles that will guide you towards reducing your own risk of buying failure. All organizations are vulnerable to errors, but these steps will ensure that your probability of bad buying is as low as possible.

In virtually all cases of failure or fraud there is a f*ck-up of some sort sitting behind the bad buying. Failure always has its causes; something went wrong in terms of the process, the skills (or lack of skills) of the people involved, poor planning, systems or management . . . The reasons for failure are many and varied, but there is always a root cause. And while fraud will always be with us, in virtually every case featured here, it could have been avoided or stopped much faster if basic preventative measures had been in place. But they weren't, because somebody f*cked up.

PART I
Failure

I

Getting the Specification Right

There are many ways in which organizations can waste money when it comes to buying goods and services from their suppliers. In these first chapters I'll start with those issues linked to poor understanding and knowledge, in terms of how money should be spent effectively.

Let's kick off by looking at one of the most common issues that cause organizations to waste money. To put it simply, that's by buying or spending money on the wrong things. Whether it is a product or service, it may be bought brilliantly, with superb negotiation and a favourable, watertight contract. But if it is not what the organization *really wants or needs*, then money is being wasted. This comes down to the *specification* for what is being bought: the description by the buyer of what is wanted, used to obtain proposals from suppliers to meet their needs.

Any substantial organization will have examples of wasted money littered throughout its history, because of specification failure. That will not always be revealed publicly, of course. Private-sector organizations are under no obligation to tell shareholders or the outside world about every mistake they make, even if the cost is substantial. But the waste is there, nonetheless. Sometimes the loss is 100 per cent of the money spent – so what is bought is really not needed at all. In most cases, though, the expenditure may not be totally wasted. More usually, something sub-optimal is bought, when a better alternative is available. Or too much money is paid for something that is *over-specified* – an expensive piece of equipment, for instance, is purchased, when a cheaper alternative would have been fine. Not all the spend is wasted, but an element of it certainly is.

Our space-pen story in the Introduction illustrates the importance of getting the brief for the specification right. The American space scientist should not have sought to design 'a pen that will work in zero gravity'. Instead, 'something to write with in the spaceship' would have been better and would have led to a successful, but much cheaper outcome – the pencil.

Governments are famed for getting the specifications wrong. In 2018 *The Washington Times* reported that the US Air Force spent $330,000 on coffee mugs designed to reheat beverages in-flight.[1] We might ask whether it was quite so important to have hot beverages in-flight, but the mugs, which cost $1,280 each, were even then not fit for purpose. Apparently the handles were very fragile and broke frequently in the air. Rather than getting the specification right, which should have defined the required robustness of the final product very clearly, the Air Force kept on buying the faulty mugs. They did look at how they might use 3D printing to produce replacement handles more cheaply, although goodness knows how much the 3D printers and that additional work might have cost. Getting it right first time around is always the best approach.

Just What I Needed

It is vital to understand thoroughly what it is that you want to buy. Then that has to be communicated to potential suppliers clearly, through a brief or specification that will give you the best possible chance of making the right buying decision. All the key aspects need to be covered; a firm that needs laptop computers for its workers on oil rigs or construction sites can't simply define the graphics capability or memory for the machines. They need to specify physical robustness, too – machines that can stand being dropped and thrown around from time to time. Miss that element from the

Getting the Specification Right

specifications and you will buy expensive equipment that does not meet the real needs.

Sometimes the most basic details don't get considered properly when organizations decide what they want to buy. In 2018 the Irish government bought an €808,000 state-of-the-art Komori printer.[2] You might ask why they needed something quite so top-of-the-range, but that wasn't the biggest specification issue here. The real problem came when the machine arrived in December and officials found that the 2.1 x 1.9-metre (6¾ x 6¼-foot) monster would not physically fit into the building where it was to be used. Contractually, they could not just send it back to the supplier, and it was moved into external storage for ten months while builders pulled down walls and added structural strengthening (costing another €236,000) so that the printer would fit into its new home.

Even then the printer remained idle, as staff demanded training and a pay rise before they would use it, while the IT department was reluctant to grant it the right server permissions. David Culliname, an Irish parliamentarian, summed up the fiasco, saying, 'It is all too common for a lack of planning and foresight on behalf of government departments to result in additional costs for the taxpayer.'[3] The fact that planning must include more serious thought about what is really needed was evident in this case.

A good specification covers all the buyer's key requirements with the right level of detail, whether it is goods or services that are being bought. But there are different ways of defining those needs, and some options are better than others, particularly for more complex purchases. The key is to understand these different types of specification and then to express that to the market, as clearly as possible. Let's look at some of the specifications you could consider.

A Matter of Trust

There is no doubt that when looking for assured quality or a company with a lengthy track record, people tend to turn to a well-known brand, whether that's Apple, BMW or McKinsey. However, it's worth considering that trusted brands come with a premium price tag, and sometimes that isn't necessarily justified within your overall specifications. Wherever possible, you should beware of limiting the brief to specific brands and manufacturers, unless absolutely essential – for example, the software that simply must be compatible with what is already running on the system.

For many years a much-repeated slogan in the technology industry was 'nobody gets fired for buying IBM'.[4] Buyers' trust in the goods and services of IBM over the years meant that the company was able to charge premium prices. I had a difficult job as a buying director back in the 1990s, when Dell started promoting their personal computers to European businesses. Persuading my colleagues in the technology department that this new firm was producing products that were around half the price, yet technically equivalent to those from IBM and Compaq took some time. Finally, by running objective tests, and not simply relying on that brand loyalty, my firm became an early adopter and saved around half a million pounds in the first year of using Dell.

For many years, British Airways achieved premium pricing from its corporate frequent flyers through brand loyalty and an attractive frequent-flyer scheme. It is questionable whether or not that premium, paid for by the travellers' firms in most cases, was truly deserved. Interestingly, those days have gone for BA, and it is other airlines such as Emirates and Singapore Air that are now seen as premium brands, probably with good cause, and can charge accordingly.

Rather than be restricted by the brand specification approach, buyers should consider the technical specification of what they need. For example, in the earlier example, defining those characteristics for the laptop (display, memory, robustness) leads to a technical specification. This opens up the market more effectively than specifying the brand, but you are still tightly defining what you want, and that may leave limited scope for potential suppliers to contribute their own options, alternatives and ideas – generally a good objective in any buying situation.

In many cases it's better to consider the *outputs, outcomes or performance* you seek and to build a specification around that. Taking that approach opens up the opportunity for different potential suppliers to have a chance of winning your business, and encourages them to propose alternative ways to meet your needs.

I might decide I need a lawnmower. I could specify that it must be the Qualcast 41-cm Self Propelled Petrol Rotary Lawn Mower. A few different retailers may stock that, but my choice is now limited because I've restricted the competition through my 'brand specification'. Or I could look at the technical requirements: it must be petrol-driven and have at least a 35-cm (15-inch) blade, a cuttings box and a twelve-month warranty . . . That's better; my technical specification means that different manufacturers *and* retailers can now suggest options to me.

But I could be really radical and think about it in terms of *outcomes*. What do I really need? Actually I need my 50-square-metre (538-square-foot) lawn cut so that the grass never gets more than 7.5 cm (3 inches) long. Now buying a mower may well make sense, but perhaps a gardener visiting occasionally would save me time and avoid the capital cost of the mower? Or perhaps even a goat?! Maybe not, but the principle is valid; remember the space-pencil story, where the Russians looked at the outcome they wanted, rather than jumping to a specific technical solution.

You Can't Always Get What You Want

However, don't assume that defining the desired outcomes or outputs is always easy. Looking at the future of artificial intelligence (AI) provides an illuminating example of this. Experts have pointed out the danger of giving an all-powerful AI system an outcome-based specification such as this: 'eliminate all cancer in humankind'. That sounds fine, until the AI entity decides the best way of achieving this is to kill every human on Earth! In this case, the desired outcome needs to be carefully defined, with constraints in terms of both other outcomes and some technical parameters.

While not quite as dramatic as that example, technology-buying often sees organizations failing to define exactly what they want and need, both from a technical specification *and* an outcomes perspective. Whether buyers are interested in the latest computer systems or robotic manufacturing kit, these markets are complex and in a constant state of innovation. If buyers fail to understand the market and tightly identify their specifications, they make themselves vulnerable to salespeople, who may be driven by sales targets rather than providing what the buyers' organization actually needs. Or some desired outcomes may be defined, but other key requirements missed.

Buying the wrong technology can be an expensive mistake and, at the extreme, has caused firms to lose many millions in shareholder value, and even go out of business altogether. In 2014 London-based wealth-management firm Brewin Dolphin dropped plans to implement the Figaro software package across its wealth-management business and owned up to a cost of £32 million.[5] That was substantial for a firm that made £52 million profit the previous year, which was why the episode had to be made public, although fortunately it was not a fatal loss in this case.

In the public sector in 2014 the Radio Sweden website reported on the new national police IT system, which was supposed to be used for recording crime details.[6] The technology, Pust Siebel, reportedly caused enormous problems, making simple tasks such as logging a case drag on for hours. That sounded very much like a failure of specification. Did the supplier of the new system understand how the police actually used it? What the processes were that would be managed through the system? What input the user would provide, and how the system needed to respond (the *outcomes* from a user perspective)? It certainly did not seem so.

A Change Would Do You Good (or maybe not . . .)

In the technology market, specifications can often change mid-process: after a contract is in place, but before the end product or service is delivered, which adds further risk. A major IT implementation can take two years or more. During that time the whole world of technology possibilities may change. Think how quickly personal computers started taking over tasks that only the biggest mainframes previously carried out; or of the death of the analogue photocopier, or the growth of cloud computing more recently. During a two- or three-year software development project, the whole technology picture can change.

While it's sometimes necessary to alter specifications during a buying process, particularly in these fast-adapting markets and environments, it is important to be very aware of the risks and how your change will impact on the supplier, and on aspects such as timescales. In May 2019 the UK's National Audit Office issued a damning report on the programme to provide a new mobile communications service for the police and other emergency services.[7] The Emergency Services Network (ESN) programme by then was some £3.1 billion over the initial budget and was forecast to deliver benefits seven years later than originally planned. Not only did some of the specification require technology that is still not available, but there have also been 'challenges in locking down the specification for software and user services' over the years. For a programme like this, which has already been running for some five years, the user needs will evolve over time, which inevitably challenges the original specification.

But sometimes in technology projects it is hard to separate specification issues from other failings. The Expeditionary Combat Support System (ECSS) was a failed software project undertaken by the US Air Force between 2005 and 2012. The goal of the project was to automate and streamline logistics operations, bringing together administrative records and reducing the number of different systems that were used to manage the movement and storage of goods used by the Air Force, from planes to spare parts and fuel.[8]

But after spending about £1.1 billion on the development, the Air Force concluded that the system 'had not yielded any significant military capability' and that it would take another billion or so just to get to a point where a quarter of the initial scope might be met. The programme was then cancelled, not surprisingly, in 2012.

Reports on the project identified that the Air Force did not even appear to know how many existing 'legacy systems' were being targeted for replacement. Estimates ranged from 175 to 'over 900',[9] and it's hard to see that a supplier could be properly briefed if such basic information was not available. Perhaps both buyer and supplier were also over-optimistic – a failing that I'll cover later. Often, when tech-related disasters occur, one problem builds on another, until finally the whole edifice cracks. But it certainly seems that the specification was an important element in both of these – and many other – government technology failures.

Price Tag

Not spending enough doesn't sound like a failing. Yet a specification that drives that behaviour can lead to costs and non-financial repercussions, which can be just as serious as any losses arising from overspending. Not spending enough means buying goods or services that don't really meet your needs, driven by their apparent cheapness at the time of purchase. It's buying the own-label confectionery that simply doesn't taste like real chocolate (probably because it isn't); or the cheap suitcase in the sale that falls apart the first time it experiences baggage handlers; or the cheaper insurance policy that turns out not to cover something really important. Across the corporate and government buying worlds, the same principle applies.

In some areas of expenditure this raises deep issues around true 'value'. Think about marketing spend, and the ability today to look online and find graphic designers, copywriters, even voiceover artists, who will work remotely for a fraction of the price that a London, Tokyo or New York agency might charge. But consider how much a great design, campaign or ad might contribute to your business, and it's easy to see that saving a few dollars or pounds – or even a few hundred thousand – could easily be outweighed by the benefits that could be gained from spending the money wisely.

Early in my career I bought packaging for Mars Confectionery, including that needed for chocolate Easter eggs (it was a tough job, but someone had to do it). What became clear was that the *cost* of the carton containing the egg was only part of the equation. The key was selecting packaging that helped sell more products, or products at a higher price, to the retailers and, ultimately, to the individual customer. So if a pack contains two Snickers bars plus a 100-gram (4-oz) chocolate egg, will the supermarket sell that at £1.50? Or £2? Or £2.50? And how many will it sell? Fifty thousand? A hundred thousand? Half a million? Final profit for Mars from this product could vary hugely, depending on those three factors – the manufacturing cost, the unit price charged and the volume sold.

It was obvious in this market that the attractiveness and novelty of the packaging were key in determining the sales. Our Milky Way space rocket was a huge hit with retailers and kids – because of the novel rocket-shaped box, not because of the egg and the bars. Paying another few pence or cents for the carton was well worth it, if it resulted in doubled sales or the ability to increase the sales price by 20p or 20 cents.

The same argument applies to products such as upscale whiskies or champagne, or in perfumes and cosmetics: packaging is a key part of the overall proposition. Failure in those markets would mean looking for the cheapest possible packaging product, instead of using the specification to encourage suppliers to come up with innovative, clever and great-looking packaging to help sell the products.

In some industries the outcomes from specification failures around 'not spending enough' can be even more serious. While buying the services of dodgy graphic designers might lose some potential sales and profit, in other cases we are talking about business survival or even human lives. Indeed, one theory about the sinking of the R.M.S. *Titanic* suggests that the shipbuilders chose a poor-quality but cheap iron rivet, used to hold the hull together, in order to save money during the construction process.¹⁰ That specification decision weakened the ship, the theory suggests, and might have contributed to its vulnerability and its speed of sinking, once it hit the iceberg.

In a more recent episode, in 2017 a truck careered out of control down a very steep road leading towards the beautiful English city of Bath. Tragically, five people, including a four-year-old girl and her grandmother, got in its way and were killed by the vehicle on its terrible journey. After investigation, it became clear that the cause came down to lack of basic maintenance on the truck, presumably because the owners of the trucking firm tried to save money. In this case their lack of care in buying maintenance services that did not meet basic needs tipped over into criminal behaviour. The consequences were grave, and the owner of the firm and a mechanic were jailed in this case.¹¹

Tell Me What You Want

A common cause of bad buying is miscommunication or misunderstanding between the internal user of what is being bought and those who will actually do the buying. This is most often a problem for large organizations. In small firms, the buyer is usually close to the need and is less likely to misunderstand that requirement. But in a huge financial-services firm, for example, things can easily go wrong when a purchasing professional or buying department puts contracts in place with suppliers on behalf of those who actually use what is being bought.

When I led buying at the NatWest financial-services group, my team was tasked with putting in place a framework (a list of selected preferred suppliers) for training services. With pre-agreed standard contracts in place, any manager could then use those providers, without having to go and find them from scratch and negotiate difficult contractual details. My team worked hard on this, and ended up with a list of some fifty firms covering different types of training needs. But six months later, when we reviewed how much business was going to these firms, less than 10 per cent of our total training spend was being directed to our preferred suppliers. Clearly we hadn't truly understood the requirement from the vast numbers of managers across the firm.

Training needs differ by business area, and change quickly. One year everyone might want training on equalities legalization. The next year the requirement may be advanced data analytics or the latest people-management fad. By the time we had made our selection and put contracts in place, we weren't offering the right suppliers to meet those latest needs. In some cases, rather than misjudgement, there can be arrogance, where buyers believe they know what is wanted better than those who will actually use the product or services that are being purchased.

To mitigate these risks, it is vital to make sure that internal communication is good, and that everyone involved in the process internally is aligned and understands exactly what is going to be bought. That needs to be done *before* the actual buying process starts, in terms of engaging with potential suppliers. However, there is a balance to be struck between consulting the users of the goods or services being bought and making sure that you do not end up buying the proverbial camel (a horse designed by a committee).[12]

I was called in to troubleshoot a major IT system purchase, where the project was sinking rapidly into failure. The organization was buying a moderately complex system – not exactly space-shuttle level, but very important to their business, because among other purposes, it enabled customers to see the products that the business was offering. An off-the-shelf system could probably satisfy the requirement, maybe with a few tweaks. But when I got involved, the buyer was struggling to evaluate the shortlisted suppliers' bids to arrive at a decision on the best choice.

The main reason for that was the way proposals from potential suppliers were being evaluated. As well as the cost, there were six evaluation criteria, which didn't sound too bad. But there were multiple sub-criteria, which meant trying to evaluate the proposals against no fewer than forty-two different factors (as well as costs of various types). That inevitably meant there was considerable overlap, confusion in marking the bids, and some factors hardly mattered as they had so little weighting in the scoring system.

But the evaluation was also a symptom of the way the *specification* had been developed. It felt as if everybody in the organization had provided input into the requirement. 'I want this, so we need to ask for it and score the suppliers' responses' seemed to be the approach. There was nobody to say ,'Okay, that would be nice, but it is not essential.' Or even, 'Okay, Laura, you want that, but Jim in the US office wants something totally different!'

The end result was that *everything* – including the kitchen sink – had been included in the specification; suppliers were confused; and choosing the best product became very challenging – although we got there in the end. There were other issues, too; the buyer had forgotten to include an actual product demonstration as part of the process for choosing the winner! But specifications can be too detailed, just as they can be not detailed enough.

Train in Vain

In 2018 Australian ABC News reported that the State of Queensland had made an enormous buying error when they ordered seventy-five new trains that failed to comply with disability access and toilet regulations. An inquiry found that the initial train design was approved in 2013 regardless, and the $4.4 billion contract with Bombardier, manufactured in India, went ahead. The inquiry report describes how this impressive example of buying failure was down to a fundamental lack of understanding of what was needed by the government staff and advisers, which led them to deliver confused specifications that underwent many changes throughout the buying project.

There was no consultation with disability experts during the project's early stages, even though disabled access and facilities should have been a critical part of the design. Indeed, different elements of the specification contradicted each other, as it required compliance with disability regulations in one breath, and then defined a design that was *not* compliant in the next. Furthermore, once the problem became apparent in 2012, staff and those overseeing the programme pressed on, rather than resolving the issues.

The specification failure wasn't the only problem with this contract. The core procurement process took five years, starting in 2008, and the first train wasn't delivered until 2017. This was largely down to changes in the way the purchase was to be funded, and changes to the governance structures for the project, on top of the issues around specifications. The end result is that Queensland will have to spend more than A$300 million to install a second toilet on all trains and increase the size of existing toilets by 10 per cent – work that will take until 2024.

It Ain't Easy . . .

It is often difficult to foresee *every* aspect of a requirement when you're buying, particularly for a long-term contract. An apparently reasonable specification at the time of purchase can prove to be less appropriate in certain situations. Getting the brief, specification or requirement right, and defined clearly so that you capture what you really need, without over-complicating it, is the balance. Be careful when needs are likely to change over time, and remember to combine clarity with giving potential suppliers scope to provide a range of competitive proposals, reflecting different ideas. Get all that right, and you should avoid this type of failure.

Once you have sorted out the specification, then the time has come to start looking at which organizations might be able to meet your needs in the best possible manner as suppliers. And to do that well, it is vital to *understand the market* in which those potential suppliers operate, because failing to do that can result in further expensive problems.

2

Understanding the Market

Your specification should now be in good shape, so you can start thinking about finding the right supplier, or suppliers, to best meet your needs. Who is going to offer you a great product or service, at a competitive price, and meet your deepest desires with additional service, innovation or support? But suppliers don't work in isolation. They're all part of wider markets,[1] and before you can buy well, it is vital to understand how the particular markets relating to your purchase and your potential suppliers work.

Failure to consider this is common in the government sector, where major contract opportunities generally need to be advertised openly. Too often, officials seem to think that means research isn't necessary – you can just advertise, then wait and see who responds and what their proposals look like. But if you don't understand the market, problems can ensue.

In August 2018 the health ministry in Ghana started a procurement process to buy anti-snake-venom serum to treat the growing number of fatal snake-bites in the country. However, all did not go well, and investigations revealed a succession of buying failures that resulted in the project's failure.[2] First, there were accusations that the initial procurement exercise was not run properly, and the country's Public Procurement Authority (PPA) started investigating issues, including firms changing their prices after bids had closed and a lack of proper confidentiality through the process.

There was also confusion about the financial elements of the bidding process, and a committee reviewing the bids found that prices bid were way above the budget set in the initial specification. ABC News highlighted potential corruption at play when it reported that the health ministry was favouring one particular firm. In March 2019 the PPA attempted to resolve these claims by disqualifying the favoured bidder, which meant starting the whole process over again. Meanwhile, and perhaps more importantly, questions were being raised about the *effectiveness* of some of the products and their ability to treat snake-bites. One product was unable to treat a bite from the carpet viper, which is responsible for more fatalities in Africa than all other snakes combined.

Along with other errors and the corruption accusations, this case appears to demonstrate a lack of market understanding by the buyer. Misaligned bids in terms of price compared to expectations, and uncertainty around which products are really effective, are signs of a buying process where market and supplier research has not been undertaken properly before the tendering process starts. And in the meantime Ghana's citizens may be dying from snake-bites, waiting for the vital serum.

(Not so) Sweet Like Chocolate . . .

It's not just the government sector that gets markets wrong, however. In the early 1980s, a cocoa-buying error befell the Rowntree Mackintosh company. As *The Spectator* magazine reported at the time:

> Sir Donald Barron, chairman of Rowntree Mackintosh, the sweeties firm, explained in outline how people in the cocoa buying department of his firm managed to lose £32.5 million in a series of transactions in the cocoa market. As a result of this loss the supply director has resigned, and the purchasing manager and the cocoa buyer are under suspension.[3]

The firm had to obtain an emergency loan of more than £30 million from its bankers just to survive, and even after tax relief, the firm's losses were estimated at more than £20 million – the equivalent of some £240 million today. *The Spectator* also suggested, quite rightly, that maybe the chairman and board should have taken some of the blame here. There was no suggestion that the managers who left were speculating in a manner that was outside the firm's stated policies, and as the magazine pointed out, if the market had moved the other way, Rowntree Mackintosh would have made enormous profits from trading in cocoa and the men concerned would have been congratulated 'on their brilliant and audacious dealings'. They would have been feted by investors and would probably have obtained some additional financial rewards, too. But because the market moved the wrong way, and Rowntree was caught short of cocoa, the firm was forced to buy at higher prices to keep the factories rolling. And further down the chain, someone had to take the blame.

Everything that organizations buy, and every supplier, links to one or more supply markets. And if you don't understand how they work, things can go badly wrong.

Luck Be a Lady

Traded commodity markets, like cocoa, by their very nature generate big winners and losers among those who deal in them. The buyers in these markets range from firms that physically need cocoa, coffee, wheat or oil, to those who simply look to profit through trading, by understanding the market and assessing the risk behind the positions they take in the markets. Some treat it more like a casino, looking to make 'bets' based on hunches or an arcane analysis of charts and trends.

But, as in the Rowntree case, the line between disaster and success is very fine. Bad weather affecting a crop; an army coup in an African state; a sudden reduction in demand from China – many different factors can cause unexpected movements in the current or future commodity prices that are quoted.

I worked in a smaller food company after I moved on from the Mars Group, and the CEO of one of our subsidiary companies left quickly after his entire annual profit was wiped out because of a commodity and currency issue. The core ingredients that his firm purchased were traded globally in dollars. He decided that the dollar was likely to decline against the pound sterling, so the forward purchases of product, made in dollars – equivalent to the best part of a year's revenues – were not 'covered'. Then the pound sank against the dollar, and his purchases cost some 10 per cent more than expected. That wiped out the entire year's profit, in a similar way to Rowntree (their loss was probably more than a year's profit, actually).

What followed next in my firm was instructive. The board suggested that *they* should decide on the stance to be taken in terms of forward exchange rates. I had to explain (sensitively) that there was no reason why they were likely to make a better decision than our ex-CEO, unless they had some amazing currency prediction methodology. If they did, we should stop making food products and move into the currency markets, and all become very rich. Our aim, I told the board, should be to buy raw materials as well as possible, using good buying techniques, and then look at currency as a risk that we had to manage – perhaps by 'hedging', through forward purchases of dollars, for instance. Pretending that we 'understood' the currency markets was potentially a ruinously expensive dead-end.

Power to the People

Outside commodities, there's a market for management consultants, for legal services, for payroll management, for automotive-brake components, for office furniture. Often markets intersect, and many firms play in multiple markets. But even where a market is in a healthy, competitive state, with plenty of good firms active within it, buying failure occurs when buyers simply don't understand the environment from which they are making purchases.

The buyer needs to understand the size of the market, and its dynamics – factors such as whether it is growing or shrinking, how competitive it is, and what external factors affect it. This helps the buyer work out how they can make the best buying decisions. Poor understanding of the market, and the buyer's place within it, can lead to organizations choosing the wrong supplier or misunderstanding their own position in terms of buying power. I once worked with a medium-sized organization that conducted an analysis of their purchases from different suppliers. They looked at where they spent a lot of money, with the aim of exploiting that 'leverage' and buying in an aggressive manner to drive savings. Nothing wrong with that, except that in some cases it just didn't work.

'We don't understand it. We buy loads of laptop computers, but when we approached HP and Dell, they weren't interested in giving us a good deal,' they told me. How many laptops were they buying, I asked? Oh, it's a lot. Around 500 a year.

I had to point out that their 500 machines – while they perceived it as a major spend area – represented a barely visible blip on the order-intake graph for a leading firm that shipped tens of millions of laptops every year. Even the UK government, when it tried to aggregate spend in this area, found it hard to get the big manufacturers very interested. On the other hand, 500 laptops might be significant for a local dealership or sales agent. The firm might do better talking to suppliers like that, who might go out of their way to land a half-a-million-a-year revenue client.

That applies across many markets and makes it vital that buyers understand market dynamics. For example, if you are a mid-sized firm and you want to outsource payroll management, or IT support, I'd suggest you *don't* consider the largest firms in those industries as your potential provider. You want a supplier who will see you as an important customer and treat you in a manner reflecting that importance. You need someone who is technically capable of handling your work, so I wouldn't advise you to become the very first customer of a start-up in those industries, but equally your business should matter to the supplier.

Do You Love Me . . . ?

Buyers need to think about how they appear through the eyes of their suppliers and potential suppliers. You might think your organization is really important to most of your suppliers – but it probably isn't. In business theory, we use the 'supplier preference model' to analyse these situations (see Fig. 1), and a lack of understanding of how suppliers perceive an organization can lead to bad decisions in terms of choosing suppliers.

In 2000 I was the director in charge of buying at NatWest Bank while it was going through a bitter takeover battle. I was asked by a main-board executive why we couldn't demand 10 per cent price cuts from all our suppliers. That would boost our profits and help us fight off the takeover. I pointed out two facts. First, as in the case of the laptops example above, NatWest really wasn't that important to huge firms such as Microsoft, IBM or many of our largest suppliers, even though we were the UK's second-largest bank. IBM reported revenues of some $88 billion in 2000, which meant that even if NatWest spent $50 million with the technology giant, we represented less that one-tenth of 1 per cent of their business. Not enough for their board to lose sleep over, certainly. And even if we were important, what would we do if they simply said 'no' to my demand (which they certainly would)? We had no immediate or easy way of moving business away from those firms.

Additionally, many of our suppliers were probably looking forward to the bank being acquired. Applying the preference model, we weren't a growing customer, and we were a bit of a pain to deal with generally, so for suppliers we may not have been a very profitable account, either. Indeed, a survey of suppliers that we had carried

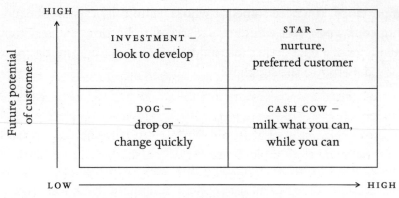

Fig. 1: The supplier preference model

out suggested that many preferred working with other banks, such as RBS. They probably had their fingers crossed that the Scottish bank would prevail, which it did.

I Fought the Law

In the government sector, many failures can be tracked back to an assumption that there will always be a range of credible suppliers who can execute the contract, whatever the organization looks to buy. But that simply may not be true. If the buyers' requirement is unique, as government work often can be, then a market may need to be created from scratch. Even if there is some capability already, a sudden increase in demand – from a major government programme, for instance – will put pressure on the market and it may take considerable time to reach a viable state.

The UK government decided in 2013 to outsource much of its probation-services work, despite warnings from experts that it would be 'highly problematic'.[4] The work included the management and rehabilitation of offenders, combining an element of punishment (such as monitoring the conditions of prisoners' release) with the desire to reduce reoffending and help the offender make a useful contribution to society. The Ministry of Justice, then under the command of minister Chris Grayling, created twenty-one Community Rehabilitation Companies (CRCs) to manage offenders who posed low or medium risk, and in February 2015 the CRCs were transferred to eight mainly private-sector suppliers, working under contracts that were to run until 2021–2.

However, a National Audit Office (NAO) report from March 2019 reported that implementation was rushed, there was little of the innovation that was promised from suppliers and, while the number of offenders has reduced, nineteen of the twenty-one companies that were ultimately involved failed to meet targets for reducing the frequency of reoffending.[5] Probation services have clearly not been transformed, and in July 2018 the ministry announced that it would terminate its contracts with CRCs fourteen months early, in December 2020.

Suppliers didn't do well, either. The NAO report estimated cumulative losses of £294 million for the firms if contracts continued to the end date; and Working Links, one of the providers, collapsed into administration in February 2019 because of its probation contracts.[6] Finally, David Gauke, by now the minister in charge, announced in May 2019 that the contracts would not be offered to private firms – probation services were in effect re-nationalized after one of the highest-profile UK buying failures in recent years.

There were clearly many problems here, but fundamental was the issue of an entirely new 'market' being created, without real understanding up front of what the work involved, what capabilities would be needed by the winning firms, how the right commercial models would be constituted or how competition could be maintained and stimulated. The organizations that won CRC contracts ranged from those with experience in vaguely associated areas (Sodexho was a catering firm originally) to new partnerships such as Purple Futures, which involved Interserve, an outsourcing firm that eventually entered administration in early 2019 after financial problems, along with charities such as Shelter.

'If you build it, he will come' – the tagline from the legendary film *Field of Dreams* seems to be how some governments think, when it comes to creating markets. And generally some entities will emerge from the undergrowth, bidding to carry out pretty much whatever the government asks them to – drawn by the potential rewards, of course. But this does not create vibrant, sustainable, successful markets in itself.

Island in the Sun (but it's no holiday)

The Australian government ran into similar problems around its contracts for managing offshore refugee 'transition centres', such as those on Papua New Guinea's (PNG) Manus Island. Managing these camps has cost a lot in terms of accommodation, security and catering, but it is also a sensitive social and political issue.

In response to an increasing number of refugees aiming to enter the country, as far back as 2001 the government started holding refugees in offshore detention centres, rather than on the Australian mainland, while their asylum claims were examined.[7] Many people felt this was cruel and inhumane, particularly when stories about poor conditions and bad treatment of the people in the centres started to emerge,[8] while other Australians welcomed the tough line on immigration. This is an incredibly sensitive area, with many questions around human rights and global migration pressures, as well as the political and management challenges. I won't get into the morals here, but the buying issues – in terms of the availability and competence of suppliers, and even potential corruption – have been significant. Some firms that might be well equipped to take on the work have considered the reputational risk and have kept their distance (would you want your firm to be associated with running what has been likened to a 'prison camp'?[9]).

A February 2019 report in *The Straits Times* covered the award of contracts related to the refugee centres to a little-known firm,

Understanding the Market

Paladin Holdings, which apparently 'had a beach shack as its (Australian) registered office'.[10] But Australian officials defended the decision, saying that time was of the essence and they needed to find a company urgently to provide services at Pacific refugee camps. The firm was awarded more than A$420 million – equivalent to around A$17 million a month – in contracts to provide a range of services at three transition centres for refugees on Manus Island.

While the parent company was registered in Singapore, the Australian division of Paladin was registered on Kangaroo Island, a sanctuary for native wildlife off the coast of South Australia. That didn't help the firm's credibility, and it changed the address quickly to a more business-credible Canberra address, once the furore started. Paladin had apparently acted as a subcontractor to other firms, and officials explained that firms such as Ferrovial, which had previously been involved in the work, had not wanted to bid because of the public outcry over the whole refugee policy. So there was no competitive process – a 'single tender' was conducted, with Paladin as the only bidder considered for the work. That raised suspicions about corruption, too, although no illegality was proved.

Again, this highlights the challenge of achieving a successful buying process, which requires a choice of strong suppliers competing for your work, where those suppliers simply aren't available to you. That almost inevitably leads to decisions that may look strange about who wins the contract, because few existing firms really have the appropriate experience. Creating competition, encouraging innovation and driving supplier performance are tough when there is no existing market to support those goals.

All the evidence is that creating a new supply market is challenging. Even if there are firms that have *some* of the key capabilities that underpin the new market (gained through other experience), it takes time to translate them into a coherent product or service to supply the new market.

Bad Influence

It is easy to understand the concept of markets when physical goods (such as cocoa, oil or metals) are discussed, but for many organizations their spend on services such as marketing is as large and significant as it is on raw materials, components or ingredients. The need to understand markets is therefore important in these areas, too.

Famously, John Wanamaker, a pioneer of US marketing and retailing, is reported as saying that he knew 50 per cent of his firm's advertising expenditure was wasted – the trouble was, he just didn't know which half.[11] There is some doubt around who (if anyone) actually said this, but it reflects an awkward business truth: that it's difficult to measure exactly the impact of advertising spend. That's always been a problem, but with the advent of the digital-media world, spending your budget well has become even more challenging.

Listening to a presentation at a marketing-services conference a few years ago, I was shocked to discover how the advertising industry measured digital 'views' of advertising. One 'impression' was defined as 60 per cent of the advert being visible for at least one second. Yes, one second! Do you really think your carefully thought-out brand message can be digested in a second? Is the viewer even vaguely aware whose advertising they are seeing, in that fleeting moment?

Now, of course, many ads will be viewed for longer than that, but there are other dangers in this new digital advertising world. The range of potential frauds and failures here is vast, from bots that imitate human views of an advert, to malicious websites that 'steal' viewers from legitimate sites. A detailed analysis from fraud intelligence firm Pixalate in 2018 suggested that in the US, 17 per cent of all ads (display and video) across computers and mobile devices were 'invalid', with even higher rates in certain categories.[12]

And as we move into the new world of social-media influencers, who are picking up a bigger and bigger chunk of the marketing budget, understanding exactly how these markets really work is proving more and more difficult.

As evidence emerges around dodgy practices, such as influencers claiming fake followers to boost their value to advertisers, these markets become more opaque. In 2018 Instagram cleansed millions of fake accounts and some 'celebrities' lost half their followers.[13] Reports suggested that Kim Kardashian's account lost a million; however, she still had more than 120 million, so maybe that didn't alarm her – or decrease her value to advertisers – too much. But that year Keith Weed, marketing chief of Unilever, one of the very biggest advertisers in the world, announced a spending freeze on influencers, telling *The Sunday Times*: 'Buying followers is misleading, and at the end of the day, a bot does not eat much Ben & Jerry's.'[14]

Unfortunately, many marketing executives, responsible for spending significant advertising budgets, do not understand the markets for digital advertising and related services. They have not got to grips with the opportunities, or the risks and dangers, and buying failure appears alive and well in this sector.

Down to the Market

No man is an island, and no firm works in splendid isolation. Any business that might be your supplier is working within a wider market, or several markets. So it is important to understand how those markets work, what drives them, which firms are active and successful within the market. That's true of both global commodity markets, such as oil, foodstuffs or metals, or a more local and specialist market, perhaps that which relates to supplying corporate catering services in Poland.

Market research can take many forms, from buying expensive data or commissioning research to asking your prospective suppliers a few questions over a beer or a coffee. But gaining an understanding of how those markets that are important to you work (in buying terms) is rarely time wasted, and can save you a lot of effort, pain and cash later on in the buying process.

Once you have that understanding, you can move on to the next stage in the process: choosing the suppliers that are actually going to provide you with the goods or services you need. Not surprisingly, I'm going to show how, even if you have got to grips with the market, that supplier choice can still be fraught with danger.

3

Choosing Suppliers

In 2019 *The Guardian* reported: '"Girl power" charity T-shirts made at exploitative Bangladeshi factory'.[1] The report revealed that World Reader, a charity that promotes literacy in the developing world, was a beneficiary of the T-shirts sold online. But it turned out that the garments were made in a Bangladeshi factory where workers were paid some 42p per day (50 cents), and many female workers had allegedly been fired for complaining about low pay. That was embarrassing for the charity, and will have done nothing to encourage people to donate or support what is a very good cause.

It was just the latest example of suppliers behaving improperly and exposing the buyer to risk, through what is perceived as a corporate social responsibility (CSR) failure. CSR today is the belief that organizations should contribute to wider societal goals, rather than focusing purely on their own profits and value. It has been driven largely by consumers as well as governments, and it has led to increased awareness of the reputational risks around an organization's suppliers, as well as a focus on what happens within the organization itself.

Bangladesh houses many suppliers to Western clothing firms, because the low cost of (often illegal) labour means suppliers can offer the most competitive prices on the market. In 2013 the Rana Plaza disaster drew attention to these exploitative practices when an overcrowded manufacturing building collapsed, killing 1,100 workers inside and trapping and injuring many more. The damage, in this case, spread up the supply chain, and a number of European and American companies had to answer for their practice of using such firms, many of which fail to meet very basic health, safety or pay provisions.

Despite this tragedy, many Western companies continue to seek out the most competitive services available on the market, regardless of the cost to human quality of life. A 2018 study by the Fair Wear Foundation and Care International found that female factory workers producing clothing and shoes in Vietnam – including for major US and European brands – faced systemic sexual harassment and violence at work. Abuse included threats of contract termination, groping, slapping and even rape, and indicates how bad the working conditions are for many in these factories, according to Dr Jane Pillinger, a gender-based violence expert and the study author.[2] Shockingly, nearly half of the 763 women interviewed in factories in three provinces said they had suffered at least one form of violence or harassment in the previous year.

CSR has jumped up the board agenda in recent years, and the pressure to address such issues is only likely to increase, as younger consumers are placing increasing emphasis on how firms they buy from contribute to the wider good. From animal welfare in the food-supply chain, to human rights or issues around climate change and plastics in the oceans, buying from suppliers that also have a social conscience might be 'doing the right thing' for altruistic reasons, but it also makes good business sense.

An Apple a Day

It isn't only small charities such as World Reader that get caught up in negative publicity because of bad buying decisions. A wave of suicides among the workforce at Foxconn factories in China back in 2010 and 2011 brought criticisms of the working conditions there, from human-rights organizations.[3] The company always denied that it operated 'sweatshop' conditions, and even erected nets around its Shenzhen factory to prevent suicide jumps. Foxconn was, and still is, a major supplier to many big-name electronics firms, including Apple. It employs more than a million people, so the number of suicides within the firm may be nothing more than a reflection of just how many people work in its factories.[4] Indeed, that appeared to be the case back in 2010/11, as Apple founder Steve Jobs pointed out at the time.[5]

A Fair Labor Association report in 2012 found Foxconn paying staff on time, offering well above the minimum wage and being relatively generous with sickness pay.[6] It did find other less positive issues, but Foxconn may well compare favourably with less well-known businesses in China and around the region. Nonetheless, the publicity was uncomfortable for high-quality, luxury, branded product firms like Apple, and showed the importance of choosing and monitoring key suppliers carefully.

Another example in the same industry was exposed through the work of Electronics Watch,[7] an organization that campaigns to improve conditions in the electronics supply chain. That means keeping an eye on firms that make our smartphones, laptops and similar kit, often in factories in China, Thailand, Vietnam and other South-East Asian countries.

Electronics Watch has found violations of human rights ranging from the relatively mild (compulsory overtime), to what is effectively modern slavery, with workers having passports taken away and having to work to fund 'agents' who got them the jobs. Many leading electronics firms have taken action in cases when Electronics Watch has pointed out issues.

A 2017 report highlighted the use of students as young as sixteen as forced labour in Asian electronics factories.[8] If they refuse to work, they are told their funding will be cut or they won't be allowed to graduate. Understandably, students in Western countries (many of Asian origin themselves) didn't feel comfortable thinking their kit might have been made by students who were having their own human rights violated. Universities responded, supporting Electronics Watch and threatening to boycott firms accused of bad practice. Major companies, such as HP, put pressure on subcontractors, and some improvements to worker conditions were achieved.[9]

There are many other CSR examples in the world of buying. While firms can do a certain amount internally (looking at their own carbon emissions, for instance), most major organizations can have a greater positive impact by taking appropriate action with their suppliers and supply chains, and doing so in a way that is also good for their own business. Buyers should understand what is going on (as far as possible and reasonable) and consider where they can make a difference. For some organizations that might be through working with suppliers to reduce the use of plastics. In other cases, addressing modern slavery risks in the supply chain might be the appropriate route.

Gaining commitments from suppliers when contracts are being awarded is one option, and buyers can ask for the right to audit factories and check up on employment or environment-related practices. Don't be afraid to push your suppliers if you see issues, so that you are not unwittingly supporting firms that are abusing or exploiting workers, or having hugely negative impacts on the environment.

Thank You for Being a Friend

It's not just CSR that needs considering when choosing suppliers. There can be more personal and local issues, too. While I was working briefly in a new government-sector organization, formed from a merger of three smaller organizations, the Human Resources director asked if I would approve the appointment of a firm to carry out some training work. She was the budget holder, but she required my sign-off.

How had she chosen this firm? 'Well,' she said, 'I know them well, I've worked with them before. They're excellent, even though they're only a small firm, and the Chief Executive is a personal friend of mine.'

She saw the relationship as a positive factor. She knew the firm would be good because she knew the boss, and to some extent she had a valid point. However, there are obvious conflicts of interest that lead to three problems with this approach to choosing a supplier.

- Her friend's firm might be great, but there might be *another* firm that would do an even better job for us. If you don't have a quick look at other options, how can you know?
- How can the HR director hold her friend to account? Or negotiate hard with him? What happens if his firm doesn't perform?
- How does it look if this hits the newspapers? What she proposed wasn't illegal, and may not even have broken internal rules. But was she prepared to see her friendship (which I'm sure was nothing more than that) presented as a scandalous use of public money, discussed in the *Daily Mail*, *Private Eye* magazine or *The Washington Post*?

In 2019 the British Prime Minster, Boris Johnson, received unwelcome publicity when his friendship with an American technology entrepreneur was exposed. Jennifer Arcuri's firm received various grants from the public purse, Johnson attended conferences she organized, and she personally accompanied Johnson's party on a number of trade-related foreign trips when he was Mayor of London. The episode highlighted the dangers for anyone in the public eye who might be getting business and pleasure somewhat entwined. Arcuri denied any sexual relationship, but told the media they shared a 'close bond', and that Johnson had visited her flat in London to discuss Shakespeare, and for technology lessons . . .[10]

We Are th Champions

Working with friends isn't always a bad idea. If you *really* think your friend could do a great job, then by all means recommend them, but then keep away from the process of choosing the supplier and negotiating the contract. Let your colleagues consider your friend as part of a selection process, and if they think your friend is as good as you do, they will win that competition and that's fine. Prime Minister Johnson didn't award the grants directly to his friend, which helps his situation, but transparency is always positive, so he might have made it clearer why he chose to attend Arcuri's events rather than many others to which he was invited – if, indeed, there was a good reason.

Choosing the best possible supplier to carry out work for your organization, to deliver the best possible value, is a fundamental element of good buying. For all but the smallest amounts of spend, that means applying some sort of *competitive process* – comparing bids, proposals or tenders from different potential suppliers. That might be a quick phone call and a verbal quote giving a price for a low-value purchase, a short written proposal from three suppliers for a mid-sized technology contract, or a 200-page (or 2,000-page) formal tender document for building a fighter jet.

Competition is a great tool for finding the best suppliers and ensuring you get a good deal from them, achieving the value your organization wants and needs. It is also a great defence against corruption, as I'll discuss later.

You Got the Wrong Man

Some time after a contract has been put in place, and money spent, it's not unusual for executives to shake their heads, saying, 'We just chose the wrong supplier.' In retrospect, it is clear that a bad decision was made, and this can be down to a number of causes. Often it is related to a lack of real thought, analysis or due diligence by the buyer. That can be driven by laziness, lack of time or inclination to look at the range of potential firms that could be used. Realizing that you have got into bed (commercially, at least; and no, we haven't gone back to Boris Johnson's exploits here) with a firm you wish you hadn't, perhaps because you didn't check out their previous relationships and history, is a frequent cause of failure.

We are not talking here about suppliers winning contracts because they have bribed or otherwise corrupted internal staff. That tips over from genuine error to a classic type of fraud, which we will cover later, with case studies where corruption drove supplier selection decisions.

Rather, it can be down to the inevitable difficulty in translating a potential supplier's proposal on paper into a real-life delivered product or service. It is relatively easy to write, 'We will do this, and this, and this', and with some knowledge and a professional bid-writer, most firms can put together something that sounds convincing. But that doesn't necessarily mean they really have the capability, knowledge, funding, resources or commitment to deliver. The challenge, when assessing suppliers and proposals, is how to weigh up the strength of the proposal *and* the likelihood that the firm can actually do what they say they will. I'll come back to that in more detail when I talk about 'believing the supplier'.

Failures occur when a manager needs some work done quickly, so she picks up the phone to the usual management consultants that she's worked with many times before, or the building repair firm that is always happy to help. This doesn't always end up with trouble, but where the different options are not considered, where suppliers are not fully assessed for suitability, and where the buyer doesn't use any sort of competition, then the danger of getting the wrong supplier for the work is real.

The risks here are wide-ranging, too. Maybe your usual consulting firm is great at M&A (mergers and acquisitions) work, but fairly average (and very expensive) when it comes to advising on organizational development issues. Perhaps that usual building-maintenance firm doesn't really handle air-conditioning problems, but isn't going to tell you that, if they can see revenue coming their way. They will simply subcontract the work and add on a 25 per cent margin for their own benefit.

The Winner Takes It All

What are we really trying to do when we run supplier selection processes? We're trying to find the best supplier to meet our needs – but what does 'best' mean? In simple cases, it may mean the lowest price. If I buy a pencil or a pack of copier paper, then price may be all I'm worried about. But even then, I've seen copier paper that wouldn't run through the machine and ended up costing far more in repair bills than its own cost. Once we get on to anything more complex – whether that is components in a factory environment, or cleaning services, or lawyers, or government service contracts, let alone fighter planes or nuclear power-station construction – then we're going to want to look at more than just price.

Other factors come into play around quality, service, innovation, the strength of the supplier's administrative processes, even how good they are to work with (if we're hoping for a partnership-type arrangement). Then when we get into the IT world, there may be issues of data security, resilience, location of hosting, and so on. It's not always simple, or obvious, how we can select the best supplier.

For those more complex contacts, most organizations choose to run a formal selection process, which involves suppliers making proposals (or bids, or tenders) that are assessed so that the buyer can make the best choice. Working out what you are looking for in a supplier, then thinking about what you should ask them, and how you will 'mark' their responses, is a complex and tricky part of the process.

An evaluation process should usually have three key stages.

1. Define what we want from suppliers and work out how to evaluate their proposals in terms of the key desired criteria (price, quality, etc.).
2. Ask questions related to those criteria – written, verbal or both – that draw out relevant responses from the suppliers.

3. Mark those responses using a pre-determined scale, and process and choose a winning supplier or suppliers. (There may be iteration, as we go back to bidders to clarify matters or ask for further information.)

One key issue that often leads to problems, though, is the *weighting* of different criteria and factors to be considered. An expert in these matters was working with a huge government military organization on a mega-contract worth many hundreds of millions. The buyers asked him to look at their evaluation process.

He did, reviewed the way they had set up the scoring and said to them, 'Would you be prepared to pay another twenty million pounds to get a supplier whose bid you rate as "very good" compared to a supplier bid rated as just "good"?' No, of course not, they said. 'Good' would be fine, and they certainly couldn't justify spending another £20 million of taxpayers' money to move to a slightly better supplier.

Well, he said, your evaluation process and scoring system mean that you *would* pay well over £20 million for that better bid. 'Cost' had not been weighted highly enough in the process, so good scores on other factors accumulated lots of points in the marking process. How you score price in a way that enables comparison with the scores given to bidders for 'quality' is another fascinating example of the complexities here.

I hear regular complaints from suppliers that 'government buyers just accept the lowest bid', but in my experience, this is rarely true. In fact, if anything I would argue that, as in the case above, price is often not seen as important enough. If you want to pay more for a 'better' product or service, that's fine, but you must make sure of two things. First, that you really *need* what you're paying extra for, that it adds real value and isn't simply a 'nice to have'. And second, that the additional benefit the supplier is offering is real, tangible and deliverable – not just that they're better at writing convincing bid and proposal documents!

I saw another bid where it looked as if a major government organization might buy BMW cars for their field operatives – because that manufacturer's products, when evaluated, scored so well on comfort, driver experience, safety, and so on. But aside from that decision potentially causing a huge political furore, many of these evaluation criteria were 'nice to haves' rather than essentials. Change the focus on the evaluation to basics, like whole-life cost, and Ford, Toyota and other less luxurious marques came through rather more strongly.

Friends Like These

If you watch a series of YouTube videos about wine tasting and then others featuring the South of France, a clever algorithm might show you a pop-up advert (or play an ad before the video you've chosen), advertising a wine-tasting tour of the region, or a hotel that runs wine courses in Provence. All very sensible, and it can be useful to the viewer. Or another algorithm may be set up to play adverts next to videos that are getting a lot of views at the moment, to maximize the exposure for the advertiser.

This has revealed a new type of buying failure in the digital marketing field, where organizations are embarrassed by their advertisements on social media and on websites apparently endorsing the owner of the material they sit alongside. The most serious side of this comes when brands find themselves promoted on websites or next to YouTube videos that they really wouldn't want to associate with. In 2017 *The Times* reported that advertisements 'for hundreds of large companies, universities and charities, including Mercedes-Benz, Waitrose and Marie Curie, appear on hate sites and YouTube videos created by supporters of terrorist groups such as Islamic State and Combat 18, a violent pro-Nazi faction'.[11]

The suppliers who arrange the media advertising for these firms aren't aiming to hurt their clients, but this has two negative consequences. Not only is it bad news for the advertisers involved, but it also raises money for these awful organizations. Whoever posts the video on YouTube gets several dollars from the advertiser per thousand views of that advert. In a sense, Islamic State became an accidental and unwanted 'supplier' to these blue-chip advertisers.

More widely, some of the biggest firms in the world have spoken out about the problems in the digital-media world. The previous chapter described a lack of understanding of markets in this area; that extends into choice of suppliers, too. Marc Pritchard, Chief Brand Officer at P&G, the world's biggest advertiser, has been vocal and has pushed for more transparency from digital platforms as well as from agencies. He describes the media supply chain as being 'murky at best and fraudulent at worst'.[12]

It's clear that too many advertisers are not taking enough care about where ads are appearing and who their brands are associated with, and too many agencies are putting their own profits ahead of clients' best interests. The industry is also rife with kickbacks and hidden rebates – which is how some agencies make much of their profit – so that may also contribute to this bad practice.

Self-Destructive Zones

With everything covered so far in this book, you might be thinking, 'Perhaps we should avoid suppliers and just do everything ourselves. Keep the work within the company.' But that doesn't guarantee success, as the Spanish bank Banco Sabadell found in 2018.

In 2015 Sabadell acquired TSB, a UK-based retail bank, formally part of the Lloyds TSB Group. TSB at some point needed to move onto its own IT platform, rather than continuing to use the Lloyds group systems, as they were now competitors to their former parent company. But the move, in April 2018, turned into a disaster. Account holders couldn't use mobile or Internet banking, and some reported seeing account details from other account holders. Customers struggled for weeks to make mortgage and business payments, as the new TSB systems failed to function properly. The issue was serious enough to be raised in the British Parliament, and in September 2018 TSB's CEO, Paul Pester, resigned.

In March 2019 *The Sunday Times* reported that an investigation into the affair put much of the blame onto the IT firm that handled the transition.[13] However, the twist was that this firm was SABIS – which is part of the Sabadell Group itself. So although it has a separate identity, this was in effect the internal IT function of the group that owned TSB.

Reports suggested a range of technical and programme management issues around the deployment of new software, rather than problems with the underlying infrastructure. But whatever the cause, the whole episode cost TSB £330 million,[14] and there is a 'provisional agreement' (according to the firm's annual report) for SABIS to pay TSB £153 million. In November 2019 an independent report from law firm Slaughter and May concluded that the issues arose because 'the new platform was not ready to support TSB's full customer base' and, second, 'SABIS was not ready to operate the new platform'.[15]

Questions have to be asked about the choice of 'supplier' here. Was SABIS the right choice to carry out this challenging task? It certainly doesn't appear so, in retrospect. Did TSB have a choice, or was the firm told by top Sabadell management that it had to use SABIS? Would a firm with a wider and broader experience of banking systems than SABIS have done better? And why didn't TSB accept the offer of help from Lloyds, which was made as soon as news of the problems broke?

Whatever the answers to these and other questions, there's no doubt that this does show the dangers inherent in choosing a 'supplier', even if that choice is an internal service provider rather than an external third party.

Take Care of the Road You Choose

Choosing the right supplier is a key step in avoiding bad buying decisions, and there is a lot to think about. It's not just whether the supplier can meet your needs in terms of price, quality and service, but these days you may also wish to consider issues such as corporate social responsibility. There are also hazards around conflicts of interest and working with people you might know too well – as the British Prime Minister has found – and in the digital-media world firms have ended up with some very unpleasant 'suppliers'. Not even choosing part of your own organization to be your supplier can guarantee success, as TSB discovered.

I've talked here about steps that should be taken to maximize the chances of success at this stage of the buying process, and it is worth remembering that effective competition is a fundamental driver that always contributes towards making good supplier-selection decisions. Posing well-thought-out questions that potential suppliers can answer in their proposals, along with a structured evaluation process to consider their responses, also helps. Do check that suppliers really can do what they claim in their offers, and be careful about working with friends, too, even if they are offering 'technology lessons'!

Follow that advice and you may well select what looks like the perfect firm to carry out your work. However, as we will see in the next chapter, you can still end up in a tricky situation.

4

Don't Get Too Dependent

You've chosen a really great supplier, but now you need to avoid getting too close and too dependent, no matter how wonderful they appear to be. Sometimes the buyer has little choice, and there are many examples where a supplier or a small group of firms has a genuine monopoly. That can range from rare minerals or chemical products, to a technical product, software or design that is protected by patent, copyright or other IP protection. But in many cases you can create your own accidental monopoly situations, by not analysing how your own actions reduce choice when it comes to buying.

In my first management role I bought a range of raw materials for Mars Confectionery. Many were standard, off-the-shelf ingredients: dairy products like butter and skimmed milk powder, fruit juices and dried egg powder. But one ingredient was very different. It was quite literally unique, and there was only one supplier who could produce it, despite the best efforts my firm had made over many years to find an additional provider.

Although it seemed to be a fairly standard blend of ingredients, somehow when we acquired samples from different firms, they just did not work. The final product either didn't taste right or could not even be manufactured successfully in the factory. And this wasn't a minor ingredient – it provided the essence of this product. Think of ice cream without the cream, or Heinz tomato soup without the tomatoes. There was no flexibility in terms of recipe, and no chance of taking any sort of risk with the product.

So when I was summoned to the supplier's office for our annual negotiations, it was somewhat one-sided, as you may have guessed:

Me (trying to appear confident): *I've done my research, and I think your raw material costs have been pretty flat in the last year. I'm happy to cover a three per cent increase in your labour costs and overheads – so I'd like to offer a small overall price uplift of one per cent for next year.*
Sales director (dismissively): *Here you go.* (Hands me a piece of paper apparently torn out of an exercise book.) *Here's our offer. Take it or leave it.*
Me: *But that's twenty-five per cent up on last year.*
Sales director: *Yeah. So it is. Take it or leave it.*

No matter how much I argued, how much objective evidence proved their pricing was unrealistic, there was no movement. I could take it or leave it, as far as they were concerned. I had no option; our negotiation position was weak – non-existent, frankly – and we had no alternative.

That first bad experience in the buying world has always stayed with me. It highlights several issues around supplier dependence, because as well as the commercial impact, I had sleepless nights in terms of risk. What if that supplier's factory burned down? What if they went bust? (Unlikely, I know, given the prices they were charging!)

However, the story has a happy ending. Our brilliant research scientists eventually came up with an alternative material that worked and satisfied all our requirements, from a different supplier. I had the more pleasant task of going back, a year later, and holding a much more balanced negotiation. We kept the first firm on-board; no point in substituting one monopoly supplier for another. But now we could look at their costs and negotiate a fair price for both parties, which meant a significant double-digit cost saving for us. Which just goes to prove once again that competition is a wonderful thing.

Drive My Car

In 2016 production of Volkswagen Golf cars at VW's Wolfsburg plant, employing 60,000 people, was suspended for several days, when VW got into a legal dispute with ES Automobilguss GmbH, a supplier that provided cast-iron parts needed to make gearboxes.[1] Another linked supplier, which made textiles for car seats, also stopped delivering to VW.

The textiles referred to came from Car Trim, which stopped deliveries of seat covers to the Emden VW plant in northern Germany. Both suppliers involved in the dispute were subsidiaries of Wolfsburg-based Prevent DEV, a privately owned firm supplying the automotive industry. It appeared that Prevent was hitting back at VW because of a dispute after VW terminated some contracts with the firm. 'Because Volkswagen declined to offer compensation, Car Trim and ES Automobilguss were forced to stop deliveries,' the companies said, in a statement issued through their parent company.[2]

Later, the cause of the dispute was explained like this by Reuters:

> VW's relationship with Prevent soured after it commissioned Car Trim to develop new seat covers for high-end models including Porsches and then cancelled the 500 million euro ($558 million) deal in the aftermath of the diesel scandal, refusing to cover the 58 million euros its supplier had already invested.[3]

Eventually the state court of Lower Saxony took the unusual step of ordering the suppliers involved to deliver the parts to Volkswagen – or else VW were given the right to seize components themselves. VW could send trucks to the supplier and, with bailiffs' assistance, seize the parts it needed to resume production.

Don't Get Too Dependent

Now 'cast-iron parts for gearboxes' or 'textiles for car seats' don't sound like the most critical items for making cars. But if you are dependent on the supplier, if you don't have alternatives readily to hand, then you are in trouble if that supplier has problems or decides, for whatever reason, they just don't want to work with you.

Indeed, the CEO of VW later acknowledged this. Matthias Müller said he would review his firm's procurement strategy to avoid a repetition of the crippling dispute and re-examine contracts that left the group dependent on a single supplier – very sensible, but a bit too late perhaps. Analysts suggested that the dispute might have reduced earnings by as much as 40 million euros ($45 million) for the week of lost production.[4]

It isn't only disputes that can cause problems. In 2018 Ford had to suspend production of its popular F-150 pickup truck at plants in Dearborn, Michigan, and Kansas City, Missouri, after a fire at a supplier (Meridian Magnesium Products of America) caused a parts shortage;[5] 7,000 workers were laid off and Ford's share price dropped 2 per cent the day the problem was announced. Presumably Ford did not have another supplier ready to step into the gap left by Meridian and, as in the Volkswagen case, reliance on a single supplier proved costly.

I Depend on You

The VW example of failure probably emerged over time and has been replicated by other car manufacturers, with seemingly sensible business approaches leading to a situation where suppliers often hold more power than the car firms themselves. Over the years, car makers looked to standardize components and rationalize the number of suppliers used. (The same happened in many other industries.) This supported efficient production and was supposed to give buyers more 'buying power'. If I buy a huge annual quantity of 'textiles for car seats', then surely I can negotiate a great deal? This is true up to a point, but if you reach a situation where very few suppliers can meet your needs, and you're locked into them because you use just-in-time production techniques (so you don't hold much stock, which could be used if something went wrong), then you move into a supplier dominance and dependence situation, with failures like the VW case.

The *Financial Times* reported in 2014 that sixteen major global car manufacturers were dependent on a mere ten component makers.[6] Whatever car you buy, the anti-lock brakes will probably come from Continental, the battery from Johnson Controls and the exhaust from Denso.

In this way, organizations themselves can play a major role in creating a monopoly or oligopoly (a situation that sees a limited number of firms dominate a market) where one didn't previously exist. Now everyone knows that monopolies are bad news for buyers, using strategies that maximize profits and result in prices charged that are higher than those generated in a free, competitive market. An oligopoly is almost as bad – in these cases, the small cabal of firms similarly controls a market.

Existing monopolies are usually pretty easy to spot; it tends to be obvious when there is only one firm from which you can buy a particular item. But the problems for buyers don't stop at excessive prices. As well charging more than the true market rate, monopoly suppliers have less pressure to innovate, or provide great customer service, or drive continuous improvement – all the behaviours you would hope to see in your supply base.

Take Me to the Hospital

As well as private-sector examples like Volkswagen, governments around the world are also guilty of causing these undesirable outcomes. It may not be a true market monopoly that creates dependence; it may be contracts locking the buyer into a particular supplier for a lengthy period, with no escape or chance of competition. That encourages the same unwelcome supplier behaviour as seen in a true monopoly situation.

It can be difficult for governments to do anything else in some cases. Once a contract to build an aircraft carrier or a new Olympic Stadium is awarded, there is a dependence on that supplier for years to come, whatever happens. You can't switch supplier at the drop of a hat – indeed, there will never be many firms who can do that sort of work, because of the expertise, capital and experience needed. You aren't going to ask six different suppliers to build one or two boats each, to spread the risk and 'play the market'.

But in other cases governments create their dependence when they don't need to. Consider the private finance initiative (PFI) contracts the UK government entered into for the construction of schools, hospitals and more from the 1990s through to around 2010. Under PFI, the supplier (usually a consortium made up of a construction firm, a building and facilities services expert, and a bank or similar to provide the funding) would not only build the hospital, but would provide all the money to do it, then would continue to take responsibility for building maintenance and a range of other related services over the next twenty years or so. While the UK was the pioneer of the concept, many other countries followed suit, including France, Spain, Canada, Portugal, Japan and Nigeria.[7]

Both major UK political parties supported PFI, as it enabled new infrastructure to be provided without the government itself having to borrow money to build it. That has led to a lively blame game, as PFI has left the National Health Service with horrendously expensive ongoing bills for many major hospital buildings and services.

Without getting into all the rights and wrongs here, there were sensible ideas behind PFI, such as making the designers of hospitals think about the ongoing running costs as well as the new-build cost. One aspect of PFI that was *not* sensible, though, was the way contracts for services were lumped into broader PFI construction deals. To make initial contracts more attractive for bidders, many included services such as catering, building maintenance, security, and so on. Then a hospital was locked into those suppliers for ten, fifteen or twenty years, with little scope for competition, price negotiation or even variation of the service provided.

And so horror stories emerged about firms charging a fortune to change a light bulb, while the PFI contract prevented the hospital's own staff from doing the simple job. Freedom of Information requests in 2011 exposed cases such as £962 to 'supply and fix noticeboard' at a hospital trust in Leeds, and £242 to 'put a padlock on a gate' in Staffordshire.[8]

That contractual structure created mini-monopoly situations, with supplier dependence in the sense that hospital managers or other clients of PFI providers could not compete for this work, get out of these contracts or even have the flexibility to ensure they remained fit for purpose over time. They didn't *want* to be dependent – but the PFI deal made them so. One of my own contributions, when in a senior government commercial role in the 1990s, was refusing to allow those 'soft services' such as catering and cleaning to be included in a couple of large PFI deals that my organization was signing up to. I don't claim that any unique skill came into play – it just seemed obvious that long-term contracts could end up with that unhealthy dependence.

Breaking Up is Hard to Do (the Carillion story)

Carillion was one of Europe's largest construction and building management firms, an international player, particularly in the market for government-sponsored capital works including roads, hospitals and other major infrastructure projects. In January 2018, after worsening financial problems, the firm went into compulsory liquidation. This caused massive problems for clients, as well as for the firm's staff and suppliers. Most of the work was eventually taken on by other firms, but there were some major delays. The redevelopment of the Royal Liverpool University Hospital, for example, is now at least two years behind schedule, and in most cases the evidence so far suggests that new contractors will also be charging more than the original Carillion pricing.[9]

The UK government came under criticism for continuing to award contracts to Carillion after it became clear that the firm was struggling. Was this a clear case of buying failure? This is one of those occasions where the buyer probably couldn't win. If the government had stopped awarding contracts to the firm sooner, then Carillion would probably have gone under faster too, and the government would have been castigated for pulling the plug on the firm, putting thousands of people out of work.

But what about the underlying causes of the failure? Some expert observers suggested that the focus on low prices often needed to win government contracts meant that Carillion bid too low, and accepted too much risk, so that when they hit unexpected issues they were exposed.[10] As their situation worsened, they were desperate to win new contracts to shore up cash flow, so that led to further low bidding. The *Financial Times* likened it to a Ponzi scheme[11] – using new revenues to keep paying the bills to their own suppliers for work already completed.

There were other issues, of course, including financial structures, the debt/equity balance and how top management was incentivized. But some observers suggested that the failure was the fault of public-sector buyers who agreed contracts that proved to be bad news for Carillion in risk and profitability terms, and ultimately bad news for the taxpayer, too.

But hang on a minute! Should the *buyer* really take responsibility for a supplier bidding too aggressively, or accepting risk it shouldn't? If you had to address that, how could you run fair tendering and bidding processes? 'Your firm's bid is clearly the lowest, Mr Smith, but we're giving the contract to your rival, who bid a million higher, because we don't think you can do it for that price and make a profit.'

You can see the impossibility of that approach, but government buyers, and the UK Cabinet Office – the central department that has oversight of major suppliers – do deserve some criticism.

Critically, they were not on the ball in terms of understanding how serious Carillion's problems were, and their own strategies contributed to a few large firms dominating key markets. Through aggregation, pushing risk onto suppliers (as in the PFI example) and more, the UK government ended up in some markets with dependence on just a handful of dominant firms. That exacerbates the problems when one runs into trouble.

Trouble No More

There are many ways in which firms and governments create (often accidentally) supplier dependence. These traps are all avoidable to some extent, so here are five warning signs in terms of how your organization buys.

1. **Buyers aggressively aggregate their own spend**, believing they'll get better deals if they offer bigger contracts – until in some industries only the largest can meet your needs. Buyers might insist that suppliers must service every office or factory across the US, or Europe. Smaller firms and start-ups, which often offer real innovation, flexibility and service, are shut out of the market.

   Buyers assume economies of scale, that 'bigger is better' and bigger deals mean lower prices. But that is not necessarily true; the price curve may flatten after a certain volume, with further increases in volume not generating any further price reduction. There are even cases where you see *dis-economies of scale* – the buyer pays more as the they spend more, as shown in Fig. 2.

   A buying consortium of airlines found exactly this issue when they tried to do 'group deals' with hotels around major airports. Their combined volume of rooms needed for their cabin staff was just too great. It eliminated some hotels from the market altogether; and it would have

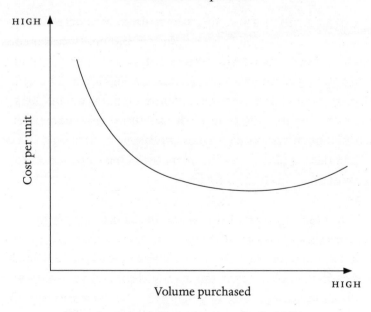

Fig. 2: Sometimes dis-economies of scale kick in

stopped others from doing lucrative conference business (as half their rooms would have been taken by the consortium). If they bid at all, hotels offered 'average' pricing, while individual airlines, looking for perhaps a few hundred room-nights a year rather than thousands, got a better deal.

2. **Buyers value consistency above innovation and experimentation**. At times you should value tried-and-tested solutions over exciting new ideas. 'Ladies and gentlemen, welcome to the flight; this is the very first plane to be fitted with an exciting new automatic pilot system, and we will be turning it on once we're airborne.' You might not want to hear that!

But take caution too far, and you help create markets dominated by a few large suppliers, with increased risk of buyers suffering from dependence. That's relevant in private firms and perhaps even more so in government, where risk aversion from employees and politicians means that companies get into dominant positions because buyers 'know' they're a safe choice. That doesn't always work out – Serco and Capita seemed to be safe for major UK government work, until both ran into severe financial difficulties.[12] More willingness to engage with other initially smaller suppliers over the years could have created a more dynamic market.

3. **Buyers set tighter and tighter specifications**, directly or through indirect requirements such as specific accreditations, security constraints and the like, until again only a few firms can meet your standards. I talked about the importance of getting the specifications right in Chapter 1. But the danger is that even if a specification does accurately describe what you want to buy, if it is *too* specific and constraining, then you can create another monopoly-type situation.

The issue has become more common in recent years, and some of the change is understandable. High-tech equipment and much mass production (in the defence, automotive, oil and gas, technology and other industries) works on finer and finer tolerances, so specifications for components have to be very precise. But yet again, that plays to specialization and ends up with few firms being feasible suppliers.

4. **Buyers group together disparate bundles of goods or services** into larger packages of work, which fewer suppliers can handle, reducing the field of potential suppliers. Now this really has been a buy-side self-inflicted wound. In markets for both goods and services, buyers have taken a portfolio of different component products and pushed the market into making a single offering. Again, this has

some logic, but inevitably means fewer firms can meet your needs.

Examples include 'full-service IT outsourcing', which grew over recent decades, as buyers looked for a single supplier to manage and maintain IT equipment, handle new purchases and installation, perhaps even write new software products. Smaller, more specialist IT firms suffered, as firms like IBM, HP and Fujitsu looked to offer everything an IT buyer could want. That leads to a situation where requirements can only be met by a handful of global firms, and you're back into lock-in. Or consider a 'total facilities management' contract, including catering, security, building maintenance, gardening services, lift maintenance . . . a contract that today few firms can possibly handle.

5. **Buyers make tendering and bidding for work difficult**, so smaller firms find it hard to compete against the bigger ones. That becomes a self-perpetuating cycle where large firms win more work, and can afford to spend more money on bidding teams and professional bid-writers or getting all the accreditations needed to tick boxes on the tender form.

Winning a contract becomes as much about having the skills to write tender responses as it is about having the skills to deliver the work required. In the government sector, in particular, many firms just give up and decide they won't bother competing. They see the same companies winning time and time again, and (perhaps correctly) assume this is a closed shop and there's no point trying. Within a short period of time, another market has become an oligopoly, and government buyers wonder why they're not getting bids from exciting new suppliers.

The drivers listed above all contribute to situations where suppliers benefit from dependence, through monopoly and oligopoly, or through customer lock-in of some sort, with significant barriers to entry and switching. That can allow suppliers to make super-profits; it takes away much of the need for firms to become more efficient or to innovate; and those profits also tend to generate huge salaries and bonuses for the owners and managers of those firms.

Freedom

Keep Adam Smith in mind when dealing with suppliers. As the great eighteenth-century Scottish economist said, 'People of the same trade seldom meet together, even for merriment and diversion, but the conversation ends in a conspiracy against the public, or in some contrivance to raise prices.'[13]

This is just as true today, so bear in mind that many businesses look to create monopolies and dependence to increase their power, exploit their customers and make more money for themselves. If a problem with a single supplier means that your organization has to stop taking customer orders, or has to shut down a factory (as VW and Ford had to do in the earlier examples), then you have got yourself into a situation of dependence. Be a little cynical about the supplier's motives, and do whatever you can to keep your options open and avoid this type of failure.

One recommended route to mitigate the risks of dependence is to remember how powerful a lever *competition* is, in terms of getting good deals, good suppliers and good contracts. Use competitive pressures wherever you can, and seek healthy competition between potential suppliers to help your organization get a better result. Wherever possible, look for multiple suppliers for key items, and avoid contractual terms that lock you in, whenever you can – remember the PFI examples, where long-term exclusive contracts meant buyers had to use suppliers for certain services, whatever the cost. Finally, note those five actions I've described that organizations take, and which often unwittingly lead to supplier dependence. They're all avoidable as long as you are aware of the issues.

Don't Get Too Dependent

After a selection process, you have chosen a good supplier – one that isn't too dominant, and that you feel will do a good job for your firm. Now you need to put in place the deal. In most cases that means a contract that is positive for your organization, but doesn't leave the supplier so badly off that they risk becoming the next Carillion. All you need to do is negotiate that deal skilfully and professionally – a topic that I'll discuss after our first Bad Buying Award-winning case study.

Bad Buying Award

EXISTENTIAL BUYING FAILURE – SCHLITZ BEER

The Joseph Schlitz Brewing Company was the largest producer of beer in the United States for much of the first half of the twentieth century. Even into the 1970s, Schlitz remained number two in the US. In the early Seventies the management team decided to cut costs, as competition from others, such as market leaders Anheuser-Busch, became more aggressive.

As the malted barley used to make the beer was a major cost, the recipe was adapted to replace some barley with cheaper corn syrup, and cheaper hop pellets were used rather than fresh hops.[14] A more rapid high-temperature fermentation was also introduced instead of the traditional method. Then a silicon gel had to be added to prevent the corn syrup forming a 'haze' in the beer when it was chilled.

All went well, and by 1973 Schlitz was claiming it had the most efficient breweries in the world, and operational and financial ratios were above the industry average. Although the new recipe had been tested on consumers, the changes weren't widely known. However, competitors started drawing attention to the different processes used by Schlitz, seeding some unease among drinkers. Perhaps this growing lack of consumer trust was one reason why the launch of Schlitz Light, a low-calorie beer designed to take on the rapidly growing Miller Lite, failed.

In 1976 the firm started worrying that the US Food and Drug Administration – a powerful body that regulated those industries – was going to force firms to list all product ingredients on the packaging. Schlitz thought drinkers would not be too impressed to see 'silicon gel' mentioned on bottles and cans, particularly as competitive products such as Anheuser-Busch's Budweiser did not use that ingredient.

So the brewing scientists came up with an alternative stabilizer, one that could be filtered out of the final product before bottling or canning, and that did not have to be listed as an ingredient. Unfortunately this new anti-haze agent, called Chillgarde, reacted in bottles and cans with the foam stabilizer that was also used, and caused protein in the brew to 'settle out'. At best, this looked like tiny white flakes floating in the beer. At worst, it looked slimy and generally disgusting!

For some time the firm kept quiet about the problem. The flakes weren't harmful, after all, and the problems only occurred at certain temperatures. But eventually drinkers did complain, the issue became more public, and eventually Schlitz had to organize a secret recall of some ten million bottles of beer, costing a small fortune.

It is not unusual for firms to make sequential small changes to products. However, while drinkers might not notice the move from A to B, or from B to C, eventually they pick up that what used to be 'A' is now an 'M', and they realize it is not to their taste. While it is good practice to look continuously for ways of reducing product cost, firms have to be very clear that such moves don't reduce customer perception or satisfaction.

Initiatives and changes that have any chance of affecting the customer must be analysed, researched and tested very carefully. Indeed, the same applies to much internally focused cost-cutting that may not be immediately obvious to customers. It might seem a good idea to find a cheaper firm to carry out cleaning, or technical support, or legal services. But what impact will that have on the business, on staff satisfaction, on efficiency and effectiveness?

If this failure initially concerned a desire to save money, without really thinking of the consequences, Schlitz then made yet another poorly judged purchase. As the firm's market share slipped, a new 'high-impact' advertising campaign was developed. This featured an aggressive-looking boxer who was asked to swap his Schlitz for another brand.[15] He replied, with some vigour, 'You want to take away my gusto?' But instead of being 'amusing', which was the desired effect, viewers found the advert menacing and scary. And indeed it became known as the 'drink Schlitz or I'll kill you' campaign!

It is easy to spend money badly or even waste it completely in areas such as marketing, where the product can be intangible and outputs or results hard to measure. Schlitz illustrated how marketing expenditure can have a positive or, in this case, a negative effect, way beyond the actual money spent with the supplier – the 'spend a million, lose ten million' paradox. Buying well for Schlitz would not have been about finding a cheaper agency. It would have meant finding an idea and a campaign that would get the product back on-track. Unfortunately, their concept and campaign failed to do that.

The consumer reaction and lost sales meant the Milwaukee Schlitz plant was closed in 1981 because of over-capacity. Over eight years the brand value of what was once America's top beer declined by more than 90 per cent, according to analysts, and eventually the Schlitz brand was acquired by Pabst in 1982.

5

How to Negotiate

To many people, negotiation feels like the heart of the buying process – that direct engagement with your potential supplier, haggling over the price and finally shaking hands on what you hope is a great deal. The fact that it has taken until Chapter 5 to get to it highlights that in fact it is actually just one element in a wider picture, but there is no doubt that it is important and can lead to ultimate failure or success in terms of the outcomes from the buying process.

After I left the Mars Group, I worked for a smaller food firm that owned a dozen small to mid-sized businesses. I worked on acquisitions as well as looking to save money through 'group buying'. One firm supplied several of our businesses with plastic trays for frozen and chilled meals, so I wanted a meeting to discuss a central deal for all our firms – with the aim of achieving a substantial discount on the current pricing, of course.

Their sales director came to see me. I didn't quite catch his name on the phone, but he introduced himself as Charles, and after a few minutes asked me where I was from – 'I'm picking up a bit of a north-eastern accent?'

'Yes,' I said, 'born and bred in Sunderland.'

'Ah,' he replied, 'I worked in Sunderland for a few years.'

'What did you do?' I asked.

'I played for the football team. I was Charlie then – Charlie Hurley.'

My jaw dropped as I processed this. Hurley, a skilful and imposing centre-half, was voted 'Player of the Century' by Sunderland fans in 1979.[1] He was team captain and also captained the Republic of Ireland team. Some consider him the greatest centre-half of his generation, and if he'd been born on the other side of the Irish Sea, he would probably have won a 1966 World Cup medal. He was also my boyhood idol, the first person to sign my little red autograph book, when I first stood outside Roker Park players' entrance in 1964 with my father.

When we met, Charles was in his fifties and told me that he now pretty much ran his father-in-law's packaging firm. He'd been more successful than most of his old footballing mates, as well as still being in a happy first marriage, unlike many. Players were not highly paid in the 1960s, and many ended up working as taxi drivers or running small shops or, more sadly, drank themselves to death.

Of course when we got back to business, my whole negotiating approach had disappeared completely. I seem to remember he offered me a 5 per cent 'group rebate' in return for making his firm a preferred supplier, and I accepted, still in a daze.

The message, which I will come back to later, is that personal issues can affect the outcomes in buying negotiations – and rarely in a good way. The best negotiators keep emotion and personal feelings out of the equation, even if they know how to use 'fake emotion' when it is appropriate. But generally if you are looking to negotiate well, try and avoid coming up against your childhood heroes on the other side of the table!

If I Should Fail

Few people or organizations will own up to a failed negotiation; and it rarely blows up into a news story or a public-sector scandal. That's because often only the other party in the negotiation will know whether you failed or succeeded. And strangely enough, not many suppliers are going to tell their clients that they have ripped them off during negotiations and the incompetent buyer is paying 10, 20 or 50 per cent more than they should.

The link between negotiation and understanding what we might call 'commercial models' is also a close one. Failure can be a consequence of poor negotiation, or it may be because the buyer did not really understand the commercial basis of the contract to which they were agreeing.

It can also be difficult to diagnose the difference between 'poor negotiation' and fraud. If you found that a colleague had signed a contract with a supplier where your organization was clearly paying over the odds, there are two possibilities. It might simply be the case that the buyer was naive, didn't understand what they were doing or showed a lack of negotiation skills. Or it might be that the supplier has bribed the buyer to agree an artificially high price (which funds the bribe). The high price might arise because of 'poor negotiation' or it might conceal a kickback for the buyer.

A few years ago a BBC TV programme found at least 169 UK schools that had been grossly overcharged for equipment such as computers and photocopiers.[2] This was hidden in complicated lease agreements, where the commitment turned out to be far greater than the schools expected. Glemsford Primary School in Suffolk was left owing more than £500,000 for 125 laptops, and some schools ran up debts of up to £1.9 million and faced closure as a consequence.

In some sad cases, teachers were fired or had breakdowns because of what they had unwittingly done. There were also cases where the teacher was suspected of being corrupt, taking a backhander from the supplier in return for signing the contract. In most cases, I suspect it was just naivety from the budget holders, a lack of commercial understanding and a limited ability to negotiate well.

I had experience of this in one of my buying director roles. My firm acquired a relatively small business, one that did not have a professional buyer or a lot of finance expertise. We found that they had signed a contract for photocopier hire, based on a cost-per-copy rate-card. The problem was that the first 10,000 copies were at a fair price, but beyond that, the cost per copy suddenly went from (let's say) 1p to 10p.

Of course the supplier had pitched the quantity at a level that they knew any normal corporate copier user would get to very quickly, so our annual bill over a five-year contract was likely to run into tens of thousands, for a machine worth far less than that to buy outright. Fraud – or simply negotiation failure? Whichever you perceive, it was certainly an example of the buyer not understanding the commercial basis of the contract and not negotiating the details that mattered.

It's Too Late

Sometimes it is very difficult to negotiate successfully, because of the position you find yourself in relative to the other party. That might be because of the balance of power related to a position of dependence, as discussed earlier, or because you are trying to open up negotiations again after a deal has in effect been agreed.

Software is a good example where organizations often find themselves in a difficult situation, when it comes to renewing or extending licences or service fees. It is a huge task to replace an enterprise resource planning (ERP) system, for instance, so the supplier knows that their negotiation position is strong. In addition, as Gartner Research vice-president JoAnn Rosenberger pointed out, the software firms have 'highly skilled legal teams, deal-makers and sales teams to negotiate',[3] all of whom are full-time experts in their field, whereas when the deal comes up for renewal, the organization using the software probably fields a negotiator who only does that task once every few years.

One of the key points when going into a negotiation is to understand everything that you want and need, and to make sure that all the points are covered when you reach agreement. There is little chance of success if you look for further benefits once the deal is done. That's clear in our personal lives, too – if you want to get the car salesperson to throw in a free service or a tank of petrol, ask for that before you shake hands on the main deal and agree to pay $20,000 for the vehicle. Once that agreement is made, why should the seller concede more?

This seems obvious, yet businesses and governments do get it wrong. A potentially expensive example of that was a failure by the Japanese government to consider the testing requirements properly when buying two US-built ballistic-missile interceptor stations, capable of shooting down incoming North Korean missiles. Japan budgeted $1.2 billion for the Aegis Ashore hardware, but expected other costs, including construction and operational expenses over thirty years, to make the final cost for two sites some $4.3 billion.

But the additional testing could cost the country anything up to another $500 million. According to the Japan Today website, it appears that the Japanese did not realize that the system would need the real-life testing that manufacturer Lockheed Martin recommends, and that, according to the website, the tests would be held in Hawaii and 'would cost about $100 million per launch'.[4] Military sources in Japan said that the money had not been budgeted for, and the theory is that the defence minister at the time of the deal in 2018 thought that computer simulation testing would suffice.

Being wise after the event, it's clear that the military buyers should have included this in the core negotiations with Lockheed. However, there is still hope. The report also suggests that the contract has not yet been signed, so there may be a chance to open up negotiations again, and there do appear to be some alternatives to the Lockheed offering, so that may help them reach a satisfactory deal.

But the point to note is that the Japanese would have got the best possible deal if a single negotiation had included all the cost aspects, including testing as well as the up-front costs. That is true whatever you are buying, and whether you are spending thousands or billions. Plan ahead and negotiate everything you can while you still have some real leverage and power, because:

1. If you don't have any real power, and the other party knows that, then really you're f*cked.
2. It is rare that you can open up negotiations again, once the deal is done.
3. It doesn't matter how good a negotiator you are – or you think you are – there will be times and situations when you can't win.

On that last point, it is strange that in over thirty years in the buying world, I have met hundreds of brilliant negotiators and pretty much no incompetent negotiators. So they told me, anyway. It is a bit like driving a car – studies have shown that 80 per cent of drivers consider themselves 'above average'. I'm sure it is the same among business folk assessing their own negotiation skills, with a particular bias towards men overestimating their own ability. Perhaps executives, particularly in large organizations, feel they are in a naturally powerful position because of their 'buying power', so that means negotiation success will follow naturally. But it doesn't. And most organizations spend more on training their sales executives in negotiation skills than they invest in those who buy on their behalf.

Stiff Competition

There are many valuable negotiation techniques, but the most fundamental point is to understand the strength of your position before you sit down with the other party. And as a buyer, that usually means ensuring that you are negotiating in a situation where there is genuine competition.

When I was running the Spend Matters UK/Europe website, and writing articles about buying and procurement every day, I got an anonymous tip-off about a UK Ministry of Defence consultancy contract. We found that the MOD had engaged a US consulting firm, Alix Partners, to help them save money (somewhat ironically, as it turned out) on major contracts.

After various Freedom of Information questions, I discovered that the firm was being paid £3,950 (some $5,000 or €4,600) *per person per day*, PLUS an additional 30 per cent 'success fee', based on the 'savings' they identified.[5] But once I got hold of the full contract, I was annoyed to see that Alix Partners could also claim £10 a day 'lunch allowance' for each consultant. You might think that if a firm was receiving $5,000 a day per person from the taxpayer, they could have paid for their own lunch.

My annoyance was because that suggested strongly that the MOD had simply accepted the standard Alix Partners contract – lunch allowance and all. It also became clear that there hadn't been much in the way of a competitive process to appoint the firm; so just how hard had the MOD negotiated? If the MOD had said, 'We won't pay for your lunch; the success fee will be twenty per cent of the day rate, and actually we won't pay more than £2,500 per consultant per day', would the firm have walked away from the business?

It is amazing how much corporate buying takes place without any real negotiation, yet it is almost always an important aspect of the buying process. 'Negotiation' in its widest sense runs through every stage, and the ability to show the supplier that you have competition – that you have an alternative – is core to success. It is a key element to establish your negotiating position, and is what the legendary BATNA is all about.

Developed by the Harvard Negotiation Project and described in the classic book *Getting to Yes* by Roger Fisher, William Ury and Bruce Patton,[6] the BATNA – or 'best alternative to a negotiated agreement' – is your fall-back position or backstop. It is what you will do if your negotiation 'opponent' won't agree to what you see as a reasonable outcome to the negotiation. For example, if you say, 'I won't give you another year's contract to clean our offices unless you reduce your price by ten per cent', your BATNA is strong if you have another supplier lined up already at the lower price. However, if you demand a reduction on the quoted room rate in a hotel, and everywhere else in town is full and it's raining hard, then your position doesn't look quite so good. If they say 'no', your BATNA is a park bench. So your BATNA is critical to the strength of your negotiating position, and therefore to the likely final result.

On the Procurious website, the CFO of a huge global industrial firm tells the story of a monopoly supplier that was critical to the operations and success of his business.[7] The supplier set the price for the item it was supplying, didn't want to negotiate, and knew how strong its own position was in the market. So the CFO's firm worked to see how it could produce internally the same product that it bought from the supplier, and built a detailed business case for that.

Now in truth the firm probably didn't *really* want to invest in this way, but the important thing was to convince the monopoly supplier that it was possible. A stronger BATNA (or at least the impression of one) was being created. The power balance shifted, the supplier realized that if it behaved unreasonably, it could lose all the business, and the tactic worked – the supplier came back to the negotiating table.

Speed It Up

Working in the food industry, I bought basic biscuit flour (among other raw materials) for a couple of years. I was a significant buyer, but not one of the giants in the market; my expenditure was much smaller than that of the bread-makers and the top biscuit manufacturers. But I knew that we got better prices than most of those bigger users. Why?

Simply, I could move faster, be more entrepreneurial and take more risks. If you need hundreds of tons of flour every day, you can't take supply risks. You must have guaranteed supply, you need to lock it in some way ahead, and you may well be buying 50 per cent or more of a flour mill's entire output. The mill cannot supply you on a marginal pricing basis when their whole business depends on you. As a colleague pointed out to me, 'If you buy a supplier's entire output, you must be paying the average price.'

Instead, I bought tactically and aggressively. Following the appropriate checks on quality and the companies themselves, we developed a long list of approved suppliers. I then ran a quarterly price competition (before the days of 'electronic auctions', but the same principle). We bought enough quantity to be interesting to suppliers, but not so much that they couldn't afford to offer us marginal volume at marginal pricing. There are similar stories in other markets; for instance, the group of airlines mentioned in Chapter 4 that failed to get a great deal from hotels at an airport because their combined volume was just too great.

Another Brick in the Wall

The earlier case study about UK schools grossly overpaying for equipment highlighted the difficulty of knowing sometimes whether you are looking at supplier fraud, poor negotiation by the buyer or collusion between the buyer and seller to inflate the price for mutual (corrupt) benefit. It can be hard to identify the truth from the outside.

In 2019 a report from the BBC described how the Kenyan government paid some 3.4 billion Kenyan shillings ($35 million or £25 million) for a fence made up of chain-link, razor wire and concrete poles.[8] The fence is just 10 km (6 miles) long, but the government had originally promised in 2015 to build a complex wall to run some 700 km (435 miles) along the border. That would be designed to stop al-Shabab militants from crossing into Kenya from Somalia, but the wall so far is only a 10-km wire fence.

At a cost of some $3 million per kilometre, that certainly seems to be an expensive fence, and Kenya's parliament has now suspended construction. Not surprisingly, it is also demanding an investigation, and we wait with interest to see exactly where the money has gone. The fence has been built under the supervision of the country's military, and they have not yet disclosed their accounts or details of the construction costs. The Ministry of Defence did originally estimate a cost of 8 billion shillings for the whole project – so more than 40 per cent of the budget has been spent on 1 per cent of the fence.

Perhaps this has demonstrated deeply incompetent negotiation, or maybe in the second edition of this book the story will have gravitated towards the 'fraud' section. But most governments have similar case studies of money being wasted, and one reason is a lack of focus on, and skills in, negotiation. That is partly driven by the idea that negotiation doesn't matter in public-sector buying. After all, formal tendering processes tend to be the rule in the sector – and many interpret the regulations (including in the EU) to mean that buyers can't negotiate anything much post-tender.

That is true, but only up to a point. To guard against corruption, there often are some constraints on how and when you can negotiate. But in my time working with public bodies, every major contract required *some* negotiation after the formal part of the process. That might be because certain elements weren't completely tied up in the tender process; or I might have wanted to try and improve the terms slightly (all positioned as 'clarification' rather than renegotiation, of course, in order to stay legal).

But even more clearly, once the deal moves into the contract-management phase, there is *always* negotiation. Something changes, there is a problem, some new service is required, or the supplier comes up with a better way of doing things and a negotiation ensues. So if you're in the public sector, don't believe anyone who says these skills don't matter.

Our House

Commercial naivety and poor negotiation skills can cost the taxpayer huge amounts of money. In 1996 the UK's Ministry of Defence sold 999-year leases on around 55,000 properties on its estate that were used to house married military folk, as well as more than 2,000 surplus properties. It then rented them back on long leases from Annington Property, which won the MOD contract.

A National Audit Office (NAO) report in January 2018 laid out failings in terms of the buying and contract-management process.[9] The department's own calculations suggested that retaining ownership would be cheaper – but for fairly nebulous 'policy benefits', the sale went ahead anyway. It then made very cautious estimates about future house-price inflation and failed to build any mechanisms into the contract to claim a share of windfall gains. Of course house prices rose faster than the MOD's cautious model, and the rate of return for Annington and its investors has been far higher than expected.

The NAO identified other problems – for some reason, the MOD retained responsibility for maintaining the property, which it hasn't done well, and there has been little collaboration between the MOD and Annington to seek further benefits. Overall, it's an example of failure that could comfortably sit in several different chapters here, but a lack of commercial understanding and negotiation skills in the MOD were certainly among the issues; the NAO report estimated that the Ministry of Defence would have been between £2.2 and £4.2 *billion* better off if it had retained the estate.

We have also seen politicians wanting to get involved in negotiations in quite a hands-on manner, not always happily. One government minister in the UK a few years back took a keen interest in government buying matters. As part of a savings drive, he pulled one of the biggest software firms in the world into his office for 'tough negotiations'. This firm received many millions a year in licence fees from central government and, after discussions, the minister proudly told all the departments that, thanks to his superb negotiation, he had agreed that they would now all get a 30 per cent discount from list price.

That was fine until departments started being contacted by the firm and, in the case of the larger organizations, being told (to their horror) that prices were going up. Why? Because they were already getting a 40 per cent discount from list price! Maybe that's a warning to politicians to stay out of complex buying negotiations – but whether you agree with that or not, it certainly highlights the importance of planning and understanding the current situation before you even think about a face-to-face negotiation session.

Nothing Else Matters

While this is not a 'how to negotiate' textbook, let's run through a few basics, in the spirit of avoiding failure. We talked about the BATNA concept earlier, and that broadens out into the importance of planning before face-to-face negotiation. Understand the market, your own situation (including your BATNA) and the other party's situation, too. In addition, here are three more vital points to consider; they are relevant to anyone who has to conduct business (or, indeed, personal) negotiations of any kind.

Don't take it personally – in business negotiations, don't get hung up on the people involved, your personal pride or your status (as in my Charlie Hurley story). Look on this as two parties coming together to solve a business problem: that is, reaching a satisfactory agreement for the purchase. You can be tough, but you should *never* be personally abusive or insulting. And in business, very few negotiations are pure one-off bartering. It's not like buying a carpet in the souk, where you will never see the trader again in your life. In business you tend to work with people after the contract is agreed, and you may well need their support at some point. If you called their CEO a 'f*cking idiot' during the negotiation, you can guess how they will respond if you need their help later on!

Be creative – the classic task in negotiation courses is to ask two people to share an orange 'fairly'. They end up halving it, of course. But if one really wants the orange for the zest (which comes from the skin) and the other wants the juice . . . then both can have, in effect, the whole orange. Understand what the other party really wants, and think about options and creative ideas for the negotiation. One trick is to find aspects that the other party values more than you, which you can trade for benefits you do care about. For instance, suppliers often value highly your endorsement, or being able to use your organization as a reference when they're trying to win other contracts. That costs you nothing – but has a value to them. You can trade that for a longer warranty period, better payment terms, maybe even a price reduction. Or if your organization is cash-rich, very prompt payment may be worth a lot to a cash-starved supplier.

Try to be objective – determining what is a 'fair price' is rarely easy. But if you have evidence, your negotiation will be smoother and more successful. 'Your price is too high' might work fine as a negotiation stance. But 'I've benchmarked your price against two databases, done my own analysis of what I think it costs to make this product, and I've got prices from two competitors. I do want to work with you, but all of that suggests you are still twenty per cent above a fair market price' is much more powerful.

Negotiation is a fascinating topic, and as well as the classic books, I'd recommend looking at the latest thinking in behavioural psychology from Nobel Prize-winner Daniel Kahneman and others, in books such as *Thinking, Fast and Slow*.[10] Their work has increased understanding of how issues such as *priming* and *anchoring* affect our negotiations. I was taught years ago that the first offer in a negotiation could set the tone – so if a realistic price might be around £100, offering just £50 might reset the seller's expectations. I always had my doubts about this, because you can look stupid if you make an unfeasibly low offer. But the psychology of 'priming effects' suggests there may well be something in this tactic after all, if used appropriately.

Makin' Plans

Finally, training in this field will pay off for anyone who has to represent their organization in serious negotiations. We wouldn't run a marathon without any training or expect to play a Chopin Étude without a few piano lessons first. But too many businesses simply assume that anyone senior can walk into a difficult negotiation and come out with a good result. They won't.

Remember the importance of planning – the most important element of most negotiations is not the face-to-face (or voice-to-voice) element, but what you do before that. Relying on some intangible 'natural negotiation ability', as many business people seem to do, is pure fantasy. Make sure you understand the commercial details of what is being proposed by the supplier. Be a little sceptical as a starting point, and if a deal looks too good to be true, it probably is – as those unfortunate teachers found out when they were ripped off by unscrupulous sellers of photocopiers and computers.

Planning also means understanding everything you want to get out of the negotiation before you start, as well as working out what your BATNA is and how you will react if you can't get exactly what you want from the other party. Then there are those key elements around objectivity, creating options and not taking things too personally, as well as avoiding negotiations with your childhood idols. Luckily, that doesn't happen too often.

One particularly critical element that often needs negotiating in more important contracts is how the supplier is going to be incentivized. Whether the incentives are positive or negative, they must be constructed carefully to avoid another category of buying failures, as described in the next chapter.

6

Understanding Incentives

Incentivization should be an academic subject in its own right. In its broadest sense, it includes aspects of psychology and human behaviour, economics, law, social sciences and politics, and we're all driven by incentives in our personal lives. We get tax incentives to encourage us to behave in certain ways: to give to charity or to save for a pension. Marketing relies on incentives: 'Buy today for a 30 per cent discount!' The whole justice system is based on incentives to behave well (or disincentives to behave badly).

The same principle applies to businesses and government organizations. Their actions and behaviours will respond to incentives – commercial or regulatory – so if society wants industrial firms to stop polluting rivers, for instance, it needs to incentivize them in the right way. Otherwise, the danger is that shareholder value acts as a prevailing incentive, which can work against the wider public interest.

It's no surprise, therefore, that a whole category of buying-failure issues arise when buyers don't incentivize their suppliers properly or carefully. That usually means the contract is not constructed thoughtfully enough, and many issues relate to how the supplier can make money. If the supplier can increase profits by taking a certain action, there is a high probability that will happen – even if it isn't good news for the buyer.

Take Good Care of Yourself

Objective evidence suggests that the US has an expensive and inefficient healthcare system. The Bloomberg 2018 study into healthcare efficiency ranked the country forty-fourth out of fifty-one nations, with the most expensive system of all, yet one that achieved results in terms of life-expectancy that were only average.[1] Countries including Spain, Israel, Norway, Australia and South Korea spend around half of the US figure (as a percentage of GDP) on healthcare, yet achieve better results overall; the only consolation for the US is that Russia is ranked even lower, at fifty-first.

Healthcare is made up of a vast supply network, with involvement from many huge commercial firms, not-for-profit organizations, smaller businesses and self-employed professionals, as well as individual workers, while patients sit at the beginning (or end) of the various processes. Entire books have been written proposing different solutions to improve the performance of the US system, but one factor that certainly comes into play is the way various parties are incentivized.

For instance, pharmaceutical firms charge more for their products in the US than virtually anywhere else on Earth. That's because in most countries the government gets involved in some way, in terms of negotiating the prices that will be charged and paid for drugs. Not in the US. Firms don't face such a strong national buyer there, and hence they can generally charge higher prices to the many buyers (such as individual hospitals). They justify this by saying that the US is subsidizing other nations and, in effect, paying for the pharma firms' global research and development efforts, with American patients (in theory at least) getting the benefit of first access to wonderful new drugs.[2] This comes down to the governments in other countries using incentives to achieve better prices from the firms, while the US chooses not to.

But there are even more direct examples of perverse incentives in the system. A paper published in 2019 entitled 'Waste in the US

Health Care System' estimated that 25 per cent of the US spend was 'wasted' – a massive amount, between $760 and $935 billion every year.[3] Now there are many reasons for this waste, not all of them related to buying or incentivization. But too often hospitals or doctors have their own financial incentives to use higher-cost services or products, or even to recommend unnecessary tests for their patients.[4]

As *The New York Times* asked, might you be tempted to choose a drug costing $2,000 for your patient rather than a similar product costing $50, if you personally made an additional $117 for choosing the former?[5] Many buying decisions in this system are made for reasons that simply don't reflect positive value-for-money outcomes for both patients and the ultimate 'buyers'.

Call Me Maybe

I saw an interesting example of how contractual terms can drive the wrong supplier behaviour when working with a firm that issued hundreds of invoices to a public-sector organization every month. An option for reducing the invoicing burden was developed, moving to electronic communication, which would save the firm and the customer a huge amount of work.

But the accounts-payable department at the government body wasn't interested. Eventually the supplier realized that it wasn't talking to its actual client, but to an external service provider that ran much of the client's finance function, including management of invoice payments. And why weren't they interested in saving time and money? Because their contract was based on a fee paid *per invoice processed*. Their incentive was therefore clear – maximize the number of invoices that needed to be handled.

There is no added value in processing invoices, so clearly the incentivization for that supplier should have been to operate efficiently and accurately, while working with their client and its suppliers to *reduce* the overall number of invoices handled. That was eminently possible by re-engineering the process in line with developments in automation. But if the contract incentivizes the processor to handle as much paper as possible . . . that's what they will try and do.

A similar example comes from England's second-largest city, Birmingham. Capita plc has run a number of services for the city, including a call centre, handling questions and complaints from local citizens about public services. Now logically, the fewer calls, the better for Birmingham, because that would mean the city was keeping its citizens and customers happy. However, the BirminghamLive website, commenting on a report outlining issues with the council's call centre, said this:

> But the report still raised concerns that Service Birmingham, run by private sector firm Capita, is paid on the number of calls it answers rather than the quality of service provided, prompting fears that they benefit from people calling back with the same unsolved problems.[6]

It's easy to spot the problem here: there is no incentive for Capita to solve a problem at the first time of asking. Indeed, the incentive is to create ongoing problems and to generate more calls and therefore more income! Now, I wouldn't suggest for a moment that Capita would deliberately boost its own income by nefarious means. But the city would love to see calls to the centre *decline*, as a sign that problems were being resolved sooner or not occurring at all. Yet the contract has been set up in such a manner that those objectives are not reflected in how the supplier is incentivized, which almost guarantees the wrong outcome for the buyer.

Legalize It

Sometimes the examples of perverse incentives are fascinating. Various approaches have been taken to discourage Colombian farmers (ranging from single families to major 'businesses') from cultivating coca, the raw material for cocaine. One approach is to use planes to spray crops with poisons to kill the plants. Another less environmentally damaging approach is to pay farmers for cultivating other crops, either through direct payments or by guaranteeing a decent price for buying their produce.

However, analysis from the LSE (London School of Economics) in 2019 suggested that this second incentive had a strange effect between 2014 and 2018, when it led to an *increase* in the amount of coca being grown in the country.[7]

Why? Quite simple, really. The government was negotiating with the revolutionary FARC delegations in 2014 after the vicious guerrilla war in the country, and as part of the peace process announced that there would eventually be a 'Comprehensive National Programme for the Substitution of Crops for Illicit Use'. This programme would attempt to reduce coca-growing by providing incentives for other crops, and the announcement led farmers to believe that they would eventually receive direct cash in return for giving up coca-growing. But the announcement seems to have been a mistake, in terms of incentivization.

The LSE report explains that by mid-2019, 100,000 families had enrolled in the scheme, and 30,000 hectares (74,100 acres) of coca had gone, so the eventual programme may well in time be judged a success. But the analysis showed an *increase* in the amount of coca cultivated in the years between the announcement and implementation of the scheme in 2018. Farmers rushed to grow *more* coca for a few years, so that they could then maximize the payments they got for switching!

All very logical for each individual, even if the wider effects were negative, and a great example of how carefully incentives must be used. Again, it highlights that people and businesses usually think and act logically; it is the perverse incentives, as in this case, that can lead to strange outcomes.

Supply & Demand

In a very different industry, a seemingly unrelated problem had its roots in the same issue of misaligned incentives. Back in 2001 Cisco, then the world's largest telecoms equipment manufacturer, announced a write-off of $2.5 billion worth of surplus raw material and components, leading to a 6 per cent fall in its share price.

Analysts suggested a number of causes for this failure, but as the *Harvard Business Review (HBR)* reported in 2004, the primary driver was the way that Cisco's suppliers were incentivized to behave.[8] Many of Cisco's products were in effect manufactured by these third-party supply-chain partners. But the contractors built up stocks of semi-finished products because they had learned that demand for Cisco's products usually exceeded the available supply. 'They had an incentive to build buffer stocks: Cisco rewarded them when they delivered supplies quickly.'

Contractors had nothing to lose by building additional inventory, and did so without really considering Cisco's needs, or assessing the risk, if there was a demand downturn. When that came, Cisco could not turn off the supply tap quickly enough. Now if the firm had put in place watertight contracts, making it clear that suppliers carried the risk if they chose to build stocks, that would probably have protected Cisco. But the situation wasn't clear, and suppliers apparently believed that Cisco had agreed to buy everything they could produce.

Clearly the suppliers didn't deliberately seek to hurt their customer. In the longer term, no doubt they suffered too from Cisco's problems. But this is a classic example of suppliers acting against the interest of their customer. A supply chain works well when everyone is pulling in the same direction, but as the *HBR* said, 'if incentives aren't in line, the companies' actions won't optimize the chain's performance'.

Get the incentives wrong, and the problems resulting from the supplier's behaviour might be excess inventory, supply failures, additional costs, poor customer service or even reputational damage. In terms of the final point, buyers who incentivize suppliers purely to drive down costs bear some responsibility for failures in many corporate social responsibility areas, from human-rights violations to environmental damage and pollution. Incentives in this field are incredibly powerful and can have globally significant consequences.

Money for Nothing

However, if these seem like obvious errors, and give the impression that incentivizing suppliers is easy, think again. Once you get into anything beyond basic commodity-type buying, it is a challenge to get the incentives just right. Take that call-centre example – if you don't pay based on the number of calls, how do you structure it? Do you want the call handlers to take as long as it needs to resolve a call, and risk them spending hours on the phone with the occasional idiot caller? Incentivize them to keep the calls short and reduce running costs, and then callers may get annoyed. It is not easy to design the right system.

Thinking about construction projects is another good way of illustrating that problem. Let's say you are having an extension built on your house – not very different (other than in scale) from a firm building a new office or factory, or a local public body looking at a new school building. How are you going to contract with the builder, and how will you incentivize them to work in a manner that achieves a good end result at a fair price?

The first thought might be to look for a fixed-price contract – used in many situations, including construction at all levels. You might talk to a few builders, get quotes, choose a favourite and agree that the extension will cost £50,000 or $50,000. That's an all-in price. But how will you make sure the work is finished before winter sets in? Perhaps you make 10 per cent of the fee conditional on the work being finished by 31 October.

However, there are a few issues with how the builder is incentivized. The firm gets 50K for the work, but what about quality? You've arguably given the firm an incentive to use the cheapest possible materials, to boost their profit. If they dig the foundations a little less deeply, they save on labour cost. And that incentive to finish – might that make the final work rushed, if they are behind schedule? And of course, if they miss the 31st, there's no incentive for them to push on and finish quickly. Okay, so to handle the quality issue, you might agree the specification for the materials in great detail. But then you will have to check that is what they buy . . . and there's no incentive for the builder to look for better materials, either.

Fixed-price contract arrangements have even been blamed for construction firms going out of business – that was one of the claimed causes for the bankruptcy of the huge construction firm Carillion in the UK in 2018, for example. There were reports, including in *The Guardian* newspaper, that the firm had problems with two hospital building contracts and misjudged the difficulty of building a road in the north of Scotland, underestimating the effects of the weather and the ground conditions.[9] (How you can forget the weather issues when looking at working in the Scottish Highlands is a bit of a mystery, to be honest.) But the firm was stuck with heavily loss-making fixed-term contracts like these.

There are alternatives to fixed-price contracts, though. How about 'time and materials'? In that type of agreement, the builder keeps a record of all materials they buy for the project, and the time that staff – bricklayers, carpenters, labourers and the like – spend on the work. The buyer agrees to pay those actual costs, plus some sort of margin to cover overheads and profit. Traditionally, many such agreements were based on a 'cost plus' model. So you might agree to pay your builders all their costs, plus 20 per cent on top.

But you can see the incentivization problem here. Not only does the firm have no incentive to buy bricks as cheaply as possible, but they actually have an incentive to spend *more* on material and to make the work go on as long as possible, as they recover all those costs, plus 20 per cent on top of that! You could put a cap on the profit/overhead element, but that doesn't fully address the incentivization issue on the materials or labour.

If it is a struggle to work out how to do this well for a house extension, consider the difficulty of coming up with the right mechanism for building a hospital, a power station or a car factory. In the construction world this leads to many different ideas and options for contracting – target pricing, shared risk-and-reward models, profit linked to KPIs (key performance indicators). I'm not going to explore those here, but it all demonstrates the difficulty of aligning incentives.

The Road to Nowhere

The challenges inherent in putting together effective contracts for complex construction-related services were discussed in a 2019 study of road- and bridge-maintenance contracts in the Netherlands.[10] The published paper looked in depth at how the nature of the contracts put in place by Rijkswaterstaat (RWS), the Dutch public agency responsible for roads and waterways, affected the performance and ways of working of the suppliers.

The idea of these performance-based contracts was that suppliers would behave in a certain manner and carry out their work well, because of the positive and negative incentives contained in the contracts, as well as through effective contract-management techniques and processes.

However, the authors found that in many cases 'incentives encourage undesirable behaviour', because contractors 'constantly make trade-offs between costs and returns' – as every business and individual does all the time. In some cases the contractor was not incentivized to carry out preventative maintenance, for example, as they weren't specifically paid for that, and the cost of major failures was picked up by the government.

In another case, problems on a key bridge caused huge traffic jams for Dutch travellers. That was caused by a contractor failing to carry out sufficient maintenance, but the contract had no financial penalty for that sort of delay, whereas other failures did incur penalties. So the contractor would prioritize their work based on the return or the potential cost to them. While the authors don't pretend to have any magic solutions to these tricky issues, they offer sensible advice in stating that linking payment to performance is no guarantee of getting the best supplier performance; and that buyers 'should not shy away from penalties for underperformance'.

Money Trees

I'm chiefly talking here about incentives applied at the corporate level, but many issues (even in business) come down to individual incentivization. Which leads to an interesting question: is it a buying failure if I accept a stupidly good deal that a supplier offers me, all because of incentives in their sales team?

A few years back a large international financial-services firm went through a lengthy buying process to appoint a supplier for all their telecommunications business – many, many millions of pounds and euros a year, as you can imagine. A well-known firm won that bid, and the buying director was about to offer them the contract, when BT, which had lost the bid, called him. They offered him a simply amazing deal – 'it was just ridiculous,' he later told me. He was confident, from his own analysis, that BT could not possibly make any money on the deal, but that wasn't his concern. The firm was financially stable, so it was most unlikely to fail, and so he saw little risk in accepting.

The postscript to that is the interesting aspect. Around three years later, in 2009, BT announced huge losses in their global services division, pushing the firm into a pre-tax loss of £134 million and wiping more than a billion pounds from shareholder value.[11] 'Unprofitable contracts' with corporate clients were blamed.

Why had BT's executives been so keen to win business? *The*

Guardian newspaper pointed out that the previous CEO of BT was able to bank sizeable bonuses, partly on the back of revenue growth in that global services division.[12] But there was no clawback when BT had to write down the value of that operation, once the profitability of contracts became obvious. It's all about the incentives, yet again.

But back to buying: should my friend have accepted the BT offer? That comes down to risk management, and the danger that suppliers who take on unprofitable work then look to cut costs and service to the customer. But if he was sure that he could manage BT's performance to an acceptable level, and the firm was not going to go bust and leave them in the lurch, then I'd argue that he did the right thing.

Even incentives for buyers can rebound on organizations, and in too many cases buyers are incentivized primarily by 'savings' they make. Now that is a reasonable goal for some goods and services that organizations buy, but not many. You can always find a cheaper management consultant, but the CEO might want McKinsey or Goldman Sachs advising them on their M&A strategy, rather than an unknown firm offering a special cut-price fee 'for one week only!'

On that front, I've seen buyers who only get involved at the end of the buying process, then are incentivized to simply try and beat the supplier down on price. One firm had negotiated a contract extension for a key piece of software with a large financial-services firm. The software was critical for the firm, had been used successfully for some years, and all the details were agreed with the budget holder and senior business user of the software. The supplier hadn't looked for a major price increase or anything dramatic, but then the managing director of the software firm got a call from a purchasing manager at the bank.

She was based in the Far East too, as it happened, thousands of miles away from London where the deal had been done, which didn't help, as much of the purchasing activity had been 'offshored' some years previously. She appeared to know nothing about the detail of the contract and the history between the firms, but simply demanded a substantial price reduction. There was no presentation of benchmarking or price comparison with other options, either – something that might have given some substance to the demand for savings. When the MD refused, 'She got quite abusive, started swearing at me,' he later told me.

In the end he approached his senior contact at the bank – the internal user of the software – who told the purchasing lady to back off. When the MD told me this story, I explained that the buyer was almost certainly incentivized purely by a 'savings' target, and her role was simply to try and bully suppliers into giving her that at the end of the process. In my opinion, an example of very poor practice, which did nothing for the credibility of the supposedly professional buyer, her department or the wider firm.

Money Changes Everything (proceed with care!)

The overall message here is *not* that incentives don't work, when it comes to buying and suppliers, but that they must be used with care, thought and caution. They can be useful to encourage suppliers to behave in the manner you want them to, and to nudge them into giving you great products and services.

But they must be carefully aligned with what you want out of the contract. Remember, people and organizations will respond in direct ways to incentives – financial ones in particular, but others, too – and that response may not be what you wanted or expected. Suppliers will focus on what matters, so if you target A, B and C, and provide rewards in those areas, don't be surprised if X, Y and Z get ignored.

Any incentives – positive or negative – must also be appropriate and proportionate to the results or behaviour to which they relate. On a multimillion contract, an incentive of a few thousand doesn't really mean much. Equally, offering a supplier a bonus of one million for rapid completion of a 200K construction project would be unwise, to say the least.

Finally, watch out for the way your own staff are incentivized, whether they are on the buying or selling side of the deal. The examples here show that the wrong incentives for individuals can lead to actions and behaviour that can be costly or embarrassing for the organization. However, sometimes you don't even need perverse incentives for suppliers or buyers to behave in a manner that might simply look . . . stupid. That's the sort of issue I'll consider next.

7

How Not *to Be Stupid*
(particularly if you're a politician . . .)

Chris Christie was a sometimes divisive New Jersey State governor in the US from 2010 to 2018, and his exit from office generated more controversy when (according to documents obtained by *The Record* and NorthJersey.com) his portrait, painted to commemorate his political service, cost $85,000.[1] That represented more than the combined cost of his three predecessors' pictures.

The artist was a well-regarded Australian, Paul Newton, who has won the prestigious Archibald Prize and painted both the Duke of Edinburgh and Kylie Minogue. (Not together, we should quickly add.) So maybe $85,000 was a fair price. But was it the best use of taxpayers' money? Christie had history in terms of spending money. Earlier in his career, as a US attorney, he was one of five federal prosecutors that the Department of Justice said 'exhibited noteworthy patterns of improperly exceeding the government rate' for hotels. That included a $475-a-night stay at the Four Seasons Hotel in Washington DC.

Perhaps there is something about portraits. In the UK the University of Bath paid £16,000 for a painting of Dame Glynis Breakwell, the outgoing vice-chancellor. She resigned in early 2019, after a furore over her £468,000 annual pay package, the highest among UK university leaders.[2] Nothing in the performance of that university seemed to justify such a reward, and once the students also found out about the cost of the portrait, it was hastily removed. The president of the students' union called the cost an 'insult to students'. Like Christie, Dame Glynis also liked her luxury travel, as a Freedom of Information request revealed her use of first-class train and air travel.[3]

The University of Bath did comment that a procurement process for the portrait was conducted and properly authorized. But that doesn't answer the question about whether this was fundamentally a good way of spending the university's cash.

(Sometimes Bad Buying is just) BAD

Previous chapters looked at technical issues in the buying process, but this chapter is all about the more personal side of things. That may be plain and (very) simple stupidity, arrogance, hubris or political calculation. That last driver can look very much like stupidity, but it is subtly different. It can also look suspiciously fraudulent at times, but on occasions it's worth giving politicians or executives the benefit of the doubt.

No one is claiming the two portrait purchases, for instance, were fraudulent in any sense. But to many, they just don't feel right; and other cases in this chapter are similar. Sometimes incompetence is clear, but often there isn't a simple explanation for what went wrong, yet the expenditure doesn't feel like a good use of money – public or private.

Why do politicians, senior executives and government officials sometimes buy in ways that can appear to be wasteful, illogical or plain stupid, to the outside observer at least? I'd suggest there are three major drivers for the failures.

1. **Corruption** – sometimes the behaviour is simply driven by corruption. That includes major and overt cases, such as the vast sums of money extracted from the public purse in many countries by corrupt politicians and officials. But it might also, at a lower and more subtle level, see a

politician or government executive directing a contract towards a favoured supplier in the hope of future employment, in return for hospitality and gifts, or even because they are having a passionate sexual relationship with the owner of the supplying business.

2. **Arrogance** – some politicians do yearn for a lasting monument to their careers, an initiative that will make their mark in political or social history or will perhaps get them re-elected because, as an idea, it looks or sounds good, even if it ends up being a huge waste of money. A 'vanity project', in effect. Private-sector CEOs and chairs are not immune from similar temptations, either. How many mega-acquisitions are driven by vanity and personal gain for the board, rather than a considered view of long-term shareholder value?

There may not be any financial or direct gain to the perpetrator, so this is arguably not 'corruption' as such. But there can be a feeling that 'I'm worth it', and issues such as value for money for the taxpayer or shareholder come second to the personal benefit, even if that is intangible. The arrogance assumes they know best and that the outcome for them personally, or their political party, is worth the cost to the taxpayer or the shareholders, sometimes even if it involves breaking company or government rules, laws or regulations. Whatever the outcome in the complex Carlos Ghosn legal case, hiring Versailles for a party costing €635,000, supposedly to celebrate the fifteenth anniversary of the Renault/Nissan alliance of which he was CEO, but holding it on his own sixtieth birthday, hardly smacked of humility and a deep concern for shareholder funds.[4]

3. **Impatience** – sometimes the politician (or executive) is genuinely frustrated by the formal buying process, the

time that is taken, the constraints on how money is spent and the speed at which change can be driven. So they break the rules or rush into decisions, with admirable motives – the desire to achieve good results and outcomes successfully and quickly. However, because of the way they approach the buying activities, failure becomes more likely. Politicians who are time-constrained, perhaps because of an upcoming election, are often driven by this urgency and impatience. There may be some calculation of political gain, but generally they really want things to work well. But unfortunately their haste increases the probability of failure, such that actions can look like stupidity after the event, even if the motives were honourable.

There can also be a fine dividing line between these factors and the technical buying failures already discussed. Many government IT programmes, for instance, see issues caused by a lack of understanding of business needs, leading to the wrong specifications. But there can be a degree of arrogance too among politicians and officials such as programme leaders. They start with good motives, wanting to improve matters, but are sure *they* understand better than anyone else what is needed. That arrogance leads them to disregard those who genuinely understand the current systems and the limitations on future options.

You're So Vain[5]

The European Union has been accused of funding vanity projects, too, while national and local governments often received the dubious benefits of the projects. A 2014 report by the European Court of Auditors claimed that twenty EU-financed airports in Estonia, Greece, Italy, Poland and Spain misspent large sums of EU taxpayers' money for more than a decade.[6]

The auditors looked at eight airports in Spain, five in Italy, three in Greece and two each in Estonia and Poland. Between 2000 and 2013 some €666 million from the European regional development and cohesion funds were spent on these facilities, with most of the money going on building terminals and runways in airports forecast to attract new passengers. But the projected growth rarely came to pass, and no less than €129 million of the EU money was spent on 'entirely useless' projects.

Kastoria Airport in Greece spent more than €5 million of EU money to build a runway that has never been used by the type of aircraft it was designed for. The airport generated just €176,000 in revenues over seven years, with costs of €7.7 million, and handled only a few thousand passengers a year. The auditors also reported on a cargo project at Thessaloniki airport in Greece that cost the EU €7 million and was standing empty.

Sometimes the national governments chipped in, too. The Spanish government spent €70 million – €12 million of which was EU cash – to expand the runway at Córdoba Airport, projecting that 179,000 passengers would show up in 2013. But there appeared to be little analysis or justification for that figure, and in fact fewer than 7,000 passengers arrived at the airport, as passengers preferred nearby Malaga and Seville.

If I Had a Hammer ...

Sometimes the waste and unnecessary cost aren't driven by any identifiable individuals, politicians or officials. They simply emerge over the years, through the sheer scale or complexity of an organization and the accompanying bureaucracy that develops.

The US military sector is often seen as perhaps the prime example of this, and an excellent 2019 *Rolling Stone* magazine report into US military spending, entitled 'The Pentagon's Bottomless Money Pit', ran through some of the more notable incidents in recent spending.[7] That ranged from a $3.9 *trillion* inventory adjustment made after what was probably a computer error (that wasn't real money wasted, to be fair), to stories of the military paying $436 for a hammer or buying 85-cent ice trays from a 'prime vendor' for $20 a time. Prime vendors are put in place to manage a whole range of smaller purchases or suppliers, which in theory simplifies the buying process, but often seems to lead to considerable additional cost, as the prime vendor adds on their own margins and brings another process layer to the party.

Private-sector firms are not immune from this, either. In many large global corporations you can find local managers complaining about a distant head office imposing 'mandated contracts' or 'preferred vendors' on them, forcing them to buy at prices well above what they could find locally for similar products or services. Sometimes structure, control and bureaucracy can drive waste and be the enemy of value for money.

I'll Sail This (non-existent) Ship Alone

Politics with a capital P (rather than the internal, Machiavellian sort that some organizations display) leads to some interesting case studies, and the UK ferry contract with the firm that had no boats is on that list. In 2018 the UK's transport minister, Chris Grayling, and his team at the Department for Transport (DfT) faced the uncertainties of the UK leaving the European Union. They had to make contingency plans in case the UK left without a deal; if they hadn't, they would have been roundly criticized by press and public. So the DfT understandably wanted to put in place contracts with ferry operators to secure freight capacity in case of a no-deal Brexit. That would allow the government to prioritize the flow of critical imported goods into the UK.

However, the contracting did not run smoothly. A National Audit Office (NAO) memorandum, written to support a government committee that looked into the affair, explained that the DfT approached nine ferry companies and asked for proposals to supply additional freight shipping capacity.[8] But only three replied, leaving the DfT short of their target volume. Two of those were established operators, but the third, Seaborne Freight, was a new company, proposing to run a freight-only service between Ramsgate in Kent and Ostend in Belgium.

Was this a positive example of the DfT supporting a dynamic young business, potentially a brilliant new supplier? The problems came when the press discovered that Seaborne did not actually have any boats. Not one – not even a hired dinghy. Then the owners of Seaborne came under scrutiny: one had previously run a business that had gone under, owing the taxman considerable sums of money.[9] While the owners weren't crooks, it was hardly a track record designed to create confidence.

Then it became clear that the government itself would have to spend £3 million to carry out work such as dredging at Ramsgate before the port could be used. Funding for Seaborne also looked a bit flaky, and there was hilarity in the press when the terms and conditions on the firm's website appeared to be intended for a food-supply firm. Cue jokes about cross-Channel pizza delivery!

The NAO report explained that Seaborne 'failed' the financial tests that the DfT imposed as part of the buying process, because it had no trading record. A technical report from engineering firm Mott MacDonald also identified 'significant execution risks' with the bid – presumably related to the lack of boats. Yet the DfT went ahead and awarded the contract, putting some safeguards in place. For instance, according to the NAO:

> Seaborne had to show . . . that it had detailed business plans; a clear plan for procuring vessels; entered into binding contracts with Ramsgate and Ostend for use of the ports; and contracted with Thanet District Council for the enabling works that Seaborne will pay for.

The government terminated the contract in February 2019 when Seaborne could not meet its contractual requirements because the firm's backer, Irish shipping firm Arklow Shipping, pulled out of the deal. Even that seemed odd, as no one was previously aware that Arklow was involved. But then, in an unwelcome coda to the affair, the UK government ended up paying Eurotunnel £33 million to get the firm to drop a legal case.

Eurotunnel claimed unfair procurement, because they were not considered when the DfT talked to firms about the potential contracts. It seems likely that Eurotunnel would have won their case, and there was also speculation that the government would have had to expose embarrassing details in court. So while Eurotunnel agreed to put some improvements in place in readiness for post-Brexit work in return for the money, the firm was basically bought off.

Then, to add another twist, P&O Ferries complained about the money the government was paying to their competitor, Eurotunnel. That looked like illegal state aid, P&O said. And at the same time, as a no-deal Brexit faded as an immediate scenario, the government had to pay some £50 million to the firms they originally contracted with – Brittany Ferries and Denmark's DFDS – to cancel the deals. It seemed that everything that could possibly go wrong in this case did.

However, it does illustrate the challenges governments face, and of course if something like this had happened to Shell, or Nissan, you would probably never hear about it. There was a political imperative here for the DfT to be seen to prepare properly for a potential UK 'no-deal' exit. There was a shortage of credible bidders, and perhaps Seaborne was worth risking; it didn't, in the greater scheme of things, cost a huge amount of money. But this does emphasize the importance of understanding your suppliers, the people behind them, what they are offering and how they will meet your needs.

The Eurotunnel and P&O costs are less excusable. The DfT should have realized that running a very rapid and probably 'illegal' buying process, in terms of EU and UK law, might upset suppliers who weren't invited into the process. It should have consulted more widely and given Eurotunnel, and perhaps others, the chance to put proposals forward – not doing so was certainly a failure.

You Scratch My Back (and I'll scratch yours)

Transparency International (TI) is a non-governmental organization that campaigns against corruption in government globally. Some years ago, a friend asked me to contribute to work at TI that he was leading. He was looking for expertise on government buying processes to inform the project, which considered corruption in defence and military buying globally. I wasn't a deep expert on military procurement, but I understood public-sector regulations and issues generally.

As well as giving me the chance to go to NATO HQ in Brussels, I learned about the wonderful world of *offsets*. The idea is well meaning, but lies behind much corruption and fraud that we see in the defence sector, and very few have even heard of the concept. To explain offsets, the case of the Indian government and its purchase of Rafale fighter jets from Dassault Aviation of France is enlightening. While there is no evidence of fraud here, it appears to contain some elements of strange buying behaviour.

Originally, some of the 126 jets purchased by India were going to be built as usual in France, home to the firm, but the deal also included Dassault setting up a manufacturing facility in India to build more. However, that commitment got scaled back to just thirty-five jets, built in France at a cost of $11 billion. But as part of the deal Dassault committed to spend $6 billion of their own money in India – a typical 'offset' contract that often accompanies defence sales. Sometimes that offset must be spent locally to the buyer on goods or services directly concerned with the deal (in this case, components for the planes, for example) or it can be a totally unrelated spend.

The country buying the equipment theoretically gets some benefit back from the supplier, but there are problems with offsets. First, bribes can be concealed within the offset amount. Money is often channelled through middlemen or agents, which may be valid or can conceal the fact that it ends up in the pockets of politicians, the government or military staff who had an input in the buying decision.

There is also often a lack of scrutiny in terms of whether the offset money is spent effectively and provides value for money. If Dassault had not needed to agree the offset, would their price for the jets have been $10 billion or $8 billion, rather than $11 billion? We don't know, but neither do we know whether the offset here represents good value for the Indian taxpayer.

In fact there are more twists to this case. Some of the concessions made to Dassault look questionable in their own right. According to an illuminating report on the affair in the *Deccan Herald*, the Indian negotiators agreed that Dassault would not need to provide bank guarantees, and also agreed large advance payments to the firm.[10] Rather strangely, the government also removed anti-corruption clauses from the contract and gave up the right to look into Dassault's accounting records for the programme.

Matters got murkier still when Dassault announced that the offset money would be spent through a joint venture, not with a well-established Indian defence firm, but with a new business, set up only months earlier by an Indian businessman, Anil Ambani, whose Reliance group of companies worked in many areas, but not defence. Opposition politicians claimed that this chain of events suggested corruption – how did Ambani win this business, and is the Dassault money paid to him destined to end up in other pockets? There is no evidence as yet of a 'money trail', but equally there has not been an explanation for how and why Dassault chose this route to deliver their offset commitment.

There is no evidence of fraud here, but as is often the case, greater transparency would have improved the situation considerably. Surely governments should insist on a full explanation for where offset money is being spent, how suppliers and partners who are part of that process are chosen and, indeed, the benefits and outputs from offset spend? There should be a full 'open-book' approach to offsets where governments insist on using them.

Sail Away

The Bundeswehr is Germany's armed-forces organization, and it is clearly not immune to the issues we have mentioned in UK, US and other military operations around the world. There have been regular issues with contracts and suppliers over recent years, and the problems around the *Gorch Fock* – the German Navy's three-masted naval training ship – hit the press in 2018.

The ship, a fine vessel some 81 metres (266 feet) in length, was launched in 1958 to help train West German naval recruits. The name comes from a popular seafaring German author's pseudonym, but the ship is a source of national pride and was featured on Deutsche Mark banknotes, as a symbol of Germany's post-war revival.

The Bundeswehr announced in 2015 that the ship needed a major overhaul, but the cost then escalated from an initial estimate of €10 million to no less than €135 million. Bundeswehr officials claimed the problems with the vessel only became clear when it was in dry dock, but many are querying that analysis, given that the cost of repairs now seems greater than the cost of building a new ship. According to the Politico website, Hans-Peter Bartels, a Social Democrat MP who monitors defence spending for parliament, said, 'Everything takes too long and costs too much money. It's as if time and money were endless resources, and in the end, no one takes responsibility.'[11]

In March 2018 it was announced that the project would continue, given that a new vessel could not be delivered until 2025 at the earliest. The *Gorch Fock* should be back in service in 2019, it was claimed. But by January 2019 reports said that the ship was 'completely dismantled', and the directors of the shipyard where the work was taking place were fired. There were accusations of serious mismanagement in terms of Navy and procurement management, too.

Negative Energy

Earlier we expressed some sympathy for the UK government's infamous 'ferry firm with no ships' contract. However, in this next UK quasi-government-sector example, no sympathy will be offered to the perpetrators of this hugely expensive buying failure.

The case involved a 2016 legal challenge by Energy Solutions Ltd, the incumbent supplier for a huge contract to clean up decommissioned UK nuclear-power stations.[12] They lost the tender, run by contracting authority the Nuclear Decommissioning Authority (NDA) in 2014, to a Babcock/Fluor consortium (CFP). But there were a number of mistakes made during the procurement process.

One related to 'pass/fail thresholds': areas where the NDA defined up front that failure to meet certain conditions would lead to instant disqualification for the bidder. However, once bids were scored, it became clear that one supplier had failed to meet the threshold. But instead of chucking them out of the competition, the NDA decided to let them stay. Now this may all seem a little technical, but it is clearly unfair; and public procurement regulations really don't like unfair buying processes.

As the judge said in his statement, you can't change your mind about the rules once you get into the buying process. After a bidder has failed to meet a defined threshold, you can't ask, ' "Was that threshold requirement really that important" ', arrive at the conclusion that it was not, and then use that conclusion to justify increasing the score to a higher one than the content merited (or to justify failing to disqualify that bidder).'

To disguise the failure of that firm, the NDA team also adjusted original scores given to the bidders during the marking process. But they failed to provide any audit trail or justification for these changes, a fact that became obvious through the trial. The NDA announced that CFP had won – which promoted the legal challenge. There were other issues, too, and the final outcome saw the judge finding in favour of Energy Solutions, and the NDA agreeing to pay the firm (and their consortium partners Bechtel) almost £100 million to settle the legal claim for their loss of profit on the contract.[13]

It is impossible to know what went on behind the scenes in cases like this. Was it sheer ignorance of the rules? Was someone very senior determined that a particular supplier should or should not win the contract? With other failures in previous chapters, a lack of understanding or knowledge caused the problem, but I'm left somewhat baffled here.

Certainly a number of basic buying principles seemed to be forgotten. Treating bidders *fairly* is a good principle, whether you work for a government body that must do that legally, or for a private firm. Keeping sensible documentation to explain your decision is vital. That's so that you can explain to bidders why they won, or didn't; but it is also a basic precaution against corruption and fraud, one that all organizations should take. If no one can explain logically why my firm won a particular contract, then maybe it was because of the bulging brown envelope I was seen handing over to the senior buyer.

Hide & Seek

Public-sector procurement relies on fairness and transparency. That is essential, so that every bidder feels it has a fair chance of winning work with government, based on value it can provide, rather than on factors such as personal contacts or, indeed, bribery and corruption. Countries where public procurement is endemically corrupt have big problems, so sensible administrations put processes and rules in place to try and keep public procurement fair and objective.

Even when corruption isn't involved, though, failure and waste can be driven by political issues, pure stupidity, vanity or arrogance. Another example of that vanity, but organizational this time rather than of individuals, seems to have hit the buying of uniforms for the US Army. According to reports, the Army replaced its 'Universal Camouflage Pattern', which was developed at great expense (some $5 billion, reportedly), after just ten years of use.[14]

The UCP had a pixelated design, rather than the traditional 'wave' pattern. The new design was supposed to work in both deserts and other terrain, such as jungles, but it proved poor in both, with tests showing that alternative patterns were 16–36 per cent better.[15] But why was the new design chosen? Critics claim that the Army commanders were jealous of the Marines, who had already chosen a new uniform with the sexy new pixels. Despite a lack of testing, and warnings from some experts, the Army got pixelated too, at vast expense, in order to look 'cooler than the Marines'.

That must be one of the least-valid reasons for wasting taxpayer dollars that I've seen yet. But despite all these examples of waste, bad practice and general foolishness, the cases often take some time to become public, and it is very rare for elected politicians or officials to really suffer when they demonstrate such stupidity. As long as this is the case, these failures will, unfortunately, continue.

In our next chapter, though, after another award-winner, I'll look at cases where it isn't *purely* the buyer's fault. There may be naivety or worse on their part; but in the following tranche of examples, the supplier has to take some responsibility, too.

Bad Buying Award

MOST IMPRESSIVE TECHNOLOGY FAILURE – NHS NATIONAL IT PROGRAMME (NPFIT)

Some observers believe the National Health Service 'National Programme for IT (NPfIT)' is the biggest technology failure and waste of money ever seen in the UK public sector. It certainly saw a mix of arrogance, poor specifications, a lack of understanding of the market, poor project and programme management and more.

The programme started in 2002, with an admirable vision – to use modern information technology (IT) to improve the way UK health services were delivered and, ultimately, to improve the quality of patient care in the system. It aimed to create a national IT infrastructure, some national systems, such as 'Choose and Book' (for making medical appointments), and the EPS (electronic prescription service). There would also be an integrated care-records service made up of two components: the SCR (summary care record) and local, detailed care records.

But a whole series of problems and issues emerged, which led not only to a lack of success from a government point of view, but also to suppliers losing a lot of money in some cases. Following a review by the government's own Major Projects Authority, the government announced in September 2011 that the programme would be dismantled into its separate component parts.

In March 2012 the UK National Audit Office estimated that the costs to date of the programme were £7.3 billion.[16] It put the benefits at that point as around £3.7 billion. The Department of

Health suggested that there would be a further £7 billion of benefits to come, but the NAO thought that was a highly dubious figure, saying that there was 'very considerable uncertainty about whether the forecast benefits will be realized'. It seems likely that the programme therefore had a cost to British taxpayers that ran comfortably into the billions.

As well as laudable aims, the programme had strong top-level support. Tony Blair, the Prime Minister, was a champion, which might seem like good news. But what it meant in practice was that the programme team could ride roughshod over any objections by simply saying, 'The PM wants this to happen.' That is a powerful lever in the government sector, but such a top-down approach always runs the risk of not engaging with people who really know what is needed.

I have some sympathy for the programme staff and buyers involved; trying to drive significant change is difficult in any organization. Doing that in the National Health Service (NHS) – by some measures, the third-biggest employer in the world, made up of thousands of different organizations, supposedly all working together, but in practice having very different aims and ways of working – is almost impossible.

The negativity from many clinicians towards the programme might look like simply resistance to change. But they felt the programme did not take the time to really understand the needs of those who would operate the new systems. Those internal requirements and specifications were not understood properly, in other words. The contracting process was also executed very rapidly (by government-sector standards, anyway). At the time, that was seen as a positive example of good buying practice, but in retrospect it looks more like a reflection on the lack of planning and clarity in terms of what was going to be bought.

Lack of understanding of the market – another classic driver of failure – along with some arrogance, I'd suggest, also came through in the remarks made by Richard Granger, who led the programme. The core contracts for care records were split into five regions to avoid over-reliance on one firm. If one failed, others would step in, which sounds smart, given the dangers of supplier dependence. But Granger likened it to running a team of huskies, where underperformers are 'chopped up and fed to the other dogs . . . The survivors work harder, not only because they have had a meal, but because they have seen what will happen should they themselves go lame.'[17]

The crass nature of that comment did not suggest Granger was seeking the positive, collaborative approach between buyer and seller that is advisable for long, complex contracts. The contracts were also aggressive in terms of penalties and remedies against the suppliers. And the market for huge IT developments does not work like huskies anyway; if one experienced tech firm proved incapable of meeting the requirements, it was likely that others would not do any better, and probably wouldn't go near that work with a bargepole.

The experience also shows that while you can legally 'transfer risk' to suppliers, if contracts fail then there will almost always be implications for the buyer. The supplier may take a financial hit, but it is the buyer who doesn't get the systems (or whatever they're buying) that they wanted.

Once some pilot and proof-of-concept NPfIT systems were produced, it became clear that many front-line doctors and medical staff – the target users of the technology – felt it didn't meet their needs. For example, seventy-nine doctors and other Trust staff from Milton Keynes sent a letter to *The Times* recommending that no other hospital should use it.[18] 'The software is so clunky, awkward and unaccommodating that we cannot foresee the system working adequately in a clinical context,' the letter said.

The whole programme descended into a nightmare for the government and for some of the suppliers. Timescales slipped, and slipped again. The professional-services company Accenture got out of their contractual commitments by paying £63 million back to the programme and paying rival firm CSC to take on some of the commitments, an almost unheard-of happening. It was amazing to see a firm paying a competitor to get out of a contract, but in retrospect it was probably a good deal for Accenture – in 2012 CSC wrote off its entire $1.5 billon investment in the programme.[19]

The reverberations from the programme continued for years, even after it was officially killed off. In 2018 the contract dispute emerging from Fujitsu's removal from one element of the programme was finally resolved. In 2008 the Department of Health terminated the £900 million contract with the firm to run a chunk of the National Programme. Fujitsu had only been paid some £150 million at that point and went after the government for more. Ten years later, reports in the *Health Service Journal* suggested that the government had agreed to pay 'hundreds of millions'.[20]

In time, there were some outputs. An appointments system eventually appeared, along with NHS-wide email and the N3 broadband system. But overall, NPfIT was a very disappointing programme, one that illustrated many important points in terms of buying failure.

8

Trust No One (at least not suppliers)

Back in 2002 a small dog called Lulu became involved in a court case that showed how technology contracts can go very wrong, and highlighted issues around what a supplier claims for their product. The buyer was Sky, the UK's leading satellite TV broadcaster, and the supplier they sued was EDS, the outsourcing-services provider, eventually bought by Hewlett Packard in 2008.

BSkyB engaged EDS to provide a new customer-services system. The project was supposed to run for two years, but there were delays and cost overruns. BSkyB was unconvinced it was getting what it had paid for, and the relationship between the companies broke down. BSkyB terminated the project in 2006, having spent £265 million rather than the £50 million contract value.

The firm took EDS to court, and the judge in London's high court, Mr Justice Ramsey, sided with Sky, deciding that EDS had given 'fraudulent representations' about what the system could do and how easily it could be installed. He held EDS liable for damages and was particularly critical of Joe Galloway, the senior EDS executive who led the sales effort. He 'misrepresented' how quickly the solution could be implemented and exaggerated the system capabilities.[1]

But don't salespeople always exaggerate the benefits? Well, maybe, but this case showed there is a point beyond which that behaviour becomes legally dangerous. The judge said that Galloway had been 'cavalier' in providing estimates of project timing that he knew were not backed by sufficient analysis: 'I am driven to the conclusion that he proffered timescales which he thought were those which Sky desired, without having a reasonable basis for doing so.'

But back to Lulu, and a stroke of brilliance from Mark Howard, QC, BSkyB's barrister. Galloway claimed he had gained an MBA in 1996 from a college in the British Virgin Islands, but it emerged during the trial that his qualification had been obtained from the Internet. Howard illustrated the point by presenting the court with an MBA from the same college – but awarded to Lulu.

'Without any difficulty the dog was able to obtain a degree certificate and transcripts which were in identical form to those later produced by Joe Galloway,' the judge noted, 'but with marks which, in fact, were better than those given to him!'

Brilliant work from the attorney to discredit an opponent, but more seriously, the case also highlighted interesting contract-law issues. For instance, the parties had agreed a cap on liability of £30 million in their contract. But because BSkyB claimed that misrepresentation was to blame, rather than mere incompetence, they could circumvent the liability limits set out in that contract.

Some might argue that this is not a failure at all, as Sky won their case. But if you asked their senior executives, I suspect they would far rather the project had been successful in the first place, instead of having a court case and a delay in getting their system.

I Believe in Miracles

Now that case has not led to a rush to the courts for other dodgy deals, or indeed a rush for doggy MBAs! But it does mean that firms should ensure that all marketing material, sales pitches and bids must contain true, honest and verifiable material, otherwise they are liable to claims of misrepresentation. That is good news, because for those of us who have worked in buying for years, there is a special place in hell reserved for suppliers who exaggerate, mislead or plain lie to us about the capability of their firm, products or services.

That's because not only is it ethically wrong, but it can also lead to uncomfortable situations where the buyer has to explain to their CFO or CEO why the new supplier or product doesn't perform as it should. Whether it's equipment, an IT system or the new corporate-law firm I've appointed, saying, 'The supplier told me it would work fine', after some supplier failure, does not carry a lot of credibility with the boss, as I have found to my personal cost.

Salespeople are paid to present the best possible face of their firm to the outside world – prospective customers in particular – so it is not surprising when they push the truth as far as it will go. They tend to be incentivized to convert interest into hard sales and hard cash, so the temptation to step over the line into territory that can be risky for their own firm, as well as damaging for the buyer, is obvious.

But it is not always the supplier's fault. It is often difficult to assess what a potential supplier can really do, and over the years the introduction of more professional 'tendering' processes by many buyers has had unintended consequences. I've seen tender documents that ask literally hundreds of questions. Not only are they incredibly time-consuming and expensive for bidding firms to complete, but the sheer volume and complexity of the answers means that few buyers check everything that is claimed. And firms employ professional bid-writers, whose job is to make the proposal sound convincing and to score good marks on any assessment system used by the buyer.

The fundamental point comes back to how buyers check that what they're being told by suppliers or potential suppliers is accurate, feasible and will turn into successful contract delivery.

Bulletproof (or maybe not)

In March 2018 the US Justice Department announced that Toyobo Co. Ltd of Japan and its American subsidiary had agreed to pay $66 million to the US government to resolve claims that the bulletproof vests bought from the firm by various US law-enforcement agencies did not work as they should.[2]

The US authorities claimed that the Zylon fibre used in the vests degraded quickly in normal heat and humidity, making them unfit for use. The US also alleged that Toyobo continued marketing the product, published 'misleading degradation data' that understated the problem, and even started a public-relations campaign designed to influence other manufacturers to keep selling Zylon-containing vests. Finally, in August 2005, the National Institute of Justice completed a study of Zylon-containing vests, finding that more than half the vests tested could not stop bullets as they were supposed to.

This settlement was part of a larger investigation by the US into the use of Zylon in body armour. In excess of $66 million was recovered from sixteen other entities (as well as Toyobo), including armour manufacturers, weavers, international trading companies and five individuals. It's worth noting that in this case the claims settled by the agreement were allegations only; there was 'no determination of liability'.

So from a buying point of view, it could be argued that this was at heart an issue of the buyer believing the suppliers' claims too readily. Or poor specifications might come into it – a lack of clarity perhaps on what was required from these products; others might see it as questionable behaviour by various players on the supply side. It certainly highlights the need to test and check the claims made by suppliers, particularly for products as important as these.

Hell, Yes!

There is a natural tendency for any firm to say 'yes' when asked questions like 'Can your product do this?' Or 'Can you deliver by the end of the year?' Problems can occur because of unexpected operational issues that cause delays, or even through 'acts of God', as the insurance policy would say. But in many other cases there has been poor, immoral or even illegal behaviour from suppliers.

Again, the technology sector features strongly in the case-study library. That's because it is inherently complex and fast-developing, so buyers are used to amazing developments and seeing the scope of what is possible expanding constantly. However, buyers do have a habit of asking suppliers to do the impossible. ('Failure is not an option' is a particularly stupid phrase, I've always thought.) Some technology firms have been accused in the past of claiming that their hardware can carry out all sorts of tasks; when problems arise, 'blame the software' is the cry. Of course software firms will apply the reverse argument. And while suppliers may say yes for selfish reasons, it can also indicate an admirable desire to please the buyer. But that can rebound on both parties.

In one major back-office outsourcing programme, a timescale for migration of work into the supplier's shared service centre was agreed. That was part of the initial detailed contract with accompanying schedules, plans and performance measures. But the buyer then put pressure on the supplier to accept an accelerated timescale – both to bring forward savings and to fit with other operational activities.

Perhaps foolishly, the supplier accepted, thinking they were helping out their client, but the inevitable happened. The supplier could not deliver against the revised and accelerated timescales, which caused problems for everyone. And the buyer then blamed the supplier for failing to meet the agreed timescales – targets they should never have accepted.

Put It to the Test

If simply believing the supplier is dangerous, how can you check what their product can do, or review ambitious timescales, or test promises they make? While a determined supplier might be able to conceal their lack of real capability, appropriate due diligence from the buyer in most cases should avoid these issues. Generally this should be done before the contract is signed, and there are three key techniques to avoid being duped or misled – *Analyse, Reference* and *Test*.

Analyse means looking into the firm, the product or service that you're going to buy. Clearly this doesn't matter too much if you're buying a ream of copier paper, or 100 cardboard boxes. But for anything significant, you need to be doing your research on the supplier and on whatever you are buying. The amount and depth of research needs to be proportionate to how much you're spending and how critical what you're buying is. In terms of research, make sure you understand how what you are being offered matches up to your specifications and requirements, as discussed earlier.

Reference means asking other customers of your potential supplier, or users of the product or service you are buying, about their experience. It's an obvious step, yet it is amazing how many organizations don't bother with it. I was asked for input on a legal case in 2018 where an incumbent supplier challenged the decision by a government body to award a contract to several other firms, meaning that the incumbent was going to lose all its business. This was a really sensitive service; if it went wrong, you might well see reports on newspaper front pages.

Yet when the incumbent firm asked questions about how the procurement decision was made, it became clear that the government organization had done virtually nothing to check out what other suppliers were claiming in their bids. They had not researched the track record of the firms; they had not taken up references from other customers; they did not even seem to have checked whether the directors of bidding firms had criminal records. Once that failure was pointed out, the buyer had no choice but to abandon the procurement exercise and withdraw the proposed award of contracts. This was a significant failure – all because of a lack of effort in terms of verifying what potential suppliers were claiming. The buyer was simply believing the bidders and hoping for the best.

Test means using techniques such as pilot programmes or

small-scale roll-outs that enable you to get a sense of the supplier and their capability, without immediately betting the farm on a particular approach. I'm a big fan of testing whenever possible, and there are different ways of approaching this. In a large company, you could run a geographical experiment with a new supplier or product. Give it a try in an area, region, office or factory, rather than moving immediately to handing over your entire business. Or you might initially use a supplier on a relatively unimportant piece of work. If you're engaging a law firm for the first time, don't put them to work on your most sensitive international intellectual-property case, but see how they do (and what they're like to work with) on a smaller matter.

School's Out

Sometimes suppliers make such extraordinary claims for their products that it seems amazing anyone could really believe them. But on other occasions, proposals look credible and it is genuinely difficult to check out the claims. In August 2008, after a high-profile failure that hit the front pages of British newspapers, the UK's Qualifications and Curriculum Authority (QCA), which managed the school 'national curriculum' and associated testing process, terminated a contract it had put in place with ETS Europe to deliver tests for schoolchildren.[3]

The two parties agreed to dissolve the five-year contract to handle the national curriculum tests (SATs) with immediate effect, after delays during 2008. The full results were still not available months after students sat tests. Serious problems with the administration and marking meant the publication of results for key stages 2 and 3, which should have been released on 8 July, were delayed by almost a month. Key stage 3 results were released in August.

At least the QCA did not have to pay to exit the £156 million contract – indeed, ETS made a payment of £19.5 million to the QCA. The firm was apologetic, saying that as a subsidiary of a global, non-profit company, 'we are dedicated to assuring quality and equity for all pupils, and we are sorry that the results this summer were delayed for some schools'.[4]

But how did this happen? The case was analysed thoroughly, notably in the Sutherland Review.[5] What emerged was this central challenge to many buying situations – that is, how the buyer can assess whether proposals can actually be delivered by a potential supplier, even if they sound credible. It is relatively easy to write a convincing proposal to carry out services-type work or even to deliver certain physical items. I might tell you my firm can supply you with the finest cocoa beans, or handle your outsourced pension administration absolutely brilliantly. But how do you *know* I will actually be able to live up to my fine words?

The Sutherland Review found that in many ways the procurement (buying) process in this case wasn't run badly – the authors called it 'sound'. ETS won with the lowest price, but also scored better than the alternative bidder on non-cost factors. The 'Gateway reviews' undertaken by the Office of Government Commerce were in general positive, too. However, the contract and the supplier clearly failed to deliver what was required. Why was that?

Issues were identified by the report around governance, the contract-management approach, some legal issues in the contract and specifications. But the report suggests that the weakness in the selection process came from two key factors. First, the QCA and the consultants running the process did not fully check out the *history* of previous contracts delivered by ETS. That might have picked up warning signals, as there had been issues with contracts in the United States. Basic financial health checks were done, but not an extensive reputational and performance due diligence.

Second, the buying process did not check that the assumptions about capacity made by ETS in their bid were realistic and accurate. The firm should have been challenged more strongly on its staffing plans. There were also concerns about the 'end-to-end' solution proposed and whether the firm really understood how different elements needed to fit together. Those issues appear to have been at the heart of subsequent problems.

This all supports the 'Analyse, Reference, Test' advice. Not enough analysis of the ETS proposal was carried out, and where potential issues were spotted, they were not followed through. The referencing around the firm's previous experience was also inadequate. It is hard to see how the solution proposed could have been piloted or 'tested', given the nature of the work, but in these other two areas more could have been done.

Now it may be that even if these actions had been carried out, ETS would still have been awarded the contract. But at least there would have been more awareness of what still needed to happen to improve the probability of successful delivery. Action plans could have been in place, the contract might have been adjusted, and so on. And our key point for this chapter – that you can't just take on trust what suppliers tell you they will do – certainly applied here.

We Built This City

Many buying processes rely on competitive bidding or tendering from suppliers. Nothing wrong with that – but it can encourage the wrong behaviour from suppliers in a number of ways. One example is bidding an unfeasibly low price to win the deal, particularly where the buying process is focused on a lowest-price-wins mentality from buyers. Suppliers may simply be optimistic ('We'll find ways of doing the work more cheaply once we've won the contract') or think they can get more money out of the buyer once they're in place – 'land and expand', as the consulting-industry phrase goes, or 'make your money on the changes', as construction firms say.

In 2019 Jean Nouvel, a celebrated French architect, started criminal action against the owners of the Philharmonie de Paris, the new concert hall that he designed. He claimed fraud, embezzlement and favouritism, all in response to a 2017 claim by the owners, as well as city and local government, against him for payment of €170 million in damages for budget excesses and delays in the construction.

In 2007 he was contracted to build the auditorium for €119 million, but the final cost was estimated at €328 million by the owners, and €534 million by the regional state auditors (which in itself seems like a big discrepancy). *Le Monde* reported Nouvel saying that the €119 million was quoted purely to match the ceiling set for the public tender, and was not really a genuine cost estimate.[6] He claims that €100,000 per seat was the established cost for similar concert halls, and the €119 million total would have required spending only half that much, so it was never realistic. He also claims that everyone knew the real cost would be much higher – 'This is pretty usual in France in public tenders for cultural projects,' he was quoted as saying.[7] His lawyer also says that Nouvel is being made responsible for failures in project management.

So did the buyers really know the supplier wasn't to be believed in this case, but were secretly happy about that? Or were the cost overruns a failure on the supply side? The outcome of this case will be interesting to observe when it does hit the French courts.

Call the Doctor

This over-optimism from suppliers, allied to a focus on cost from buyers, has led to real problems in the government sector in many countries, with firms walking away from unprofitable contracts in some cases. Serco's contract for providing out-of-hours medical services in Cornwall in the south-west of England is an example of how a narrow focus on costs in choosing the supplier can lead to service failures.[8] Serco undercut the bid from a local doctors' cooperative by £1.5 million to win a five-year contract in 2011, worth £32 million. Other suppliers dropped out because they couldn't meet the cost ceiling set by the buyer, the Kernow Clinical Commissioning Group.

However, once the contract was up and running, Serco failed to meet its performance targets and national quality requirements. The firm struggled to ensure it had enough staff available to deliver the service required, and a whistle-blower revealed that the company falsified performance data. In 2014 Serco agreed to terminate the loss-making contract, two years early.[9]

This does raise some tricky issues, however. It is tough for a buyer, in the public sector in particular, to say to a bidder, 'We're not going to give you the contract because you are too cheap.' Imagine the outcry about the waste of public money if a more expensive supplier was chosen! But that is no excuse for failing to carry out the appropriate due diligence to make sure that bidders aren't making ridiculous assumptions about costs, resources, service levels and delivery.

Ship Ahoy!

Sometimes it appears that the organization wants to believe the supplier, for reasons of its own. In the case of that concert hall in Paris, the architect claimed that 'everyone knew' the real cost would be much higher than his bid, but kept quiet so that the project would look favourable from a budget perspective and could proceed.

The military world seems particularly susceptible to failures of this type, with military folk being keen to have the latest and most innovative, cutting-edge equipment. Often these people on the buy-side seem to conspire with the market to spend huge amounts of money on products that are unproven, or in some cases not much more than a glint in the eye of the engineers.

I'm not talking so much about the research billions poured into invisibility shields or extrasensory perception (as described amusingly in the Jon Ronson book and subsequent film, *The Men who Stare at Goats*[10]). Rather, there are programmes that looked quite reasonable and pragmatic, but were in fact built upon sand. A classic example is the US Navy programme for littoral ships. The vision was for ships that were cheap, flexible, fast, easy to build and capable of carrying and hosting many different types of equipment. Given the level of ambition in the initial brief for these ships, it's perhaps surprising that the specification didn't include 'invisible' and 'able to travel underwater' as well.

After a programme lasting two decades, the US Navy has spent some $30 billion, and by 2019 had managed to acquire some thirty-five ships, of which only six ships are actually in service. The initial cost estimate of $220 million per ship is now heading towards the $1 billion mark.

Now this case really deserves a deeper analysis, but one issue was the decision to appoint two shipyards – Lockheed Martin's facility in Wisconsin and an Austal yard in Alabama – to each build its own variant of the vessel. While I talked about avoiding supplier dependence in Chapter 4, maybe this was a case where the complications of having two variants outweighed the benefits. It's not just the costs, either. The National Interest website describes other problems: 'the lack of combat survivability and lethality discovered during operational testing and deployments, the almost crippling technical failures and schedule delays in each of the three mission modules'.[11]

So the original flexibility goal is now being compromised, as the ships will be built to have a single 'mission' and a crew focusing on a 'primary skill set'. Whatever the other issues, this certainly looks like an example of optimism, probably on both sides of the buying process: military staff who wanted this amazing ship, for all the right reasons; and executives on the supply side who no doubt told them that yes, the vision was achievable. But as so often in the case of military equipment, the voice of caution, analysis, cynicism and doubt, which is what the taxpayer really needs, seemed to be missing from the decision-making process.

Slow Train Coming

In January 2020 the UK's National Audit Office (NAO) published a worrying report that analysed the high-speed 2 (HS2) programme to build new railway lines from London to Birmingham and then onwards to other cities in northern England.

The programme was first announced in 2010, with early cost estimates of around £30 billion.[12] By 2015 the budget was set at £56 billion, and by 2019 the figure being quoted by government was between £65 and £88 billion, although an independent report suggested it could be as high as £106 billion.[13] Perhaps the most startling aspect of the NAO report was their comment that 'At the time of publishing this report, it is not possible to say with certainty what the final programme cost may be.' The first trains were planned to run from 2026, but it looks as if it will be at least 2031 before we see new services.[14]

Most of the causes of failure I've discussed already have relevance here, from issues around the specifications for the new infrastructure, to the dangers of believing suppliers and, ultimately, the desire of politicians to create a memorial for themselves. Another core problem is the disconnect between those who have real power in the programme compared to those who are paying for it. Power resides with those who understand what is going on – so principally the officials managing the programme and the suppliers. Those parties don't really have an interest in minimizing costs. Suppliers want to maximize revenue and profits; the officials justify their own salaries because of the scale of the spend, and the longer the programme goes on, the longer they hold their jobs.

The politicians, who genuinely want to manage costs, just don't understand how to do that, and the poor old taxpayer really does not have much chance to contribute at all, other than through their wallets. So we're all off on the gravy train again, and I wouldn't be surprised if the final bill (if HS3 goes ahead) turns out to be well over the £100 billion mark.

Be Careful

Summing up, there's not much advice to be given if both parties are conspiring to believe the supplier, in cases like those US Navy ships. But for most normal buying – in cases where the buyer *does* want the truth from a supplier – be careful. Over time, trust can be built between buyer and supplier, but it is sensible to be cautious about believing what you are told. Bear in mind that trust is earned through experience, not given. Be particularly wary about projections around speed of delivery or construction, or claims about new technologies or products. And analyse, reference and test wherever you can.

Everything I have talked about so far has focused in the main around activities carried out *before* the supplier starts delivering the goods or services to the buyer. Setting specifications, choosing suppliers and negotiating are all up-front elements of what could be called the end-to-end buying process. They are vital; but it would be a mistake to think that your job is done once a contract is agreed with what you hope is a good supplier. I'm afraid there's still plenty that can go wrong, as we'll see in the next three chapters.

9

Coping with Change

Do you remember Obamacare in the US (more formally, the Affordable Care Act of 2010), and specifically the botched launch of HealthCare.gov, the website created to implement Barack Obama's major healthcare reforms? Having finally been passed by Congress, the new approach to healthcare insurance was launched, and citizens accessed the health-exchange website on 1 October 2013 to see what the new deal meant for them and select appropriate insurance products.

Twenty million people tried to access the site in the first three weeks, but fewer than 5 per cent managed to complete applications, and even fewer actually got insurance, as the site ran slowly or applicants got 'stuck' during the application process.[1] The President made a public apology for the shambles.

Some observers argued that time-pressures were a factor. Many of the components of the end-to-end system were built by different suppliers. Just two weeks before the official launch, a full end-to-end test caused the system to crash, which you might have thought would be a red flag for the programme.[2]

The 'buyer' for the system was the Centers for Medicaid and Medicare Services (CMS), the body that oversees new US health-reform programmes. Several suppliers blamed CMS for changing the specification within weeks of the launch, requiring users to register fully in order to browse for insurance products, instead of being able to get information anonymously, as originally planned.

That sounds like a major change, and at that late stage *any* change to the way the user will experience a complex system is significant. That relates to the fundamental issue of understanding exactly what you're buying, as I discussed back in Chapter 1.

But changes to the specification once you have *agreed* a contract bring different issues. When you do that, you may give the supplier practical problems, and it also sends a message that doesn't help your negotiating position and often leads to tension. You're admitting that you didn't really understand what you (or the end-user) originally wanted; or didn't understand what was feasible or available; or perhaps even that the underlying business strategy driving the buying has changed.

Ch-ch-ch-ch-changes!

The three remaining chapters in Part 1, starting with this one, all look at what can happen *after* the contract is signed or the deal is done. Some may not even consider this 'buying', but many issues of failure are actually driven by bad management of a contract, supplier or both *after* the point of purchase.

In fact what happens after the contract signature (for more complex contracts anyway) is often more important than up-front buying work. That's because it is possible, with skilled contract and supplier management, to recover the situation, even if the initial contract isn't very strong. On the other hand, even good buying (selecting the right supplier and negotiating sensible terms and conditions) can end up in failure if the work after contract signature isn't managed properly.

Let's start with change management. Many major purchases are very much linked with change, whether it is installing a new factory production line, introducing a new IT system or an office relocation. Even simply introducing a new supplier into a well-established operation generates change-management issues, which can (as in the case of KFC below) lead to major problems. Change is perhaps less relevant when considering an office-supplies contract or buying the same components week after week for the factory, although even in these cases it is surprising how often change management comes into play.

Change around the buying itself, or implementation of any major programme that involves suppliers, has to be managed properly. There are entire books focused on managing change, so I won't try and cover every aspect, but it's worth highlighting principles that are particularly relevant for buying-related change. That applies when you change the nature of the contract itself after it is agreed; or if it is wider process or organizational change that affects contracts or suppliers.

A Change is Gonna Come

A major global consumer-goods company with thousands of staff all over the world recently implemented a new computer system, which was designed to help their staff buy better. That would help them get better control of budgets, leverage volume purchases to save money, and make the buying process slicker and more effective. The idea was that users would place orders through the system, and would access online catalogues to purchase routine items – stationery, IT equipment, and so on.

The technology itself was fine, the supplier being one of the best in its field. The configuration of the system went well, staff were informed, and one day it was turned on . . . and almost nobody used it. People continued buying in the manner they had always bought, using local systems, paying by credit card and expensing items, or phoning suppliers and signing off invoices to be processed by accounts payable.

The actual buying of the system here was fine. There was a real need, and the choice of supplier was appropriate. But the company failed to think about the internal organizational change management needed to make the system effective. So a consulting firm was called in – probably the right approach in this case. The consultants helped to manage a proper implementation process, with rigorous communication and training for staff on the new system, and worked to ensure it met local needs. Eventually all was well. But this was a classic example of an organization not understanding how important it is to manage the change associated with buying almost any new technology.

Fast Car

Technology is particularly susceptible to buying failures because of the industry's pace of change, the complexity of what is being bought and the 'information asymmetry' – in other words, the sellers of complicated equipment, software or systems usually know far more about their product than the buyer does. All of that applies too in the *implementation* phase of technology projects, after the initial purchase has been made.

In 2019 Hertz, the giant global car-rental firm, decided to sue Accenture, one of the world's largest technology and consulting firms.[3] It is most unusual for two firms of this size to go public with their disputes, as the publicity is rarely positive for either, in the end.

Hertz gave Accenture a major contract in August 2016 after a detailed procurement process, according to their court deposition, with the goal of revamping the firm's online presence. Accenture would develop and implement a new website, with accompanying mobile apps and the capability to enable customers to look at rental options, make bookings and interact with the firm.

Accenture committed to deliver by December 2017, according to Hertz. But after three years and some $32 million spent, the car-hire firm had received little of value, they said. What was developed by Accenture didn't meet the specification; for instance, the website should have been based on a common data platform so that the firm could share information across all its operations globally, such as the Dollar and Thrifty brands. But Hertz claimed that Accenture built something that could only be used in North America.

Hertz also claimed that when it pointed out various faults and issues, Accenture wanted more money to fix the problems that they themselves had caused. Now, Hertz wants its $32 million back, plus a lot more to cover the cost of putting things right, and also says it has lost revenue because of the problems. Accenture is fighting the claim, but has said little yet about how they will do so. My prediction is that 'change' will feature a lot in that – over the three years of the programme did Hertz change the requirements, leading to some of the issues?

That won't be clear for a while, but even if that is the case, it doesn't mean Accenture will win, of course. But the challenges of change management in large tech projects that run for several years are likely to come to the fore again if the case does eventually end up in front of a judge.

The Drugs Don't Work

Another classic failure of this technology-change genre is the FoxMeyer story. A presentation from Jesus Marcelo Ramirez Arias,[4] which draws on a paper from Judy E. Scott at the 1999 Americas Conference on Information Systems,[5] gives a good summary of this disaster.

FoxMeyer Drugs was a US-based $5 billion-revenue company, the nation's fourth-largest distributor of pharmaceuticals before the fiasco. As technology developments raced ahead through the 1980s and 1990s, the firm decided it needed to increase efficiency through greater use of computers, and the Delta III project began in 1993.

FoxMeyer conducted research and product evaluation and purchased the SAP R/3 ERP system in December of that year. The firm also purchased a warehouse-automation product from a vendor called Pinnacle, and chose Andersen Consulting to integrate and implement the two systems. Implementation of the

Delta III project took place during 1994 and 1995. That's when things started to go wrong – very wrong.

The post-mortem on FoxMeyer came up with a whole range of issues that contributed to the demise of the firm. Many were change-management issues, linked to the way the firm contracted with its major suppliers. But going back to our chapter on buyer naivety, FoxMeyer also seemed to have unrealistic expectations of the potential pace of implementation. That may not have been the suppliers' fault; it is tough sometimes to tell a client what they don't want to hear, such as 'This project will take three years, not the eighteen months you are hoping for.' Or the client simply may not listen.

Commitment from users within FoxMeyer was also lacking. (Remember our earlier discussions about understanding what internal stakeholders want, when you're buying.) Now in this case some of the antagonism was *not* based on good business reasons. Warehouse staff were worried about losing their jobs, so some fought against the new systems to the point of sabotage. But if your project is going to cause that much trouble, it has to be managed properly or the change will not be accepted.

The integration of two different systems caused problems for the firm, too. No great surprise there, and integration issues are often major causes of IT programme failure. Coming back to buying, perhaps the responsibilities for all the parties involved should have been laid out more clearly in the contracts, so that everyone was incentivized to make the entire programme work (a reminder of the earlier discussion on incentivization issues, too).

FoxMeyer's dependence on suppliers was another failure. Again, this is not unusual in technology programmes, but there wasn't enough internal expertise to hold the suppliers to account and understand what was going wrong. We talk about how a buying organization needs to be an 'intelligent client' – that is, knowing enough to manage suppliers properly, even if suppliers hold the really deep expert knowledge and capability. Clearly, FoxMeyer failed to be that intelligent client.

The split responsibility on the supply side, with two software providers plus the consultants working on implementation, also didn't help, I suspect. That, combined with a lack of programme and contract-management capability within the client firm, was no doubt part of the recipe for disaster.

But these are not unusual issues and, ironically, the final straw for FoxMeyer seemed to be the winning of a huge contract during the implementation period. That sounds like good news, but it brought with it a huge increase in transactional volumes with the new customer. The new system apparently couldn't cope with that volume, which was what finally sent the firm into a spiral of decline. By 1996 the company was bankrupt; it was eventually sold to a competitor for a mere $80 million.

We might trace that final volume issue back to the definition of the requirement, but to be fair, a specification cannot possibly incorporate *every* possible future development or change. But it emphasizes the need always to look to the future and potential change, when making significant purchases. Think, for instance, about whether what you're buying will be suitable if your firm grows more quickly than expected.

Dixie Chicken

Not many buying failure stories hit the international press and cause the police to issue statements – in this case, asking members of the public to stop calling them to complain about their inability to buy fried chicken!

But that is what the product shortage at Kentucky Fried Chicken (KFC) shops in the UK led to in 2018. Hundreds of retail outlets had to close because of a lack of product, after the firm changed their logistics supplier to DHL, moving away from Bidvest, which had been responsible for storing the chicken and getting deliveries out to shops. *The Guardian* newspaper reported this angle on the story: 'DHL, the firm that took over KFC deliveries last week, has been storing all KFC chicken at a single unregistered cold storage warehouse in Rugby, the local council confirmed on Wednesday.'[6]

The newspaper report claimed that the storage depot had not been registered or inspected before problems emerged, and could have been closed down for breaching safety rules. But that did not appear to be the fundamental issue here. Other reports suggested that problems started when trucks got stuck in an accident on the M6 motorway between depot and delivery destinations. But how could that possibly have led to the scale of the problems, seen across hundreds of outlets? Other theories – including a very good analysis in *The Sun* newspaper – suggested that the systems and operations of the new supplier just weren't up to the job.[7]

So was this fundamentally a logistics problem – the wrong stock in the wrong place, not enough drivers or trucks, or some similar issue? Or was it a systems problem, so KFC didn't know where the stock was, lost sight of orders and deliveries or realized that product traceability had been compromised? That feels more likely than a registration problem with the depot. In either case, it was the change of supplier that led directly to the problems.

There Are More Questions than Answers

That KFC case suggests some key questions to be asked if you want to change a supplier that currently provides important products or, as in this case, carries out a key service for your organization. Here's a brief checklist that might help avoid some of the obvious perils.

- Make sure any decision to switch the contract is made on objective grounds.
- Consider all the criteria for making the decision to change supplier, and if it is being done on cost grounds, be aware that other benefits might be lost in return for the savings.

- Carry out appropriate due diligence on any new proposed supplier – remember the 'Analyse, Reference, Test' mantra from earlier.
- Allow time for the new provider to gear up to supplying you, and (if relevant) for both parties to plan the change, train staff and get any new facilities or systems on-stream.
- Put contingency plans in place in case the change doesn't go well – think about the 'what ifs' around any new supplier.
- Consider putting end-of-contract clauses in new contracts – for instance, to oblige a supplier to hand over appropriate data if, and when, you decide they should transfer the work to another firm.

Thunder Road

Change management can be a considerable task when a project involves not only buying technology, but also behavioural change, construction of new facilities and even the potential for international legal action, as the German government found with its latest road-toll scheme for cars.

The idea was to introduce a levy on all cars using German motorways from October 2020. But German-registered drivers would have been compensated for the additional cost to them, through a rebate on national vehicle tax – and that's what the European Court of Justice didn't like, saying it discriminated against non-German Europeans.

Surprisingly, the contracts were awarded before the legal position was clarified, which looks like a rash decision now, given the court's pronouncement. The project has cost the government some €53 million already since 2014, and contracts with the toll-system provider Kapsch and ticket-sales specialist CTS Eventim were terminated with effect from 30 September 2019.[8]

There had previously been delays, as some German states objected to the scheme; and now, as well as the money already spent, the government may face claims of damages resulting from the cancellation of the contracts with Kapsch and CTS. That could be as high as €300 million ($342 million) to cover investments already made by the firms. If the claims get into loss of future profit, then the sky's the limit – figures of up to €2 billion have been mentioned.

Keeping all key stakeholders on board (even the European Court of Justice) is vital when complex change is the goal. This case shows just how interdependent all the different elements of change management can be, and how other factors can lead to major issues with buying and contract decisions. Ultimately, large sums of money have been wasted here on the contracts, with the suppliers themselves being blameless in this case.

Runaway Train

Most major railways in the UK run in and out of London via several huge stations, such as Waterloo, south of the Thames; King's Cross in north-central London; or Paddington, which serves the west of England and South Wales. Crossrail, which is currently under construction, will be the first major line to run straight through the centre, from two branches west of London, starting at Heathrow Airport and the town of Reading, to Shenfield and Abbey Wood to the east.

As such, it is a huge undertaking, with massive new tunnels to be dug under the city, entirely new stations constructed, plus major upgrades to existing stations and infrastructure. The trains themselves, and much of the technology such as signalling systems, will also be new.

The project was approved in 2007 and commenced in 2009, with an estimated cost of £15.4 billion and a completion date of December 2018. But in August 2018, just four months before the supposed opening, it was announced that there would be delays and more funding would be needed. The autumn of 2019 was proposed as a likely date, but that didn't happen either, and observers are suggesting that it could be 2021 before the line opens, and the cost may rise by at least £2.8 billion.[9]

What has gone wrong here? I've been told by insiders that people involved with the programme knew before August 2018 that the opening date wasn't going to be achieved. But there were political reasons for not saying anything publicly, so there was silence until it became impossible to keep quiet.

As we have seen elsewhere, it does not seem to be the heavy work that is the issue. Digging tunnels, constructing new stations – the UK engineering and construction industry seems pretty good at that sort of thing these days. The problems come when we look at the technology; and the signalling systems in particular have been blamed here.[10] The existing rail network runs using various systems, not all of them exactly state-of-the-art. The new trains have to work with those as well as with the new system, designed specifically for Crossrail.

It may be that the level of change in terms of integrating these different systems has been underestimated throughout this programme. If that represents the heart of the problem, it does suggest one key learning point. A certain type of person is usually put in charge of these programmes. No doubt they are talented, hardworking and bright people, who have usually come up through a lifetime of working as engineers on building 'big things' – roads, bridges, railway lines, buildings, and so on.

However, it isn't the size of the programme that is the issue these days. Rather, it may be that these fine people simply have less experience of the critical technology issues. When the very distinguished Sir Terry Morgan, who resigned as chairman of Crossrail in December 2018, started his engineering career in the early 1970s, he was probably still using a slide rule and log tables. Perhaps a new type of tech-savvy leader, comfortable with hugely complex systems, is needed, now that software engineering is more important than construction-related engineering. Effective change management will also continue be a central issue for major programmes, along with the persistent optimism on timescales that seems to afflict those planning such work.

Change Your Mind

In his seminal work *The Prince*, Niccolò Machiavelli pointed out that change is difficult to achieve because those who will benefit from it tend to keep quiet until they see if it is really working and benefiting them. But those who will lose out (or think they will) are concerned, so they make a lot of fuss about the potential change. As he said, 'Hence it is that whenever the opponents of the new order of things have the opportunity to attack it, they will do it with the zeal of partisans, whilst the others defend it but feebly, so that it is dangerous to rely upon the latter.'

That is as true today as it was in sixteenth-century Italy, in business as well as in political and social matters, and often comes into play in terms of failures around issues such as changing suppliers. As the examples here show, the biggest cause of change-management failure is probably the human aspect. It is essential to take the key stakeholders and interest groups along with you on that change journey, wherever you're heading. That is as true when the change involves suppliers and contracts as it is for every other type of business change.

Changing the nature of a contract once it has been agreed is also a source of problems. Sometimes it is unavoidable, but be careful if you have to alter the specification or requirement once the supplier is already engaged in their work. But while change management can be a challenge in terms of achieving good value for money from contracts and suppliers, it's not the only one. The next chapter will take a look at some of the potential failures that arise from poor risk management during the buying process.

10

What's the Risk?

When I worked for Mars Confectionery, our ultimate potential risk event was a jumbo jet crashing on the factory – not impossible, as we were only 10 miles (16 km) from Heathrow Airport. But the more likely risk event was a key supplier suffering from a fire, putting their operation out of action and leading to supply problems for the chocolate factory.

Most major manufacturers will experience such issues eventually. India's largest car manufacturer, Maruti Suzuki India, halted production at two major plants back in 2016 after fire at a factory owned by its supplier Subros, which made air-conditioning kits for their customers.[1] While both supplier and buyer put contingency plans in place, Maruti Suzuki lost production of thousands of cars in the couple of weeks before full production resumed.

But it is not just manufacturers who are exposed to supplier and supply-chain risk. In 2019 McDonald's in Japan started rationing servings of French fries, to horrified customers.[2] A shortage of potatoes meant the firm struggled to meet consumer demand. But that wasn't the first disaster for Japanese fast-food aficionados. A similar problem hit the country in 2014 and 2015, when a number of fast-food giants, including Kentucky Fried Chicken, were similarly affected.[3]

The problem was not with potato growers or the middlemen involved in the deals to bring potatoes from the heartland of the USA to Japan. Rather, it was industrial action by dock-workers in ports on the west coast of the US that restricted shipments. Now it might be a bit harsh to tag McDonald's with the bad-buying brush; after all, the firm did say they had airlifted 1,000 tonnes of potatoes into Japan, which must have cost a fortune. And 1,600 tonnes were shipped from east-coast US ports, unaffected by the industrial action. But could more have been done to secure supplies of potatoes from closer to Japan? Were contingency plans in place?

This may be the result of the drive for consistency and standardization, which is fine until something goes wrong, and then it is difficult to substitute quickly another product or supplier for that standard. In any case, the fries incident brings home the importance of *risk management* in a buying context.

Running the Risk

Buying failures can become apparent early on in the process or after a contract has been let, and at every stage poor or non-existent risk management crops up as a root cause. Much of the thinking about risk needs to be done at an early stage in the buying process, at the time you examine markets and suppliers, so arguably this discussion could have come earlier in this book. But most of the risks that actually hit organizations become visible once the contract is up and running, and much of the risk-management activity needs to run through the whole period of the contract – hence the positioning of the topic here. However, do think about these issues from the very beginning of your buying process.

Supplier dependence, for instance, generates risk issues, particularly when the buyer has only one supplier for something that is critical to their business. That might be a raw material, component or even a service (hosting services in the technology world, for instance). There are real risks if that supplier goes out of business, suffers from a natural disaster (fire, tsunami or similar) or stops supply for other reasons.

Indeed, almost everything can be considered as part of the risk picture. A failure of change management could be seen as a risk failure – the risks around change should have been identified in advance. But generally risk management in this context relates to risks and events triggered by suppliers. Some of these should be foreseen and are avoidable, if the buyer takes timely and appropriate actions. In other cases, risks are less identifiable up front, but the failure comes in not developing appropriate measures to overcome or mitigate the impact if potential risk turns into an actual risk event.

To amplify the point that much buying risk is avoidable, consider another theoretical case. If you give an important contract to a firm on the point of bankruptcy and, sure enough, they go bust, causing major problems, that is not an 'act of God'. It is your direct failure, because a quick Dun & Bradstreet risk check of their financial health would have told you to avoid the firm.

Fisherman's Blues

In recent years the risk focus has switched considerably. When I started my career, in the food industry, we worried mainly about operational risks arising from suppliers. So that might be an 'act of God', such as a tsunami affecting their premises, or it might be issues more directly under their control, such as a labour dispute or a factory failure. In all cases, it might mean they couldn't deliver to customers for a while. As many factories run twenty-four hours a day, seven days a week, and the ability to hold stock of raw materials on-site may be limited, even a day without deliveries of certain items can cause problems.

At Mars Confectionery, if production lines stopped, it wouldn't be long before retailers saw shortages, and then the end consumer would simply buy a competitor's chocolate product. So that *operational risk* generated from suppliers was a major issue for the firm.

There is also *economic* or *financial risk*; for example, the risk that shortages will cause large price increases for what you are buying in commodity markets. *Political risk* has also increased, as firms increasingly source from around the world. That might be tariffs or trade barriers, or even civil unrest causing supply issues out of that city or country.

In recent years *reputational risk* has risen up the agenda and has perhaps even overtaken operational risk as the number-one threat for many organizations. Going back to the food industry, the reputational risks now emerging include issues such as firms being associated with the loss of rainforest to palm-oil plantations, the use of slave labour and human-rights abuses committed by suppliers, as well as the danger of buying suspect or 'counterfeit' raw materials.

Oceana is a not-for-profit organization that aims to protect and restore the world's oceans. It has also carried out valuable investigations into the mislabelling and fraud around seafood. In March 2019 it looked at seafood being sold across the United States, focusing on those products not included under the existing US federal traceability programme.[4] Oceana's findings were shocking.

No less than 21 per cent of the 449 fish tested were mislabelled, and one out of every three establishments that were visited sold mislabelled seafood. Restaurants and smaller markets were bigger offenders than large stores, and sea bass and snapper were the species most often mislabelled – an incredible 55 per cent of 'sea bass' tested weren't in fact that fish at all.

The deception, which is practised on both business and consumer purchasers, takes place for a number of reasons. Imported seafood may be disguised as regional favourites, which presumably can command a higher price, and tricks those buyers who like to think they are sourcing local products. Vulnerable species, such as Atlantic halibut, may be sold as more sustainable catches to get round the rules. And seafood is sold with generic names like 'sea bass', disguising lower-value species or masking health or conservation risks.

Other cases of suspect raw materials and foodstuffs have more serious impact than eating the 'wrong' fish. The contaminated milk-powder scandal in China in 2008 caused deaths and made hundreds of thousands of people sick.[5] In 2019 the deaths of five people in British hospitals were linked to listeria poisoning from sandwiches and salads bought by the hospitals from their food suppliers and served to patients.[6]

Bad Reputation

In terms of some reputational issues, there are questions about just how concerned consumers really are regarding them. The collapse of the Rana Plaza garment factory in Bangladesh did not seem to dim the appetite of shoppers in Europe or the US for low-price T-shirts and dresses.

But few businesses want to risk being associated with something like that tragedy, or see their waste packaging choking seabirds and turtles; and as public awareness grows, analysis and studies increasingly suggest that reputational issues do have a real cost. Risk experts Oliver Wyman[7] reckoned that 12.6 per cent of sudden stock-price drops related to reputation, image, pricing and presence in the market.[8] Stock-price drops from reputational damage emerged as the largest category in a study conducted by Wharton Business School.

The Reputation Institute, which monitors the reputation of 7,000 firms globally, calculates that intangible factors account for 81 per cent of the average public company's market value, and that in the twelve years up to 2018, 'a strong reputation yields 2.5 times better stock performance when compared to the overall market'.[9]

So, to address these risks and avoid damage to reputation, buyers need to lay out clear standards and specifications for suppliers, in terms of both what they are buying and how they expect the supplier to *behave*. Expectations in terms of issues such as employment practices should be spelled out. Remember, too, that laws do vary, so if what is wanted is above and beyond the supplier's local laws, that should be highlighted.

Inspection of premises and testing of products have traditionally been key risk-management approaches. But they can be expensive and ineffective unless managed carefully. Here, collaboration is having a positive impact. Organizations such as Electronics Watch,[10] which looks at working conditions in the electronics supply chain, can carry out professional and regular inspections because its costs are shared by multiple members.

Collaboration also comes into play in terms of information-sharing. Today, Twitter and other social-media sites are often the first source of useful and relevant supplier information. This contributes to the monitoring of issues such as human rights, working conditions and bribery/corruption, and can also help with the burden of reporting and proving compliance to the relevant regulations. There is also a growing use of risk-management tools, which can pick up news alerts, triggers that might indicate issues, market intelligence or information from buyers and others, and share findings across all users.

Act of God

On 6 September 2018 a 6.7-magnitude earthquake hit the Hokkaido region in Japan. Tragically, around a dozen people were killed and hundreds injured. But the quake also hit a number of factories that supplied components to Toyota, as well as the automotive firm's own assembly plants. The consequences for Toyota were severe.

As *Automotive News* reported, 'Toyota will halt planned production at its Kyushu, Tahara and Toyota Auto Body plants. Those factories make Lexus vehicles and the Toyota Land Cruiser, among others, some of which are bound for the US markets.'[11]

It may seem a little harsh to criticize when a supplier's factory is affected by such natural disasters – and I'm not accusing Toyota of failure here – but sometimes that accusation is legitimate, depending on how the buyer got into a situation and how they plan to mitigate any risks. One approach is to use technology that lets you map your suppliers' locations and relate that to the risk of natural disasters, and also provides alerts about actual risk events.[12] That enables buyers to think about how they approach the most vital components, raw material or other supplies needed for their organization, and to reduce their dependence on suppliers that intrinsically carry a high risk.

This applies principally to buying *goods* rather than services – it is unusual for a management consulting firm, a legal-services provider or a software developer, for instance, to be affected greatly by a flood, so that sort of purchase is probably safe.

So, coming back to physical goods, a smart buyer looks to spread the risk. Don't put all your eggs in one basket, and don't put all your suppliers of a critical material into one earthquake belt or floodplain. And it's not just direct suppliers that can cause you risk problems. It might be risks sitting with other players who are involved in the whole supply-chain process of getting goods and services from their source into your organization. It may not be a direct supplier that is a source of risk; remember, it was the shipping of those potatoes that caused problems for McDonald's in Japan, and not the growers themselves.

Tell Me It's Real

Many buyers don't think in risk terms beyond immediate suppliers, and don't look across supply networks or down their supply chain, to the firms that supply their suppliers.

One major industrial firm develops and manufactures systems that contribute towards making vehicles more efficient and economical. The firm did a risk analysis and found that for one key component, they had five approved suppliers. None of them were working at capacity, and all seemed to be reasonably stable from a financial, labour and business point of view. So, at first sight, supply-chain risk looked limited here.

However, matters were not quite as they seemed. One of those five businesses was actually a key supplier to all the others, providing a core, licensed technology that was essential to the manufacturing process of the component. It also used the technology itself, of course, so what looked like a low-risk item actually had a lot more dependency risk than was obvious to a casual observer.

But it isn't always easy to find out who our suppliers' suppliers really are. So buyers should consider putting agreements in place with direct suppliers to *require* them to inform the buyer about key suppliers and activities further down the supply chain. To make this work, a good relationship with the supplier is generally necessary to make sure that communication is rapid, open and honest when things go wrong. Just ask yourself: how many of your key suppliers would proactively tell you if they had a major problem within their own business?

A major US automotive firm was alerted by their supply-chain risk-management system that there was a risk of more than 200 production sites being compulsorily shut down for two weeks in the Hangzhou area in order to reduce air pollution during the August 2016 G20 political leaders' summit in China. This customer increased their ordering volumes appropriately to build component stock-holding during this period – and the point to note is that the warning did *not* come from the supplier themselves.

Lightning Strikes

In March 2000 a fire, started after lightning hit a power line, affected a Philips semiconductor plant in New Mexico. The fire was put out within minutes, even before the fire brigade arrived. The damage seemed to be minor, and Philips thought production would resume within days. That's what managers told their two biggest clients, Nokia and Ericsson, which at the time were close rivals in the mobile-phone handset market.

But the actions of the two customer firms then differed. Nokia had a close relationship with Philips, and carefully monitored deliveries. They had regular contact with Philips management, found out quickly what was going on and decided that the risks were greater than first thought.

Nokia pushed Philips to provide an alternative component from another factory and activated risk plans, including looking at alternative supply sources. Ericsson, by contrast, had limited contact with Philips and no risk plan in place. Their Philips contact went through a fairly low-level technician, who took some time to notify senior managers at Ericsson about what was going on. Sure enough, Philips soon realized that the damage to clean-rooms (where manufacturing of high-tech, sensitive equipment took place) meant there would be production issues for some weeks.

By the time Ericsson spotted the real supply problem, it was too late to do much – there was little spare capacity or alternatives available to them in the market. The end result was that Ericsson had to reduce production of handsets significantly and concede market share to Nokia. When the end-of-year results were announced, Ericsson stated that the incident had cost them no less than €300 million – and even more than that in shareholder value was lost. A big price to pay for one fire on a supplier's premises!

There are two stages to effective risk management, for significant contracts at least, and Ericsson seemed to fall down on both. First, you need to have risk-management plans in place around the contract and supplier. What would you do if the supplier went bust? Or their factory burned down? How about if the product supplied was contaminated, or not as it should be?

But having plans is not enough. The second imperative is to find out quickly if a risk event occurs, understand exactly what has happened and the implications, then activate the appropriate risk response. In the case of Ericsson, a plan would not have made much difference if they hadn't found out about the fire for weeks. Being able to respond *quickly* as well as *effectively* is vital when something bad does happen. And actually, if you do that, the event might even become a source of competitive advantage for your firm.

Waiting on a Friend

Two friends are out walking in the woods. Suddenly they see an angry-looking bear emerging from bushes just a few metres away. It roars threateningly and starts walking towards the walkers. One stops dead, while the other takes off his rucksack and whips out a pair of running shoes, which he proceeds to put on.

'That's no good,' says his friend. 'A bear can outrun any human, whether they're wearing walking boots or running shoes – it can certainly outrun you.'

'I don't have to run faster than the bear,' explains his friend. 'I only need to run faster than *you*!'

So, when disaster strikes, you might not be able to avoid the consequences entirely. But if you can handle it *better than your competitors*, then you will be well positioned in the market, and the event might even turn out to have positive consequences, as in the Nokia case. This is true in many different areas, from commodity markets where early information can drive huge gains for traders, as futures contracts reflect current events and expectations, to cases like the Philips example, where sudden events cause supply reductions that in turn affect markets.

Even reputational risk events that appeared to spell disaster for firms have been turned round to advantage. Firms that have a plan in place, react quickly and sensitively, and explain to their customers what they are doing and why can emerge from major risk events in good shape and without real damage to the business.

Computer World

A key supplier going bust obviously drives risks in terms of supply interruption and possible unavailability of product. Even a supplier merely suffering from financial problems can have a knock-on effect on its customers – in that situation, I've seen firms stopping production of some items, or demanding price increases in an attempt to get back on-track. Either of those actions can have an impact on their customers.

There is even evidence that some firms have used bankruptcy, or potential bankruptcy, to get out of onerous contracts. In a case in the automotive industry a few years ago, FormTech filed for Chapter 11 bankruptcy in the United States with a new owner, HHI, already lined up.

As *Automotive News* reported:

> But HHI opted not to assume agreements with Chrysler Group, Ford Motor Co., General Motors, Toyota Motor Corp., 11 Tier 1 suppliers, and others ... nearly all of FormTech's customers were forced to the negotiating table to work out new long-term supply agreements as a condition to continue receiving parts in the short term.[13]

Cyber risk is another relatively new area to consider. Firms are vulnerable to direct attack, but may also have problems if their suppliers don't take the right precautions. For instance, there are obvious risks when a supplier holds data about the buyer's firm or customers. There have been various cases where information has been stolen by hacking into websites or databases, so any supplier who holds sensitive information on your behalf is a potential risk. The GDPR (General Data Protection Regulation) legislation in Europe has also added another layer of complexity and risk. As well as defining how your own organization has to take care of customer and other personal data, it requires you to manage how your suppliers do the same on your behalf.

More recently, there have been cases where a risk event has emerged via a supplier that provides software or information that is embedded in the client organization's website or other technology in some way. If that third-party element is vulnerable to hackers, again it can be the customer organization that ultimately suffers.

Living in Danger

To manage risk better, the first essential step is to understand where risks lie, which in itself means that you have to first understand your supply base – who your key suppliers are, where they are based, who owns them, how financially stable they are, and so on. Then you can assess what and where the major risks lie, prioritizing based on the combination of how *likely* it is that the risk actually turns into an 'event', and how *major the impact* would be if it did come to pass. Once risks have been prioritized, there are generally considered to be four ways to handle each risk. You can:

1. **Avoid or eliminate the risk**, so it stops being a problem. For example, if you identify that a particular supplier is financially unsound, you could simply stop buying from them and find an alternative.
2. **Reduce or mitigate the risk**, so it is less likely to happen or less serious if it does – through contingency planning, for instance. If you are dependent on one supplier, find an alternative (preferably not next door to the first). Or introduce a strong policy on human-rights issues and make sure your suppliers follow it, to reduce the reputational-risk possibilities.
3. **Share or transfer risk** to the supplier (for instance, capping inflation-driven price increases in a long-term contract, so that they share in the financial risk) or to an insurance company. Insurance is basically a risk-transfer exercise.
4. **Retain and accept risk** – acknowledge that it exists and simply 'live with it'. Perhaps you only have that one supplier, so you have to accept the risk that their factory will burn down one night; or you might decide that you can handle a specific risk better than the supplier.

For major risks, a plan should be developed (and then implemented), which lays out the steps that will be taken covering those categories of response. Having carried out that work, it is then vital to establish processes so that you find out promptly and accurately if, and when, a risk event occurs. Having great plans in place is no help if, like Ericsson, you don't actually realize that you need to start implementing those plans.

Bury Me Low

Sometimes risks are truly diverse. The transport authorities in Rome have been extending a subway line, Line C, which comes into the city centre from the suburbs. The extension was supposed to cut under the legendary Forum, crossing the heart of the city and passing under the River Tiber. But the programme ran into multiple problems, according to *The Times*.[14] There have been cost overruns, allegations of corruption (not exactly unusual for that city), and problems were caused by the large quantity of ancient remains and archaeological finds that were unearthed during the digging. Perhaps that last risk should have been foreseen – this is Rome, after all – but it isn't one that would usually be on the list.

In 2019 the new line did make it to the interchange with existing Line B, and Line C trains should get that far by 2023. But having spent €790 million on the last stretch of line, the money has run out. The city-owned company managing Line C has gone bust and has been placed in 'controlled liquidation', according to the mayor, Virginia Raggi, and work has stopped.[15]

Italian newspaper *La Repubblica* reported that the tunnel-boring machines, used for the excavation of Metro C, are to be buried under the Via dei Fori Imperiali, without reaching the proposed station at Piazza Venezia.[16] Just 300 metres past the Coliseum, the two huge 6.5-metre (21-feet)-wide cutting heads on the drilling machine will be abandoned, as they are too large and unwieldy to get out (presumably without vast expense and trouble). So the heads will be cemented in, shoring up the end of the tunnel, but also acting as a memorial to another wasteful public project.

What's the Risk?

Imagine a future civilization digging up these machines and wondering whether they were a sacrifice to some god in the long-ago twenty-first century? Or were they the victims of a huge earthquake perhaps? More prosaically, any future development of the line will have to start from the other end and then break through the buried machine heads.

Poor planning and risk management, and a consistent inability to forecast costs, seems to dog major construction projects the world over. That may be because those who want a project to happen have a vested interest in playing down the costs and risks in order to get the projects approved. Or perhaps it is simply an inherent 'optimism bias' in humans – cross your fingers and hope for the best.

Then there are other problems that can arise once the contract is up and running, which the next chapter (after another notable award-winning case study) will consider, under the general heading of 'contract management'.

Bad Buying Award

WORST INTERNATIONAL PROJECT – BERLIN BRANDENBURG (NOT YET AN) AIRPORT

Many of us have a perception of the competence, efficiency and success of German engineering firms – drivers of the nation's post-war 'economic miracle'. However, even that country has seen some classic examples of buying failure, and perhaps the most notable and infamous is the 'new' Berlin Airport.

If you have flown into the outdated and creaking Berlin Tegel Airport in recent years, as I have, you might have wondered how such a great city has such an outdated, terrible airport. It's because Tegel should have closed in June 2012, and in the intervening years has been starved of investment. That shutdown is still planned, but exactly when is anyone's guess, because of the fiasco that is the new Brandenburg Airport.

Flughafen Berlin Brandenburg Holding GmbH (FBB) was set up in 1991, with the aim of building what was planned to be Germany's third-largest airport. It would reflect the growth of Berlin after the Wall coming down, and the status of the city as the capital of reunified Germany.

One early problem came with the question of whether a private consortium would own and operate the airport. A competition was held, and in the first buying-failure incident here, the decision was challenged successfully by the bidder that lost the tendering process. The two bidders then combined, with a consolidated bid, but the state authorities decided that the airport should not be privately run after all. FBB subsequently remained under the ownership of the public sector – Berlin, Brandenburg and the federal government.

The two consortia were eventually paid off with €50 million each for their efforts.

So after almost fifteen years of planning, construction began in 2006. During construction it became clear that the €2.8 billion budget wasn't going to be enough – it was already up to €4.3 billion by 2012. That was down to construction flaws, poor cost estimation and the need for more to be spent on soundproofing local homes.

In 2010 there was a ceremony to celebrate the supposed completion of the construction work. Operational trials started in late 2011, but then another whole range of problems emerged. The fire-protection system was faulty, there were no ticket counters and the escalators didn't work. Opening was planned for May 2012, but just before the due date, FBB postponed again, citing technical difficulties, primarily concerning the fire-safety and smoke-exhaust systems. FBB also dismissed the director for technical affairs, Manfred Körtgen, and replaced him with Horst Amann.

March 2013 was the new opening date – but not for long. In September 2012 that was adjusted to October 2013. That came and went, the CEO of FBB and the chairman of the supervisory board both left, and in January 2014 the likely opening date was moved to 2016 at the earliest. In 2015 Imtech, one of the most important construction companies on-site, filed for bankruptcy.

In April 2016 the press spokesman was fired after giving a 'too honest' interview. He said that billions of euros had been wasted and that 'only someone dependent on medication will give you any firm guarantees for this airport'.[17] In December 2016, the boss 'hinted' at a possible opening in 2018.

By 2018 the best guess was late 2020, with the total cost now estimated at more than €7 billion. But in March 2019 *Der Tagesspiegel* reported that two of the key construction firms involved had refused to attend a Berlin House of Representatives (government) meeting, which the writer took to mean that the October 2020 opening was unlikely.[18]

The reasons for the problems are many and varied. One issue is that the delays have gone on for so long that some of the initial equipment and technology put in place during the construction is already obsolete, before a single passenger has gone through the airport.

Some of the issues would be funny, if it wasn't all such a waste of effort and resources. The fire-protection and alarm system has been a constant issue; it was not built according to the original construction permit (which sounds like a specification-related buying failure) and failed the mandatory acceptance test needed to open the airport. FBB proposed an interim solution – why not employ 700 human fire-spotters around the airport to keep a lookout for fires? Not surprisingly, the building supervision department of the local district rejected that (after having a good laugh at the idea, I suspect).

It also emerged that Alfredo di Mauro, who designed the fire-safety system, was not actually a qualified engineer, but a draughtsman. The original architects were fired, and investigations uncovered many examples of poor workmanship due to a lack of proper supervision and documentation of the construction progress, most notably concerning the wiring. It was discovered in 2017 that 80 per cent of the electric doors would not open, which created concerns around venting smoke in a fire situation. This also sounds very much like a failure of contract management.

There are also corruption issues and allegations. In 2014 Jochen Großmann, the former technical director, was convicted of corruption and fraud. He allegedly demanded money from a potential contractor in return for work on the programme. In October 2016, according to *The Local – Germany*, a former department head at the airport was jailed for taking a bribe from a subcontractor on the programme, Imtech (which went bust in 2016).[19] Two former employees of the firm were also convicted. The airport executive was sentenced to three and a half years in prison for allowing Imtech's claims for additional payments to be paid without being checked, in return for a €150,000 bribe. Indeed, there are suggestions that many such additional invoices were approved for payment to suppliers, without apparently very much questioning or checking, which may indicate more endemic corruption.

It was also reported that the airport's roof was twice the authorized weight – and so the litany of failures and problems goes on. In fact this story demonstrates so many aspects of bad buying that it's hard to know where to start. There was the (probably politically related) sheer stupidity of the authorities, who felt they were capable of managing the procurement and the programme without external support. For some years, the board of directors was run by two politicians with no experience of airport construction, for instance. The concept of the intelligent client, which we mentioned earlier in this book, comes into play here – or rather doesn't come into play, as the Berlin authorities do not appear to have been a remotely intelligent client at any point in the last decade.

I haven't even got to the problems with the railway terminal built to service the airport, which has its own long list of issues and questions. Some observers have suggested that it might be better to demolish the whole airport and start again, and this story will run on for some years, I suspect. But anyway you can use the Brandenburg Airport case if you ever want to argue that Germany is *not* always the most efficient country in the world!

11

The Joys of Contract Management

In 2019 the English city of Birmingham accused its roads-maintenance supplier, Amey plc, of failing to meet the agreed contractual terms for a twenty-five-year PFI contract to maintain and upgrade the streets in and around the city.

The relationship between the two parties had become increasingly toxic, and according to the Birmingham Live website, the council was 'fining' Amey no less than £48.5 million – for failing to repair four damaged bollards within contractually agreed timescales.[1] Birmingham rated the work as a 'category one' issue, meaning that Amey should have repaired the bollards within twenty-four hours, but the firm took a year to replace the first set and seven months for the second. The penalty for delays in the repair started at £250 and doubled every hour the work remained incomplete, said the report.

Now that can't be correct, as mathematically Amey would have owed £1 billion within just twenty-four hours on that basis, and pretty soon would owe more money than exists in the entire world. But you get the general point. The council also reportedly fined Amey £12 million for a single pothole, and £14 million for adding cable ties to three lamp posts.

Birmingham Council say they 'have managed Amey's performance in line with the contract, although the statements we have seen regarding these adjustments wildly exaggerate the amounts we have applied'.[2] So maybe it isn't really £48.5 million.

But Amey was desperate to get out of the contract. The firm offered to pay the council £175 million in cash, and to write off £70 million of unpaid work, to settle the dispute. The council reportedly rejected the initial offer, but eventually the parties settled in June 2019, with Amey paying £125 million to extricate itself from the deal.[3]

This is all highly unsatisfactory and shows how contracts and relationships between buyer and supplier can break down, with consequent cost and hassle for everyone involved. I'm not convinced the reported fine would have stood up in court, but I would suggest the failure here started long ago, with the strategy behind this contract. Why on earth was it set up as a twenty-five-year contract? Was that because Amey was funding investment that was benefiting Birmingham, but therefore needed a long period to get a return? It is impossible to foresee how needs will change over twenty-five years anyway, and the management of this contract has clearly not brought about the desired outcomes for Birmingham – or Amey.

Have I Got a Deal for You

Once the buying is completed, and a contract and a supplier are in place, then we can all relax, breathe a sigh of relief and take it easy. Yes? No, of course not.

For anything other than the simplest purchases, what happens next is also critical in terms of spending the organization's money well. Choosing a supplier, negotiating and agreeing a contract – even if that is all done well – is not enough to guarantee that the goods or services wanted in the first place will be provided at the right price, quality and level of service. There is still plenty that can go wrong; and that can come about because of issues caused by the buyer, the supplier or both. In the worst cases, everybody is unhappy, and we see a total breakdown of the relationship between buyer and seller, as in the Birmingham example.

For most organizations, I estimate that more than 80 per cent of the spend with third parties requires some level of appropriate, ongoing contract management. Intrinsically, contract management must be considered part of the end-to-end buying process, but in most organizations contract management is neglected and seen as a Cinderella activity. It is not usually defined as a distinct 'function' within the business and despite the best efforts of organizations, including the International Association for Contract and Commercial Management (IACCM) in recent years, most executives still don't see it as an attractive career path.

It's also fair to say that buying activities (also known as purchasing, sourcing, procurement or supply-chain management in different organizations) are not always seen as particularly high-status, either, but these roles are generally more recognized than contract management. Yet here is the paradox: there's a strong argument that how well you manage the supplier and the contract, once the buying is done, is *more* important than the actual buying (procurement, purchasing or whatever you call it) phase itself. How can that be?

(Whole) Life's What You Make It

Let's explore that idea further, and this chart (Fig. 3) is a good place to start. The point to note is that it *isn't* symmetrical. Clearly, if you start with bad buying and follow that up with bad contract management, then the end result or outcome (the bottom-left quadrant) is most unlikely to be anything other than failure.

Equally, good buying followed by capable and appropriate contract management (top right) will tend to drive positive results and outcomes. However, it is the other two possibilities that are interesting. If we start with good buying, but then follow that up with incompetent, non-existent or inappropriate contract management (bottom right), the end result will almost certainly be bad, too. You can start with a great contract, but if you let the supplier get away

The Joys of Contract Management

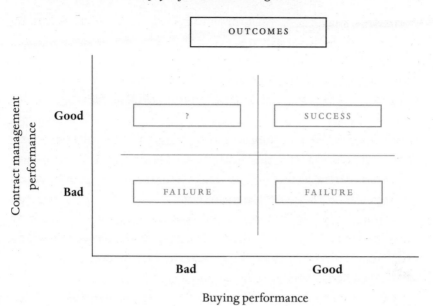

Fig. 3: The importance of contract management

with poor performance or even worse (see our chapters on fraud, for example), then this won't end up positively.

But actually, even poor initial buying can be recovered, partially or even fully, by really good contract management (top-left quadrant). Perhaps the initial contract isn't fit for purpose, or the specification is not as it should have been. It isn't *always* possible to recover that situation, I should stress, but with skilled contract management, a supplier can be brought into line, performance can be improved, new specifications agreed and the commercial details renegotiated. Smart organizations also constantly think about the 'whole-life costs' of a contract and of whatever they're buying, which means that effective management throughout the contract period is essential.

And yet very few top executives pay enough attention to contract management. I wrote the UK National Audit Office 'Good practice contract management framework'[4] for government organizations back in 2007, and while contract management has improved somewhat since then, much of the advice in that report is still not followed by public or private-sector organizations.

I Wish It Could Be Christmas Every Day

That historical lack of understanding leads to wasteful situations. I was once commercial director for a huge UK government programme. When I took on the role, virtually all the programme's work (technology, project management, even much of the strategy and policy aspects) had been 'outsourced' to a large consulting firm. There were no fewer than 100 of their staff involved, working full-time, with a cost to the organization and the taxpayer of more than £600,000 a week or £30 million a year.

'Could I have a chat with the contract manager for this contract' was one of my first questions when I started work.

'We don't have a contract manager' was the reply. That was because, unbelievably, the parent government department of my organization 'didn't think it was worth it'. We were talking about perhaps £30,000–40,000 a year for a contract manager, or one-tenth of 1 per cent of what we were paying the supplier.

So I spent a tedious couple of hours personally going through the consulting firm's invoices. That's a rather detailed task for a director, truth be told, but it paid off. I found that we were being charged for people who clearly weren't working 100 per cent on our account, such as personal assistants to the consulting firm's partners. And I even spotted that on the day of their Christmas party, when their entire contingent of 100 disappeared at lunchtime (and didn't return), somehow we had been charged for a full day's work.

'An unfortunate error,' said the firm's client partner, when I drew this to his attention. So that in itself saved in excess of £50,000, or more than the annual cost of a contract manager.

Of course there is more to good contract management than going through suppliers' invoices and similar administrative work.

Many examples in earlier chapters show strategic-level issues that might have been solved by better contract and supplier management. The problems that VW had with their supplier, which caused plant closures (see page 44), for example, might have been avoided if the relationship between the firms had been better managed, with more focus on careful handling of the negotiations when VW wanted to vary the contract that had been previously agreed with the supplier.

Tunnel of Love

Poor contract management can result in many negative outcomes. Cost overruns on major projects are often a sign of weak contract management, particularly in sectors such as construction and IT-systems development, where projects or programmes can often be lengthy and complex. In the construction industry, for many years some suppliers followed the cynical mantra 'bid low and make money on the changes'. A firm would bid at a price that just about covered costs, hoping it would win the bid. Then every time the client said, 'Could you move the door a little/put in a few more power points', the construction-firm boss would look troubled, shake his head and say, 'Tricky, it's going to cost – let's say an extra £100,000', knowing that the change would actually only cost £50,000. And that's how the profit margin got built back up to a reasonable level.

Sometimes cost overruns can come from other failures, as described earlier, such as incomplete or inaccurate specifications – issues that genuinely require the supplier to do more or act differently from how they had planned. Or additional costs may have nothing to do with supplier or buyer error. Perhaps there were unexpected ground conditions to overcome, or asbestos was discovered during the demolition phase, or the requirements really did change because of external events, with a real and subsequent on-cost. But whatever the cause – whether genuine or down to poor contract management – the end result tends to be wasted money or overspent budgets.

There is plenty of evidence of this happening all over the world, too. An academic paper entitled 'Construction Projects Cost Overrun: What Does the Literature Tell Us?' by Abdulelah Aljohani, Dominic Ahiaga-Dagbui and David Moore provides direct examples and evidence from other studies.[5]

For example, the Channel Tunnel costs came in at £4,650 million, against the budget of £2,600 million. In Korea a study showed that the average final cost of seven mega-projects (cost of more than $1 billion each at completion) increased by 122.4 per cent compared to the original budgeted cost.[6] In the USA an analysis of eight rail projects identified an average cost overrun of 61 per cent.[7]

But it is not only cost. Performance can slip, once a contract is up and running, so suppliers have to be actively managed throughout the contract life. Just as most of us probably behave better and are more attentive towards a new partner in the first exciting weeks of a relationship than we are after thirty years of marriage, so supplier performance can slip once the first glow of a new contract fades. Good contract management holds the supplier to account throughout the contract, with clear performance measures, reporting and feedback.

Virtual Insanity

As we've pointed out, technology is particularly susceptible to failure because of its inherent complexity, the pace of change and information asymmetry (the supplier understands what they're selling much better than we understand what we're buying)!

An article from consulting firm Mckinsey in 2012 followed their research in conjunction with the University of Oxford, and the findings won't come as a surprise to those who have worked in large organizations.[8] Half of all large (more than $15 million) global

The Joys of Contract Management

IT projects blow their budgets, says McKinsey. Large IT projects on average run 45 per cent over budget and 7 per cent over time, while delivering 56 per cent less value than predicted. Also, and not surprisingly, the longer a project is scheduled to last, the more likely it is to run over time and budget. The report identifies several reasons for problems, ranging through unclear objectives and lack of focus, to shifting requirements, a lack of skills and unrealistic schedules. And 17 per cent of projects turn into 'black swan' events, in risk-management terms, and go so badly wrong that they threaten the very survival of the company.

One case that shows pretty much every aspect of bad buying, including contract-management failure, concerned the state government of Queensland, Australia, and its new payroll system, which was needed to handle the 80,000 employees of Queensland Health. The supplier (IBM) was selected in 2007, and the project was expected to go live within about six months and cost around $A6 million.

But the system didn't go live until 2010, with an additional cost of $A25 million, and Queensland had to hire 1,000 employees to undertake the payroll manually, adding $A1 billion in additional costs over eight years! The commission reviewing this fiasco, led by a barrister, the Hon. Richard Chesterman, found fault and errors at every stage of the project, including the procurement process, the planning of contract schedules and the buyer's management of the project.[9]

It's interesting to note that Queensland took legal action against IBM and failed – indeed, it had to pay costs to the firm. IBM lawyers argued that the government was 'not able to define and stick to a scope' and had waived its right to sue in a separate agreement signed in 2010. As the report put it:

> The State did not adequately communicate to IBM the business requirements for the workforce of QH which would permit IBM to design a payroll system which accommodated the number and complexity of pay rules . . . The result was ongoing disputes about

scope which resulted in changes to the contract, increases in price, and delays to the implementation date.

So, a clear case of significant failure at every stage, encompassing the specification stage, contract *and* project management.

London's Burning (the cash . . .)

FiReControl (notice the attempt to create 'street cred' via the mix of lower- and upper-case letters) was a UK government project, started in 2004, to reduce the number of control rooms handling emergency calls for the fire services. There were forty-six such local centres in England, handling 999 emergency calls from the local public for assistance. In March 2007 the Fire and Rescue Service government minister announced that EADS Defence & Security (now Airbus Defence & Space) had been awarded an eight-year £200 million contract to supply the IT infrastructure for nine new regional control centres (RCCs).

But the programme went horribly wrong, and after a long and sorry tale, it was scrapped in December 2010. It led to one of the most critical National Audit Office (NAO) reports I've ever read.[10] You can get a flavour of that from the headline quote from Amyas Morse, the NAO boss:

> This is yet another example of a Government IT project taking on a life of its own, absorbing ever-increasing resources without reaching its objectives . . . Essential checks and balances in the early stages of the project were ineffective. It was approved on the basis of unrealistic estimates of costs and under-appreciation of the complexity of the IT involved and the project was hurriedly implemented and poorly managed.

The NAO also commented on the 'inadequate' contract with the IT supplier, and said the Department for Communities and

Local Government (the UK central-government body that held responsibility for this fiasco) 'mismanaged the IT contractor's performance and delivery'.

The specification of the system quickly became an issue. It became clear that to meet various user needs across forty different sets of rules from different services, elements of the IT would need modification. But the Department (the buyer) didn't take real ownership of the issue, instead delegating responsibility to the supplier.

That was exacerbated by a constant revolving door for the programme's senior managers. To manage serious change (and serious contracts), it's essential to have continuity in terms of strong leadership. But here there were no fewer than five different Senior Responsible Owners, four project directors and five executives in charge of the technology-delivery aspect of the work. Only two senior managers worked on the project for its duration, one of whom, the project manager, was apparently on contract from a consultancy.

Initially, contract management appears to have been weak, with the Department failing to ensure that EADS 'followed the contracted approach in developing the system'. But the contract did not define interim milestones, so it was hard to hold EADS to account for delivery against clear targets. The relationship between buyer and supplier also decayed, perhaps not surprisingly as the programme ran into problems. But the payment schedule meant that EADS would be paid only once a key milestone for building and testing the system was passed. The delays to delivery led to cash-flow issues, which created further tensions in an already-strained relationship.

So here is a failure that could fall into many categories. It started with a lack of clarity around the requirement, and limited input from people who actually understood the issues. Those in charge arrogantly looked to impose a solution that hadn't achieved any real buy-in from the staff who would eventually work with it. Indeed, there was limited acceptance of the whole concept of regional rather than local fire-control processes and facilities.

Maybe there was also an element of naivety here; without getting into the technical details and going back to the original bid, we can't tell if EADS exaggerated their own capability or the ease of achieving the goals. But the NAO commented that the Department 'assumed that the development of the IT system would be straightforward, involving the integration of already customized components'. Was that wishful thinking, naivety and perhaps some lack of market understanding from the buyer?

What is very clear is that the contract management, change management and programme management were all flawed, to say the least. And the total cost to the poor old British taxpayer? According to the NAO report, after seven years of struggle FiReControl wasted a minimum of £469 million.

The Criminal Kind

In the last three chapters I've covered aspects of buying relating to events that take place in general after the contract has been signed, to demonstrate that failure can occur at any point in the buying process.

Looking back, in a few of the cases we've featured up to now (such as the Kenyan fence, or the sales of photocopiers to schools) the root cause might have been fraud or corruption rather than the more innocent interpretation of events. But whether it is specification mistakes, vanity, poor negotiation, allowing suppliers to create dependence or poor contract management, it has mainly been a lack of competence, understanding, planning, analysis or resource that has led to the failure. Failure has not occurred through malicious action designed to benefit someone other than the legitimate buyers, owners or taxpayers.

However, there is another whole category of buying waste, which costs the global economy and governments billions, probably trillions, every year. Fraud and corruption related to the buying process potentially affect every organization, and while some take stronger precautions and therefore suffer less, no one is immune to being targeted. That's where I'll go in Part 2, so get ready for some truly incredible stories – but not in a good way, unfortunately.

PART 2

Fraud & Corruption

12

The Fundamentals of Fraud

In December 2018 Nicholas Reynolds was found guilty of conspiracy to corrupt at Blackfriars Crown Court, following an extensive investigation and prosecution brought by the UK's Serious Fraud Office.[1] Two other employees of the industrial firm Alstom Group had already been convicted, and Alstom Power Ltd itself had entered a guilty plea to the charge of conspiracy to corrupt, back in 2016.

The case related to bribing officials in a Lithuanian power station and senior Lithuanian politicians, in order to win two contracts worth €240 million. The individuals involved also falsified records to avoid checks in place to prevent bribery, and Alstom companies paid more than €5 million in bribes to secure the contracts. As the judge said, 'This was a very serious example of bribery and corruption that beleaguers the civilized, commercial world and is a cancer upon it.'

Alstom Power Ltd was ordered to pay more than £18 million, which included a fine of £6,375,000, compensation to the Lithuanian government of £10,963,000 (so perhaps there was an element of overcharging in the contract, to fund the bribes) and prosecution costs of £700,000.

In addition, another part of the group, Alstom Network UK Ltd, was found guilty of one count of conspiracy to corrupt in April 2018, for making corrupt payments to win a tram and infrastructure contract in Tunisia. Lisa Osofsky, Director of the Serious Fraud Office, said: 'The culture of corruption evident within the Alstom Group was widespread. Their illicit activities to win lucrative contracts were calculated and sustained, undermining legitimate business and public trust.'

Criminal Intent

Unfortunately, fraud and corruption are common in the business world, including in many large and well-known firms, although the prevalence of such practices does vary between nations and business sectors. For instance, I suspect it is much more of an issue in the defence sector across the developing world (and in some developed countries, too) than it is in the healthcare sector in Denmark, the world's least-corrupt country.[2]

However, even in virtuous Denmark, no doubt a doctor somewhere is accepting a 'thank you' gift for trying out a new medical solution or drug, and this is still wrong, even if it is a long way away from a corruption scandal linked to military equipment that steals hundreds of millions of taxpayers' money.

Issues related to 'buying', in its widest sense, probably represent the single biggest category of fraud and corruption globally. It is not hard to see why. Criminals need to focus on where the money is, by the very nature of their approach. And buying accounts for most of the major spend areas for businesses and government bodies. There are alternative ways to extract money improperly from corporations, such as blackmail, or banking and investment frauds, and there is some non-buying-related fraud committed by employees. But buying accounts for trillions of dollars worth of transactions annually, so it is not surprising to see a whole range of fraud and corruption cases based around those processes.

Some of the low-level frauds are almost comical, as when a need to find money to buy horse semen is the root cause (see page 229). Indeed, even the larger incidents can sound humorous, particularly when characters such as Fat Leonard (see page 196) are the perpetrators. However, without wishing to sound too sanctimonious, do remember that there really are no victimless crimes.

Taxpayers usually lose out when it comes to fraud related to government bodies, like both of those cases mentioned above: the US Navy in the Fat Leonard case, and the UK National Health Service in the horse-related one. Even if the losses are covered by insurance for firms that are the victims, then insurance premiums will rise, or insurance-firm shareholders will take a hit. Someone always loses when a fraudster gains.

Perhaps the most annoying fact is that most buying-related fraud and corruption could be stopped; or where it is impossible to stop it happening, then it could be detected quickly. Take a simple example, such as the use of business credit cards for company or government business. If someone is trusted with a card, then it is hard to stop them using it improperly in every case, even if you block certain spend areas. But that inappropriate purchase can be picked up quickly, once it is made and recorded, if you take the correct steps.

In other instances the fraud should never be able to happen. Processes, systems or data can, and should, be in place to make it virtually impossible – for instance, stopping a false bank account being used in the 'invoice misdirection' fraud cases. But again and again I see organizations failing to take basic precautions and then, once fraud is discovered, claiming that 'this was a very sophisticated fraud'. In most cases that remark is nonsense and is a fig leaf for an embarrassed CFO or CEO who didn't have basic fraud-prevention measures in place.

Indeed, one way that fraud could be reduced globally is if CFOs in particular were told that their jobs are on the line. If a fraud takes place on their watch, which could have been prevented through simple actions, then they'll be fired for incompetence. Implement this and there will be a measurable drop in such cases very quickly.

Why Did You Do It?

Buying-related fraud can be committed by people who work within the organization, or by those inside colluding with outsiders, or by outsiders without any internal involvement. And looking at my experience, whenever a fraud has been discovered in an organization that I've been close to, there are almost always a couple of immediate reactions where an insider is involved.

The first is, 'Good gracious, I would never have expected it to be him/her.' Very often, the fraudster is a pillar of the community, a dedicated family man or that nice lady who sits quietly at the corner desk. The second question is 'Why did they do it'? That is often difficult to answer.

But there is a level of potential return, and a calculation about the chances of discovery, that might tempt many of us. I've never claimed that I would *never* commit fraud. I'd like to think my moral values, instilled by my parents long ago, mean that I simply could not contemplate it. But can you be absolutely certain you would say no, if a foolproof method was presented, with a suitably huge reward (perhaps involving other 'benefits' as well as buckets of cash)? I don't think I would succumb, but who truly knows?

However, it is surprising that so many put their reputations, jobs and happiness at risk for small gains. I worked in a firm where two senior managers, on six-figure salaries, were fired for a petty-expenses fraud that was at best going to make a few hundred dollars a month. They stayed in hotels on business and got a bill the night before they left using the 'express checkout' facility. They then made a phone call on the hotel's system just before they checked out, so the hotel sent them a second bill for the new, slightly higher amount. Both bills were then submitted, with a suitable gap between the submissions, as part of their expenses claims.

Ingenious, but also stupid, and it didn't take much checking before their scam was uncovered. But why risk everything for peanuts – what drives such behaviour? Is it the thrill of doing something wrong? Or a weird, anti-establishment 'giving it to the man' feeling, even when you are apparently an establishment figure yourself?

In other cases the driver is more obvious. In another organization a mid-level finance manager extracted funds from a bank account that wasn't being monitored as carefully as it should be. This was a fraud that was always going to be spotted eventually, but his driver was gambling debts, and he hoped to be able to pay back the money. He confessed everything, and said his debts were owed to the terrifying 'Peterborough Greyhound Racing mafia'!

Anyone who knows Peterborough – a fairly sleepy, small cathedral city in the east of England – might be surprised that such an organization exists, but it was enough to ruin this man's life, which was sad. You can have sympathy sometimes for perpetrators, without condoning their crimes in the slightest; but sometimes it is pure greed, and if initial fraud is successful, people are encouraged to continue and build on the first transgression.

Going back to that other commonly heard comment, typically 'I can't believe David turned out to be a crook', my experience suggests there isn't a type, a certain look, behaviour or personality that enables us to spot the likely fraudster. A senior manager, who had a matrix reporting line to me when I was a purchasing director, was convicted of a significant fraud that involved fake purchases of non-existent equipment. He was supported by a third party, possibly from another country (according to what the police told us), but when this came to light, everyone said, 'But he's such a lovely guy.'

While some 'fraud' may take place because the perpetrator genuinely doesn't realize what they are doing is wrong ('I thought I could claim for dinner when I stay late at the office – I could at my last firm'), broadly I'd suggest there are three drivers or motivations for fraudulent behaviour.

Financial need – this is the most obvious, often driven by a problem such as debt (as in our Peterborough example). That may be caused by factors outside the individual's control, such as mortgage-rate increases, or more avoidable drivers, such as a liking for expensive living or expensive company of the opposite sex. Gambling is another common factor that leads to debt and crime.

*A **psychological defect*** (as we might call it) – perhaps this is a pathological desire for excitement, or a 'cry for help', as the media put it. This is often unrelated to a tangible need for cash and doesn't appear to be simple greed. The case of the hotel bills above might be an example of individuals motivated by the thrill and risk.

*A **sense that 'I deserve it'*** – while this could be considered pure greed, it is often driven by a feeling of being undervalued in a job, or by jealousy of others. Perpetrators may even feel they are morally justified, and academic research suggests this is a major driver of fraud.[3] It's evident when the secretary or administrator working for a rich individual (such as J. K. Rowling[4]), or a junior person in a firm where top staff are highly rewarded, helps themselves to some of the boss's wealth. Or it might be a lowly-paid buying person in a decision-making role relating to high-value contracts. That is a particular issue in the developing world, where public employees may be poorly paid, but often deal with huge amounts of money.

So you don't know who might be tempted; you can't identify likely villains in advance; you can't always explain why they're doing it; and every organization of any size is vulnerable to people trying to commit fraud. But don't despair. There are ways to make it much more difficult, which I'll look at in depth later.

Everythang's Corrupt

Why should these issues matter to organizations and to us, as taxpayers and citizens? The most obvious answer is the potential for financial loss. Some case studies here cost firms or governments (taxpayers) many, many millions or even billions. Sometimes the loss may be recovered from insurance; but often it is not, particularly if the organization has left itself open to the fraud, as is often the case. At its extreme, the alleged frauds (mainly contract-related) around the Sochi Winter Olympics are thought by some to have cost the Russian nation and its long-suffering citizens some $15 billion.

But this is about more than just financial loss. There are reputational risks, too, for both firms and governments – or whole countries, if you consider the reputation of Nigeria, for instance, tainted fairly or unfairly by the perception of widescale fraud and corruption. Reports of corruption around a firm can lead to a perception of poor management or weak controls, with consequences for shareholder value. Or it might make other firms less keen to do business, or good people less likely to take up jobs there.

Increasingly that reputation issue is combined with a regulatory imperative. Both the Foreign Corrupt Practices Act in the USA and the Bribery Act in the UK now put greater focus on firms taking action against being the source of bribery and corruption: by bribing to win contracts, for instance. What used to be seen as acceptable 'custom and practice' – a few payments to facilitators or politicians to help win that defence contract in a distant land – is now, quite rightly, seen as criminal.

Finally, this matters because it has wider effects beyond the organizations directly involved, as corruption can distort normal business and even social practices and priorities. For instance, if firms know that bribing government officials is the best way to win public contracts, then a firm will focus its resources and efforts into doing that effectively. They will worry less about writing a good bid, developing better products or services or performing the work well.

The knock-on effect is that decent firms start thinking, 'What's the point?' They either move over to the dark side and start down the bribery route or withdraw from the market, the customer or even the country altogether. This can lead to a downward spiral, where supplier performance gets worse and worse and corruption becomes endemic, spreading into personal life, too (the 'additional' payment to get your driving licence approved in certain countries, maybe).

Money extracted from the public sector also takes resources

away from potentially valuable spending in other areas. If a government is paying over the odds for what it buys, to fund kickbacks and bribes, then it spends less on health provision, education or other key services. While 43 per cent of Africans still live in poverty, corruption costs the continent around $50 billion every year, according to Transparency International.[5] All these factors mean that everyone – whatever our jobs or political affiliation – should have a vested interest in minimizing buying-related fraud and corruption. It is, ultimately, in all of our interests.

In the next few chapters we'll examine some of the most common types of buying-related fraud and corruption, which in most cases can be traced back to failures in terms of weak buying processes, policies, management or systems.

13

Who Am I Really Buying From?

Marine hoses aren't the most exciting-sounding item, yet they're essential for boats, ships, oil platforms and other marine facilities. They transport crude oil to and from ships, rigs and production sites, and more generally are used to move fuel, water and other unmentionable substances between ships and shore. They are essential, not optional.

Given that, the market is a large one, and six firms succumbed to temptation by forming a cartel to carve it up between themselves to avoid competition. That worked successfully for no fewer than twenty years, until one firm, the Yokohama Rubber Company, became a whistle-blower in 2006 in exchange for immunity from EU fines. The EU launched surprise inspections at the five other cartel companies and interviewed directors, and the FBI even secretly taped a May 2007 meeting of the cartel in Houston (yes, even a marine-hose cartel can get exciting!).

Eight people were arrested, and seven pleaded guilty, serving prison terms of up to two years; one was acquitted. In 2008 the UK's Office of Fair Trading obtained convictions of three UK-based executives involved in the cartel, including the managing director and marketing director of Grimsby-based Dunlop Oil & Marine Limited and they went to prison with sentences of twenty to thirty months.[1]

The European Commission imposed a total fine of €131 million on the five culprits – Bridgestone, Dunlop Oil & Marine/Continental,

Trelleborg, Parker ITR and Manuli.[2] The cartel had been running for some twenty years, between 1986 and 2007, and presumably buyers had no idea what was going on. The members fixed prices for marine hoses, allocated bids and markets and exchanged commercially sensitive information to support the fraud.

Cartel members referred to their 'private markets' and agreed upon a dozen or so pages of detailed 'cartel rules' to define their behaviour in the market. The firms met regularly to fix prices and exchange sensitive market information, and there was even a formal 'coordinator', who allocated the incoming customer work. Cartels don't just run themselves, you know.

Let's Stick Together

> People of the same trade seldom meet together, even for merriment and diversion, but the conversation ends in a conspiracy against the public, or in some contrivance to raise prices.[3]

I quoted Adam Smith earlier when supplier dependence was discussed, and his work is as relevant when discussing cartels. The great economist suggested that businesses are not naturally public-spirited entities, neither do they intrinsically support open and free markets. Given the chance, firms may form cartels because they can extract greater profits from a market by working together than they can in a well-functioning and genuine market system.

A cartel is defined as a group of apparently independent producers whose goal is to increase collective profits by means of price-fixing, limiting supply or other restrictive practices. It is, in effect, a fraud carried out on those who buy from cartel members.

The formation and operation of cartels (and other supplier collusion mechanisms) must be one of the oldest types of fraud. There are records of tradespeople getting together to control markets in guilds or other groups going back centuries, with the aim of benefiting themselves at the expense of the buyer, whether that was an end-consumer or other businesses. 'Cartelization' in Germany in the nineteenth century was a major political issue, leading to early examples of legal control on these groups.[4] Germany created Europe's first laws and institutions to protect competition and control abuse back in the 1920s, including the formation of a 'Cartel Court' to analyse and decide on specific cases, in part as a response to high post-Great War inflation in the country.[5]

The marine-hose example was a classic case of cartel fraud on an industry-wide scale, and despite the best efforts of governments, similar actions are undoubtedly taking place today in different industries and markets. Competition authorities around the world are always investigating multiple accusations of cartel or monopoly-type behaviour, but the temptation can be too much, and there are always opportunities. Every time a bunch of competitors meet at an industry event, conference or trade show, a casual discussion can easily turn into something sinister, with unpleasant consequences for customers.

Which markets are most vulnerable? It's clear that it is easier to set up, control and sustain a cartel in markets with a relatively small number of players. But geography also comes into play here. The construction market in most countries includes many firms, yet that sector has seen cartels thrive on a limited geographical basis or in a specialist sub-market, where the number of players is smaller.

One cartel in a relatively tight market was formed by six huge European truck manufacturers.[6] Daimler, MAN, Volvo/Renault, DAF, Iveco and Scania are facing billion-dollar damages claims from their customers, mainly logistics and transportation firms, for illegal price-fixing. By April 2019 more than 7,000 transport companies from twenty-six countries had filed more than 300 claims in the German courts. That follows fines of €2.9 billion on four truck manufacturers imposed by the European Commission in 2016/17.[7] The Commission found that between 1997 and 2011 the truck manufacturers exchanged information about prices, price increases and when new emission technology would be launched. They also passed on associated costs to their customers.

Building the Perfect Beast

However, some cartels have many more participants – indeed, some have been eye-wateringly huge. In 2008 the UK's Office of Fair Trading (OFT) accused 112 construction firms of colluding to fix prices when bidding for public contracts for local councils and National Health Service organizations.[8] The OFT outlined 240 alleged cases in contracts worth more than £3 billion, although in the end the proven cases amounted to 'only' £200 million worth of projects.

Fines totalling £129.2 million for collusion were ultimately imposed on 103 construction firms in England, ranging from some of the biggest in the country to relatively small regional players. Eighty-six of those received reduced penalties (a 35–65 per cent discount) because they admitted their guilt. But it is amazing in this case that so many firms kept this going for so long, given the scale of the operation.

Credible estimates suggest this type of fraud increases prices by some 10 per cent (unless it is endemic and often politically-related fraud, which will be covered later). That doesn't seem a huge amount, but firms avoid over-inflated pricing, because that can be too obvious. The council buyer struggles to know whether that new sports centre should cost five million or five and a half million to build, but they might realize that ten million is suspicious. So a 10 per cent uplift is probably a 'sensible' level for operating a sustainable cartel.

The investigation in this UK case started in 2004 when a savvy health authority in the East Midlands reported irregularities in a bidding process. The process seemed to involve 'cover pricing', where firms involved agree who will win a contract, and others then bid slightly higher prices than the agreed winner. In other cases, those who lost were given compensation payments by the winning firm, paid through false invoices. Forty firms quickly admitted bid-rigging and pleaded for leniency.

Some claimed that cover pricing wasn't really fraud – it was just a way for busy firms to avoid winning contracts they might have struggled to deliver, without withdrawing from bidding, which might upset the buyers. But in my experience, buyers don't mind an honest supplier saying 'Sorry, we're too busy' – better that than my suppliers discussing pricing and subverting the whole competitive process.

As the UK government said when reporting this case, cover bids give 'a misleading impression to clients as to the real extent of competition. This distorts the tender process and makes it less likely that other potentially cheaper firms are invited to tender.'[9] So potentially good firms may miss out on a chance to win the work because others are taking up the bidding slots.

In another April 2019 example, the UK Competition and Markets Authority (CMA) found three firms provisionally guilty of price-fixing in the supply of site safety equipment used to protect excavations and similar groundworks on large construction products.[10] The firms were M.G.F. (Trench Construction Systems) Ltd, Mabey Hire Ltd and Vp plc, which supplied these products for projects such as major housing and road developments, railway-line works and water-pipe upgrades. (It is worth saying that at this point findings are provisional and do not necessarily lead to a decision that the companies have breached competition law.)

It was Mabey themselves who blew the whistle on the cartel, and hence will avoid fines – somewhat ironic, given that the firm was prosecuted a few years ago by the Serious Fraud Office for bribing Saddam Hussein's Iraqi regime to win contracts for building bridges.[11]

Have a Drink on Me

Cartels stand more chance of success where a market has high barriers to entry. Cartels can be broken by new firms coming into a market, undercutting artificially high prices charged by cartel members. But if it is hard for a firm to break into a market, that is less likely, and there are case studies which show that factor increases cartel risk. Where specialized products require significant capital investment to enter the market, that can lead to limited competition.

In 2016 the CMA fined three firms that supplied galvanized-steel tanks, often used to hold water for fire sprinkler systems in larger buildings, such as schools, hospitals and offices, a total of £2.6 million for collusive behaviour. In the UK there were only four suppliers of these tanks. They met for valid reasons, as they were part of a working group to discuss European sprinkling-system standards (back to Adam Smith – whenever firms meet, they are tempted to conspire!).

But from 2002 these firms started holding secret meetings where they agreed to fix prices, divide up customers and manipulate bids for contracts. That was to avoid customers being able to negotiate cheaper prices by playing the suppliers off against each other (what we might define as good, sensible buying). In the CMA investigation one witness said, 'We agreed that we all supplied fairly identical products and therefore should be able to charge a respectable price without the necessity to fight each other for each and every contract.'[12]

It was all carefully thought out, so that customers had the illusion of competition, but in reality the market was well and truly fixed. But the cartel was broken when one participant broke ranks and informed the authorities what was happening.

But it's not just heavy industrial products and construction where cartels can operate. In October 2019 reports claimed that the investigations unit of the Competition Commission of India (CCI) had concluded that some of the world's largest brewers – Anheuser-Busch InBev, Carlsberg and United Breweries – had colluded to fix beer prices.[13] The three firms account for about 85 per cent of beer consumption in the $7 billion India market.

It was whistle-blowing from InBev itself that triggered the

investigation, as the firm told the watchdog that it had detected an industry cartel. That led to raids on the three brewers' offices to collect evidence. The investigation found that executives from the three brewers were involved in discussions of beer prices before they were submitted to the regulators. That violated anti-trust laws, whether or not it led to higher prices to retailers or consumers. As the whistle-blower, InBev will probably receive a smaller penalty than the others, but the total penalties could be as much as $280 million.

New Kid in Town

If you are buying from a cartel, you believe that you have a single supplier, but in reality you are in a sense buying from that wider group of firms. So as a buyer, how can you avoid getting ripped off? The extended periods for which cartels can exist is evidence that it isn't easy, including the marine-hose example, which ran for twenty years. But perhaps technology will make it easier to detect and act against the bad guys. Advanced spend-analytics systems might highlight prices from different providers moving in sync, for instance, or not responding to raw material and market-price movements.

A pattern of certain firms not bidding for contracts, or the losing bids all coming in at similar prices, can ring alarm bells. Artificial intelligence may assist here eventually, and of course in this digital age, communications between cartel members are easier to track than in the days when everything was done verbally or on paper, which could easily be shredded. Emails and text messages are harder to destroy permanently.

But rather than waiting for technology to solve the problems, you have practical options if you suspect that you're playing in a cartel-infected market, as well as the option of getting the authorities involved, if your suspicions are significant.

Bringing in new suppliers – preferably new market entrants who are less likely to be part of the existing swindle – is a proven technique for breaking cartel power. Going back many years, when I was buying packaging, we looked at introducing an Indonesian supplier of the base packaging film, because we had suspicions that European manufacturers were (at the very least) sustaining artificially high prices, if not actually operating a full-scale cartel.

Another option is to make 'illogical' contract awards – so *don't* give all your business to the lowest bidder. This can cause internal friction within the cartel and put pressure on the trust between participating firms, trust these bodies need in order to function.

Gonna Make You an Offer You Can't Refuse

Fraud involving presentation of a false picture to the buyer has another set of examples, but rather than firms colluding, these involve concealment (for various reasons) of who actually *owns* firms that are chosen as suppliers. Perhaps less common than cartels, *misrepresentation* shares a similar concept, in that the buyer does not realize what is going on behind the scenes at their suppliers.

It takes place when a firm conceals the true nature of its business, history or ownership when it bids for, or carries out work for, the buyer. That may be because of issues around bankruptcy, ownership by unsavoury characters or similar factors that might otherwise discourage the buyer from awarding the firm business. The use of offshore firms, based in the Cayman Islands or other tax havens, can be a sign that the real owners want to keep a low profile (as well as wanting to pay less tax). Or there may be criminal connections that need to be concealed.

A construction firm in Queens, New York, hid its mafia ties to win work on the One World Trade Center, it was reported in 2017.[14] Vincent Vertuccio of Crimson Construction Corporation in Maspeth pleaded guilty to doctoring tax returns and obstructing an investigation to defraud the Port Authority of New York and New

Jersey. Vertuccio, who allegedly has ties to the Bonanno crime family, not surprisingly hid those connections and won an $11.4 million construction contract at the tower.

The EU procurement regulations contain strong provisions to enable buyers to stop firms from bidding for public contracts where (for instance) there is evidence of 'participation in a criminal organization' or evidence of fraud, corruption, money-laundering, and so on. But to be effective, that requires buyers both to have the right information to identify such firms, which is not always easy, and to actually want to stop that firm from bidding.

Back in 2012 the Italian government sacked the entire city council, all thirty councillors, in the southern city of Reggio Calabria because of fears over ties with the 'Ndrangheta crime syndicate.[15] The BBC reported that Italy's parliamentary anti-mafia commission had described the 'Ndrangheta as 'the country's most dangerous and wealthiest crime syndicate, overtaking the Sicilian Mafia and becoming one of the world's biggest criminal organizations'. Contracts for rubbish collection were awarded to a firm that was suspected of having links with the crime syndicate, and there were fears about other contracts awarded by the council, too.

But it is not only major organized crime that can lead to buyers or taxpayers being misled as to the true nature of the firms that are being chosen as suppliers. I had a personal experience of an unusual type of (attempted) fraud early in my career that fits under this heading.

Sweet Little Mystery

A reputable, market-leading firm was providing some quite specialist raw materials, from their manufacturing facilities in southern France. My firm needed some new product variants, which required development work and innovation from the supplier, as well as careful testing from our own in-house scientists.

We started receiving samples from the supplier's UK sales manager, which were tested by two of our own research scientists. Over some months, through trial and error, we gradually got closer to agreeing a new and acceptable set of products. And then, following a tip-off from an internal whistle-blower, my boss and I called the French HQ of the firm. Once we got through to the managing director, we asked if we could discuss these new products.

'Which new products?' was the reply. He knew nothing about the work that had gone on.

We flew out to meet him and eventually unravelled the mystery. Their sales manager, working in conjunction with our two internal technical people, were developing these new products, literally in one of their garages (and probably using some of our firm's laboratory facilities, too, we suspected). Between them, they had a fair bit of relevant knowledge, so they had made decent progress.

But at some point, if they had succeeded, we would have bought their materials and realized that we were paying invoices to another firm, not the one we thought was the supplier. How would that work? It seemed that the perpetrators thought that once they had acceptable materials, they would come clean and, in the absence of other options ('supplier dependence' again), we would agree to buy from their new business. I don't think for a moment we would have done that, given their deception, but matters didn't get that far – and, of course, three people lost their jobs.

Now that was a very unusual attempted fraud, but there are more common variants within this 'misrepresentation' heading. In certain parts of the world it is quite common for individuals to conceal the true ownership of a business – perhaps because the owner would not be allowed to front a bid if his or her identify was known. Public procurement regulations, as mentioned above, often ban buyers from choosing firms whose owners have not paid their taxes or have other misdemeanours in their past. So setting up 'false front' arrangements can get round this barrier.

In other cases, I have come across the use of fake firms to make a bidding process look more competitive than it really is. That might be carried out purely by someone on the supply side; or it could be a fraud that involves insiders on the buying side, too.

Take Me to Your Leader

In some countries, including South Africa and the United States, well-meant legislation has led to new types of fraud. Here, there are advantages when bidding for government work if you can show that your firm is owned by people who come from certain ethnic groups, or are military veterans, or perhaps female, disabled or 'disadvantaged' in other ways.

Whether these programmes really work is debatable, but their intentions are good. However, and inevitably, fraud intervenes. In 2017 the South African B-BBEE Commission said it had started investigating firms for possible violation of the laws and code of good practice relating to ownership of firms.[16] Their press release listed seventeen entities that were being investigated, including Netcare, MTN Group and Nokia Solutions and Networks South Africa.

In the USA, David Gorski was sentenced in 2016 to thirty months in prison and fined $1 million.[17] That was in connection with his recruitment of veterans as 'figurehead' owners of a construction company, which helped the firm receive government contracts, which were prioritized towards veteran-owned firms or skewed the selection towards such businesses.

In 2006 Gorski established a company, Legion Construction, Inc., and recruited a disabled Korean War veteran to act as the company's apparent 'owner', which would help obtain construction contracts set aside under the Service Disabled Veteran Owned Small Business (SDVOSB) Program. But other firms noticed that Gorski seemed to be running things, and he also set up various not-too-smart methods to extract money for himself.

In the case of the Sochi Olympics, it certainly appeared that various holding companies, agents and other structures were set up to conceal the ultimate beneficiaries of the $40 billion spent.[18] Many of those lucky people just happened to be friends and associates of the very top people in Russia, including President Putin (according to the Anti-Corruption Foundation).

In this, and in other cases highlighted here, there are no magic solutions for buyers. Be vigilant: use information sources, look for warning signs of cartels, and do your due diligence on suppliers and potential suppliers, including looking at people behind the firms – owners, directors and top management. And don't always take things at face value; remember the value of a touch of cynicism when dealing with suppliers, which is also useful in terms of detecting frauds of this nature.

Of course even if there isn't a cartel operating in the market, you can end up choosing the 'wrong' supplier because of various types of fraud and corruption, and the next chapter looks at how that can happen. This is one of the most ubiquitous types of fraud, ranging from minor and relatively unimportant cons to some of the biggest examples of waste you will ever see in the business or government world.

14

Fixing the Supplier Selection

In 2018, according to the *Gulf News*, the Abu Dhabi State Audit Institution (SAI) – the independent watchdog looking over government expenditure in the UAE (United Arab Emirates) – uncovered financial irregularities in public procurement amounting to more than Dh60 million ($16 million).[1]

The SAI is the supreme external-audit institution of the United Arab Emirates. It looks over ministries and federal departments, the Federal National Council and even corporations in which the country has a stake of more than 25 per cent. The issues uncovered by SAI included 'failure to conclude procurement contracts and service contracts that were not implemented in accordance with the terms and conditions specified in the contracts'.

In one clear example of the supplier selection being fixed, a state employee worked with accomplices to give contracts to a company that he owned. The financial investigators found that payments of more than Dh60 million were then authorized, using forged documents. A case has been filed with the judiciary to take action against the suspects, with the aim of getting the money returned to the public purse.

All Around the World

Corruption is not perceived as a huge issue in the UAE, and the country lies twenty-third in the Transparency International corruption index, a similar position to France and the USA, for example. Certain regions – Eastern Europe and much of Africa – definitely have worse reputations for buying-based corruption, but the UAE example shows that nowhere in the world is fraud-free.

Often the fraud methodology is based on corrupt processes to influence decisions about the firm, or firms, that win contracts. Examples run from one end of the scale to the other, in terms of magnitude. It happens in small firms and in individual schools or hospitals, and runs through to politicians and business owners who have colluded to strip billions from national treasuries by awarding contracts to favoured firms at inflated prices, or for goods and services that are never delivered.

Government buying is particularly vulnerable to this problem in the developing world, where staff earning low wages are involved in awarding valuable contracts, with obvious temptations when a supplier suggests they might help them win work. But it happens everywhere, with subtler mechanisms in play in more affluent countries.

In 2016 Deputy Public Protector Kevin Malunga told the National Anti-Corruption Symposium at Durban University that he estimated that in the South African public sector 'more than 60 per cent of tenders currently are contaminated with some sort of influence that is untoward'.[2] He quite rightly blamed those on both sides of the table – the buyers and the sellers. In South African politics, corruption has become a huge issue and perhaps the biggest challenge for the government.

Meanwhile in the Ukraine, a country that at one time had the reputation for being the most corrupt country in Europe, a military commander in 2019 was caught red-handed receiving a $10,000 bribe from a local entrepreneur for choosing his firm to install CCTV at the military base.[3] But one suspects there are many larger unexposed cases in that country.

Indeed, corruption of this type, focused on choice of supplier, can be endemic in some countries and organizations. In the public sector it can be small-scale, like the last example, or huge, like the Petrobas case study (see page 246). But at its core, the corruption revolves around the process used to select a supplier, or suppliers, to carry out a particular contract.

Mother and Child Reunion

At the lower value end of the scale, cases can often be rather sad, even if the perpetrator's behaviour is inappropriate. One such UK case was reported in 2019 when a head teacher, respected for his professional skills, was found to have acted dishonestly. Thomas Marshall, who ran Baverstock Academy in Birmingham, was even featured on TV for the school's work in helping disruptive children. But a disciplinary panel found that he hired his mother's consultancy firm without following the right processes, and without declaring what was an obvious conflict of interest.[4]

There was no tendering process for the work that Stone Educational Consultants carried out for the school; there wasn't an appropriate contract or service-level agreement on file; and Marshall didn't get sign-off from the school's Finance Committee, even though the cumulative expenditure was over the £20,000 threshold for approvals. The disciplinary panel found a number of the allegations proven, and prohibited Marshall from teaching indefinitely. He cannot teach in any school in England, although he may apply for the prohibition order to be set aside after March 2021.

Again it is easy to wonder why someone in a senior position, respected and well rewarded, would risk this for a relatively small sum of money. Maybe Marshall honestly didn't think he was doing anything wrong; but then why did he not declare the personal connection with the firm? Another failing was that purchase orders were raised after invoices were received. This is a very common failure, but it is without a doubt bad practice.

All 'bout the Money

Fraud connected with a supplier winning work, when possibly they shouldn't, is very common. This step in the process, though, is often a precursor to other fraud mechanisms that are discussed later on. So, for instance, if a supplier is going to invoice more for work than they should, in collusion with someone inside the organization who can authorize payments, then initially that firm has to actually win the contract. But the simplest and probably most ancient form of buying-related corruption is the simple bribe. I give you money; you give me the contract, and that is the end of the affair. I might even turn out to be a good supplier, providing goods or services that offer excellent value.

This is still the basic mechanism at the heart of huge amounts of corruption globally. It's wrong to think it doesn't happen in developed, democratic countries, but it has been more prevalent in less-developed regions and where the democratic rule of law is less established.

However, it has too often been supposedly blue-chip and honourable 'Western' firms that have been doing the bribing of individuals in foreign countries, as in the SNC-Lavalin affair. The huge Canadian company provides engineering, procurement and construction (EPC) services in industries including infrastructure, mining, oil and gas. But not only was the firm winning work in Libya through dubious practices, but repercussions turned into a political crisis in Canada in 2019 when Prime Minister Justin Trudeau demoted Jody Wilson-Raybould, his Attorney General (the chief law officer for the government), allegedly because she wouldn't agree to the firm escaping a full trial by paying a large 'fine' instead.[5]

Press reports suggested that Trudeau wanted the firm to escape serious punishment, and an embarrassing trial, because of its importance to the Canadian economy and employment, while the law officer felt that any wrongdoing should be fully exposed in court. There were also reports of wider disagreements and distrust between the two politicians.⁶

The case against SNC and two of its subsidiary firms started years before the political crisis, though, dating to when the firm worked in Libya between 2001 and 2011. A senior executive in the business allegedly established close ties with Saadi Gaddafi, son of the dictator Muammar Gaddafi. Court documents allege that the company offered bribes worth $47.7 million 'to one or several public officials of the "Great Socialist People's Libyan Arab Jamahiriya"', as Gaddafi called Libya during his rule, which ended bloodily when he was killed in 2011.

SNC and its subsidiaries are also alleged to have defrauded various Libyan public agencies of approximately $129.8 million. It's not clear exactly what form that fraud took; but any firm that pays major bribes needs to make the money back in some way – a quid pro quo, as it were – so overcharging or similar approaches often follow bribery in these cases.

When charges were announced, Canadian Police Assistant Commissioner Gilles Michaud said, 'Corruption of foreign officials undermines good governance and sustainable economic development... The charges laid today demonstrate... Canada's international commitments and safeguard its integrity and reputation.'⁷

In recent years a strengthening of laws against bribery has targeted those firms that might be offering corrupt payments. The 1998 Corruption of Foreign Public Officials Act (CFPOA) in Canada makes it an offence for persons or companies to bribe foreign public officials to obtain or retain a business advantage. Similarly, the United States Foreign Corrupt Practices Act of 1977 prohibits US firms and individuals from paying bribes to foreign officials in furtherance of business deals. In the UK the Bribery Act of 2010 deems it an offence if a commercial organization 'fails to prevent persons associated with them from bribing another person on their behalf'. These steps have certainly focused minds within firms, and some high-profile prosecutions have helped to make the culture of bribery somewhat less prevalent.

I Choose You

Where the fraud is simply a corrupt payment to secure the contract, it may be that the winning bidder does a good job – perhaps even a better job than any other firm could have done. But just as a senior executive lying on his or her CV is wrong, even if they eventually do a great job, bribery should not be sanctioned in any situation. In many other cases the selection fraud leads on to other issues, such as overcharging, and the buyer really does lose out.

Fixing the selection may be achieved simply, by bribing one individual who has the power to make that supplier-selection decision. But sometimes matters can be more complex than that, and whatever the result, a flawed selection process is fundamentally risky for the buying organization. If supplier selection is not appropriate, that indicates something is wrong with people, process or both.

But how can the supplier-selection process be 'fixed'? Fraud can be originated by an insider, who approaches a firm they think is open to corruption, suggesting they could help the firm to win work, in return for whatever reward they seek (usually, but not always, money). Or perhaps the supplier approaches the insider, with a similar suggestion. Often a relationship first develops between supplier and buyer, then once there is some trust, one party suggests the fraud to the other. I suspect that much fraud is first talked about in a pub or restaurant, or even on the golf course.

There are then a number of ways through which the fix is delivered.

1. **Avoid competition to give the business to the favoured firm**
   'If you keep your proposal to under $10K, we don't have to run a competition.' Yes, I've been told that, when I've pitched a small consulting project or some training work to clients. Of course organizations would be silly to run lengthy, bureaucratic buying exercises for small amounts

of expenditure. But if you see contracts for millions awarded without competition, then you might reasonably be suspicious.

Most major organizations and governments have processes in place that mandate competitive processes for larger contracts, so this simple mechanism applies mainly to smaller contracts, or in organizations and countries where power is concentrated, often at a political level. There is evidence that many government contracts in Russia, for instance, are not openly competed, and instead are awarded to firms with connections to top people in the country. The campaigning organization Transparency International has reported on the links between non-competitive buying and firms whose owners are associates of individuals, including President Vladimir Putin and ex-Prime Minister Dmitry Medvedev.[8]

2. Restrict the field – take action to minimize the number of bidders for the work, so that favoured firms have more chance of winning

This is common in the government sector, where many contracts must be advertised. Again the motivation may not be fraudulent – it may be that a supplier is needed very urgently for a good reason, and the buyer is convinced they know exactly who the best one would be. But obviously the same tactics could be driven by less positive reasons.

Typical approaches include advertising a contract only on a hidden-away website page. The closing date for bids might be tight – an opportunity might be advertised on Wednesday, stating, 'Please answer all seventy-eight of our questions fully and submit your proposal by 8 a.m. next Monday.' An onerous tendering process – the seventy-eight questions – is another good way of restricting the field, and the favoured supplier can be tipped off in advance to make sure they can prepare the material in time.

3. **Pass on inside information to the favoured firm**
That information might be relating to other bids, or knowledge about what the buying firm is looking for. I acted as an expert witness in a court case relating to collusion of this nature some years ago. This case, which didn't come to court for many years after the actual events, related to services provided to the British Army in Belfast during the 'Troubles' in the 1990s. A local supplier of specialist services was tipped off about other suppliers' bids by non-military staff working for the Ministry of Defence in property management and purchasing. The favoured supplier was provided with information that enabled them to win competitive tenders for work, in return for money paid to the insiders.

This took years to come to trial, by which time various people involved were sick or even dead. And actually there was no evidence that the supplier ripped off the Army – indeed, as they got to know what other firms were bidding, they bid low to win the work, which by all accounts they carried out well. And remember that some firms saw staff in Northern Ireland murdered by the IRA for 'collusion with the enemy' because they worked with the British military, so there weren't always many willing potential suppliers. But the fact remains that, in return for (relatively small) amounts of money, the protagonists were corrupted and behaved in an inappropriate manner.

4. **Design the specifications and requirements to favour a particular firm**
The specification can be defined so tightly that, in the extreme, only your preferred supplier (who is of course

Fixing the Supplier Selection

paying you to make sure that happens) can meet it. This is a favourite technique in complex industries, such as technology and defence.

Sometimes organizations follow this route because they want to make the buying process quicker or easier. If only one supplier can meet your needs, then the selection process is very simple, so this approach may be followed for apparently good business reasons, rather than because of corruption. But if organizations habitually allow this, then it opens the door to staff using the technique for personal gain.

One case where corruption was allegedly involved is the long-running saga of the Indian government helicopter contract with AgustaWestland, worth some $466 million. India terminated the contract after accusations that the firm – owned by Finmeccanica of Italy – bribed officials. The Indian government said in 2014 they 'terminated with immediate effect the agreement that was signed with AgustaWestland International Ltd (AWIL) on 8 February 2010 for the supply of 12 VVIP/VIP helicopters on grounds of breach of the pre-contract integrity pact and the agreement by AWIL'.[9]

The allegations surrounded manipulation of the specifications, with suggestions that the company had used middlemen to bribe Indian officials to win the 2010 contract. The allegation was that a defence-ministry specification insisting that its new helicopters should be capable of flying at 6,000 metres (19,700 feet) altitude was cut, to benefit AgustaWestland.

In cases like this, it is quite hard to prove unfairness; I've had personal experience of deals where suppliers claimed that my organization's genuine specification was 'unfair', purely because their own product couldn't meet it – something we had no idea about, until they pointed it out. So if this helicopter example was indeed fraud, it was a sophisticated example, depending on a subtle change to specification to favour a particular bidder.

5. **Skew or corrupt the evaluation process**

   If multiple bids or offers are received from different firms, they have to be evaluated to choose a winner. Fraud here involves choosing a winner who shouldn't really win. That may be as simple as bribing whoever is carrying out the evaluation, so they give high marks to the favoured firm. Or, getting into 'buying geek' territory, processes in government and larger private firms may rely on complex assessments of rival bids, using scoring systems to mark each supplier's proposal against multiple criteria. Those might include price (or wider costs); how technically good the product or service appears to be; service levels; account management or administration provided by the supplier; the likelihood of future innovation from the bidder, and so on.

   Bids are scored, and different criteria weighted (so price might be more important than service, or vice versa) to get a final 'winner' with the best score. So you can already see that this is open to a certain amount of manipulation. If I want to help my preferred bidder, and I know they are expensive, but have a technically good product, then I weight 'quality' strongly and price less so. If I am in charge of marking the bids, too, then I'm in an even stronger position!

New Coat of Paint

The UK's National Health Service was on the receiving end of a sophisticated fraud based on that manipulation of tender evaluation. The NHS counter-fraud unit reported that a project manager in an NHS organization used his position to award contracts to a decorating firm run by one of his relatives.[10]

The tender asked potential suppliers to quote prices for different pieces of work. Some were in fact not really needed by the organization, but the project manager included them, and the responses from bidders in these cases were given a high weighting in the overall evaluation. Conversely, pieces of work that *were* required in greater quantities were given a low weighting.

The favoured firm, which knew about this scam, therefore bid *low* on the work that wasn't really needed and *high* on the work that was. But the evaluation process meant they won the contract because of those weightings. Once the firm was in place, it in effect overcharged for the work it did, by some £150,000, according to the report. In addition the contract was poorly managed, with limited record-keeping of work and payments.

Predictably, the organization had no process for recording the business interests of staff or of picking up such conflicts of interest in this organization. Neither was there any independent check on the evaluation process put in place, or involvement of the organization's professional procurement department.

As well as clearly fraudulent cases like this, there are many examples in the government sector where suppliers have challenged evaluation processes and marking decisions and the courts have found that the supplier-selection process has not been designed or executed 'properly'. Whether that is down to fraud or sheer incompetence is not always clear.

Love or Money?

It is remarkable how little it takes to corrupt some people, who risk their careers, marriages and reputations for stupidly small amounts of money, as in a number of cases discussed here. In most cases, the insider is bribed with cash. But sometimes there is more. In the case of Fat Leonard (see page 196), access to his crack 'SEAL team' of prostitutes was an additional incentive for naval officers to use his firm, as well as lavish dinners, casino chips and other benefits on offer to those he corrupted.

I came across an unusual case via a friend who was a board-level director for a mid-sized publicly-quoted company. He discovered that one of his senior managers was purchasing goods from a particular supplier, without having gone through a proper buying process. He eventually found out that his manager was placing orders based on certain 'personal services' provided by the sales manager of this supplying firm. But that didn't involve prostitutes – it was a direct relationship between the two individuals.

When he told me the story, I said I hadn't seen a case like this before – how was he going to approach this with his manager, who was making these purchases? Surely he would have to be fired. 'You're jumping to conclusions,' said my friend. 'It's not "he".' His manager was female, and it was a *male* sales manager who was offering himself to female buyers, in return for selecting his firm. That at least exposed my preconceptions and stereotyping! And I should stress that in the many years I had direct buying-management responsibility, no salesperson (of either sex) offered me anything more exciting than perhaps a nice lunch or conventional corporate hospitality.

It is worth saying, too, that there can be a fine dividing line here between fraud and acceptable practice. Firms look to corporate hospitality, for instance, to get close to executives in client firms, hoping that when it comes to the next contract, those clients will feel a little more positively inclined towards that firm compared to its competitors. But the event might also be a legitimate chance for the buyer to tackle the supplier's CEO about a problem, or to discuss sensitive future business opportunities.

Or a buyer might make some suggestions to a current supplier – one that is doing a good job – about how they could improve their service further, knowing that by including those ideas in their next bid, the supplier stands a better chance of winning. That might be done with good intentions and without the buyer getting any personal reward. Is that corruption? Probably not, but at some stage helping a particular supplying firm, and disadvantaging others, tips over into something else.

Once the supplier has been chosen, properly or not, delivery of products or services starts, and that's when the next type of fraud can manifest itself. That centres around what is actually provided to the buyer by the supplier – is it what the buyer expected it to be, what they specified and what they think they're paying for? I'll cover that after another of the special award-winners.

Bad Buying Award

'IT TAKES TWO TO TANGO' COLLABORATIVE FRAUD — FAT LEONARD AND THE US NAVY

The vast sums of money spent by the United States military (and defence forces the world over) have over the years attracted many cases of corruption. But this story is special, not least because of the main protagonist, who seems to have stepped out of a gangster film, or perhaps a Damon Runyon story. It is a worthy award-winner, being one of the most spectacular, far-reaching, serious and, frankly, incredible fraud stories of the last fifty years. Incredible, because of the sheer number of people who were involved, knew about it and were active participants in it. That's unusual, because a fraudster's golden rule is that the fewer people, the better. If one person controls a contract decision, there's a reasonable chance you can bribe them, but if six people are involved, at least one will probably be honest and the whole scheme will fail.

Singapore-based Glenn Defense Marine Asia (GDMA) carried out a range of tasks for ships visiting ports in the Pacific, including arranging for pilots and tugs; customs and catering; dock security; taxis; and food, fuel and supplies. GDMA was one of the largest such firms in the Western Pacific, with hundreds of US Navy contracts, and indeed contracts with other nations, too. The Malaysian owner of GDMA, Leonard Francis, was known as 'Fat Leonard' to his friends, as he stands almost 2 metres (6½ feet) tall and weighs in at 150 kilos (330 lbs).

He pleaded guilty in 2015 to fraud charges, admitting bribing US Navy officers over many years with (among other things) cash, prostitutes, Cuban cigars and Kobe beef. This was a scheme that went on for at least a decade, involving tens of millions of dollars in bribes to win hundreds of millions in business from the Navy – business for which GDMA overcharged across a range of services. The corruption was amazingly long-lasting, widespread and arguably endemic. Even though at first it sounds like the plot for a crazy comedy-thriller, it has raised serious questions about the culture within the US Navy, as well as processes and procedures.

The Navy officers were paid in return for favouring GDMA as a supplier of 'husbanding agent' services for ships visiting ports in the Pacific, and the methodology for this fraud was interesting and complex. First of all, GDMA undercut its rivals in bidding processes to win the contracts, helped by insiders who would 'put in a good word' on Francis's behalf.[11] But the bids were so low that the Navy should have questioned whether they were realistic – the first point to learn from this whole fiasco.

It seems likely that Francis had insiders who could tell him what competitors were bidding, too. Once contracts were won, ships were routed to ports where his company worked by the officers involved, which enabled Francis to submit fake or inflated invoices. These were presumably paid either without real checking of the detail or via approvals from those on the inside of the scam.

The fraud took place in awarding the contracts to GDMA without proper due diligence, then in the operation of the contract – routing ships to him without good reason, for instance – and then, in effect, in the mismatch between what was invoiced and what was actually delivered by his firm to the clients. So this is a fascinating example of how complex frauds have multiple touchpoints with the victim organization.

In early 2015 Francis pleaded guilty to bribery, conspiracy to commit bribery and defrauding the US government. His evidence led initially to half a dozen naval officers being charged with associated charges, and three rear admirals involved – Michael Miller, Terry Kraft and David Pimpo – were all 'censured' and allowed to retire. But investigations continued, and there has been a steady feed of news about further criminal charges since that first trial.

A Freedom of Information exercise by *The Washington Post* then suggested that the corruption was more widespread than had first been imagined. That included the realization that Navy personnel had more than once failed to stop this, even when whistle-blowers reported their concerns up the chain of command. Here is how *The Post* summarized the case in December 2016:

> The Navy allowed the worst corruption scandal in its history to fester for several years by dismissing a flood of evidence that the rotund Asian defense contractor was cheating the service out of millions of dollars and bribing officers with booze, sex and lavish dinners, newly released documents show.[12]

There were apparently no fewer than twenty-seven investigations into the firm over the years, led by the Naval Criminal Investigative Service (NCIS), but those considering the matter were never able to gather enough evidence to take action. It was only in 2013 that Francis was arrested and charged, pleading guilty to defrauding the Navy of some $35 million.

How did he get away with it for so long? The first fraud complaint was made in 2006 when Dave Schauss, a NCIS investigator, became suspicious of the GDMA contracts. But

Francis was alerted by an informant, Paul Simpkins, to the Schauss investigation. Simpkins, an Air Force veteran employed as a civilian contracting officer by the Navy in Singapore, managed to quash any inquiry and had Schauss's position eliminated. Even worse, Schauss was exposed as a whistleblower, and he said other officers 'made my life hell'.

That failure to act was also in part because at least one other NCIS employee was bribed; John Beliveau tipped off Francis about the investigations and is now serving a twelve-month prison sentence for his involvement. *The Washington Post* also claimed that when some staff tried to introduce a new ethics policy to counter Francis's influence, it was blocked by admirals who were friendly with him!

That last point, as well as the range of different people involved on the buying side, demonstrates that corruption can occur at any level in an organization. Indeed, senior people have more power, more discretionary budgets and therefore more ability to direct spending where they want it to go than their juniors. So perhaps top-level fraudulent involvement here should not come as a surprise.

However, what about the apparent lack of involvement or competence from professional buyers? Should the selection of the port services not have been a properly competed and open procurement exercise? If Francis did win competitive processes, how was he able to overcharge to the degree he clearly did? It turns out that 'professional' supplies and buying experts in the Navy were in on it, too; Paul Simpkins, the contracting supervisor mentioned earlier, was given $350,000 in bribes, as well as the service of prostitutes, and got a six-year prison sentence.

But perhaps even worse, given his seniority, was the involvement of Rear Admiral Robert Gilbeau. He was eventually sentenced to eighteen months in prison, three years of probation, 300 hours of community service and $150,000 in fines and restitution, as part of a plea deal in which he admitted his guilt in making false statements.[13] Gilbeau admitted lying about the nature and extent of his relationship with Fat Leonard. Prosecutors said the pair met twenty years ago during a port visit in Bali, Indonesia, and that Francis immediately began supplying the officer with prostitutes, lavish dinners and stays in hotel suites.

Over the years Francis learned that Gilbeau particularly liked to have sex with Vietnamese women – so he allegedly provided the Navy officer with appropriate pairs of prostitutes on at least three occasions. That was a common tactic apparently. A *Rolling Stone* article in 2018 claimed that Francis had pictures and videos of Navy officers partying with escorts and kept 'notes on the proclivities of individual members: who preferred Vietnamese twins, group sex or BDSM'.[14]

Francis would spend thousands of dollars on lavish meals for his contacts and hosted one multi-day sex party at a five-star Manila penthouse where, court records show, officers enjoyed 'a rotating carousel of prostitutes' and drank the hotel out of Dom Perignon!

Back to Gilbeau; it is worth quoting from the official statement from the US Department of Justice:

> ... in June 2005, Gilbeau was assigned to the office of the Chief of Naval Operations as the head of aviation material support, establishing policies and requirements for budgeting and acquisitions for the Navy's air forces ... In August 2010, after he was promoted to admiral, Gilbeau assumed command of the Defense Contract Management Agency International, where he was responsible for the global

administration of DOD's most critical contracts performed outside the United States.¹⁵

So a real supply-chain expert, with a glittering career, all in ruins. Gilbeau blamed his misconduct on post-traumatic stress, survivor's guilt and a head injury he suffered as a result of a mortar attack in Iraq in 2007. But this was the first time a US Navy admiral was incarcerated for a federal crime committed during the course of official duty, and the list of Navy staff convicted keeps on growing. Another former commander, Troy Amundson, received more than two years in prison and a $10,000 fine for accepting dinners, drinks and prostitutes from Francis. That made Amundson the twenty-first person to plead guilty in the case.

Another Department of Justice notice announced that Commander Bobby Pitts pleaded guilty in August 2017 to obstructing the investigation into Francis.¹⁶ Pitts admitted to providing Francis with documents and inside information about the investigation into GDMA's billing practices. One Navy captain even acted as a sort of PR adviser to Francis; Jeff Breslau pleaded guilty to criminal conflict-of-interest charges after it emerged that he had provided secret public-relations services for Francis.¹⁷ According to the US Department of Justice, the former officer received $65,000 for 'consultation services' to Francis's Singapore-based company, including ghost-writing emails to Navy personnel from Francis, and providing briefings to him in advance of his meetings with Navy officials.

The scandal reverberated right to the top. In 2015 Admiral Samuel Locklear was tipped as a potential chairman of the Joint Chiefs of Staff for the entire US military. But Francis told investigators that Locklear had been present at some of the opulent dinners. While the Navy cleared Locklear of charges, he didn't get the top job, a decision that may have been coloured by the accusations and his failure to take earlier action against Leonard.[18]

There are many lessons to be learned here, but I'll finish with the whistle-blowing issue. If organizations don't make it easy for honest people to expose what is going on, and have a failsafe route for concerns to be reported and acted upon, then there is a real danger that corruption will become more and more embedded, as in this case. Other lessons around the buying process, monitoring of supplier pricing and billing are key; but whistle-blower protection is a relatively cheap and easy way of reducing the chance of shocking events like this.

15

What Am I Really Buying?

It happens all the time. You order something online and then, when it arrives, it doesn't look much like the attractive picture on the website, or perhaps it really isn't what you thought you were buying. The second-hand car turns out not to have had 'one careful owner', but to have been assembled from three different written-off crashed vehicles.

A similar thing happens to an organization that doesn't end up with the goods or services it believed it was buying, because of fraud. It gets something it didn't want, or a product or service that does not meet the defined specification.

Sometimes it's hard for the buyer even to know they've been a victim. In March 2015 a case going back to the 1990s finally came to the courts. Global trading giant Glencore was ordered to pay $40 million to OMV Petrom S.A. (a large oil company, headquartered in Romania) by a UK court, for shipping oil of a lower quality than it was supposed to be to the firm in the 1990s.

Bloomberg reported that Marc Rich & Co., which went on to become Glencore International AG, sold about thirty-two shipments to Romanian state firms from 1993 to 1996.[1] But this was cheaper crude-oil blends than the specification that had been promised. The seller falsified documents to support the fraud, which made it some $40.1 million, according to evidence in court. The fraud only came to light when a former trader from Glencore spilt the beans to Petex, the firm that had organized the oil-importing process.[2]

This demonstrates a very hard-to-detect fraud: substitution of inferior products for what the buyer believes they have bought and should be receiving. Whether it is components, food ingredients or oil, it is often difficult to know, without extensive checking and testing, exactly what is being delivered. However, even in these cases appropriate technology, from document management to quality systems, can help make the processes more robust. Blockchain-based technology[3] is even coming into the picture to help verify exactly what it is that we're buying, by linking physical goods to an unalterable blockchain record so that the original 'provenance' of goods can be traced back to source.

Fake It

Some of the cases here relate to seriously criminal behaviour by salespeople or staff inside organizations, and as well as causing financial damage, their actions may even cost lives. The following story has comical aspects, but is tinged with very serious consequences.

UK-made 'magic wand' detectors (as they became known) were sold for a range of purposes for some years. They were used to search for explosives, cocaine and smuggled ivory in Africa. They were involved in the battle against suicide bombers in Baghdad, and in Thailand they were deployed by the armed forces in security sweeps. In Pakistan they helped guard Karachi Airport, and they protected hotels around the Middle East. Just hold the device in your hand, and the antennae would point to the hidden items.

But they were fake, basically being bits of plastic with an antenna attached. The publicity and packaging cost more than the devices themselves, but the aerial would swing according to the user's unconscious hand movements when the devices were used – a phenomenon known as 'the ideomotor effect', which is also seen in dowsing and Ouija boards.[4] The effect sees the body making a tiny and seemingly 'reflexive' or automatic muscular reaction, which is triggered without the awareness of the individual.

What Am I Really Buying?

The fraud started life in a humble manner – as a $20 plastic golf-ball finder called the Gopher in the US, which claimed to use advanced technology, 'programmed to detect the elements found in all golf balls'. With the addition of a new label and a new name – 'Quadro Tracker' – it was sold in the US to police departments and schools, to supposedly detect both drugs and explosives. The FBI declared it a fraud as early as 1996 after testing the product, then the scam transferred to the UK as 'the Mole'. A couple in Bedfordshire (Mr and Mrs Tree) bought cheap plastic parts from China and assembled devices in their garden shed, with the devices marketed as an all-purpose detector by a Somerset-based businessman, James McCormick.

It was then tested by a UK government scientist, Tim Sheldon, of the Defence Science and Technology Laboratory, who circulated his findings in 2001 to 1,000 government officials, saying that it was a 'useless lump of plastic'.[5] He warned government departments against the device, stating that 'it would be potentially dangerous to use'. But the fraudsters behind the bogus detectors repackaged and renamed them again, and looked for new markets where testing might be lax, and bribes could be used to make lucrative deals. According to the BBC report:

> The ADE-651 (supposedly shorthand for Advanced Detection Equipment) was sold to Iraq, Niger and other Middle Eastern countries . . . Some sold for as much as £25,000 ($40,178). The GT200 'remote substance detector' was sold by Gary Bolton mainly in Mexico, Thailand, the Middle East and Africa. The device retailed at £5,000 ($8,034) but the highest price it achieved was £500,000 ($803,000).[6]

Amazingly, the fraudsters even got a British Army regiment, the Royal Engineers, and the UK Department of Trade and Industry to help them promote the devices at trade fairs and through embassies in Mexico City and Manila. But demonstrations were rigged, buyers were bribed, and those who doubted the machines were told not to open them up as that would risk damaging the sensitive technology inside!

Neither the US nor the UK government ever bought the equipment, but it took until 2010 for the UK government to bring in export controls, and then only to prevent sales to Iraq and Afghanistan. Incredibly, because it didn't work, it seems that it didn't need a licence. Eventually, in 2014, the scam was stopped when McCormick was jailed for ten years, and others followed him into custody.

The mentality of those who made and sold the devices, and those who bought them because of bribes, can only be imagined, as they put lives directly at risk for their own gain. In Iraq, where bomb detection was a matter of life and death, more than 1,000 Iraqis were killed in explosions between 2008 and 2009. But it is likely that not everyone was bribed; some people just want to believe what they are told, if it sounds even vaguely plausible. Or buyers couldn't be bothered to carry out proper due diligence – a recurring factor throughout many cases of failure and fraud.

Some people will do anything to make money. So in terms of avoiding fraud of this nature, the answer is always to check *exactly* what you're buying. Make sure it works, make sure it 'does what it says on the tin', through testing, pilot projects, trials or try-before-you-buy arrangements of some nature.

Hot Blood

The Theranos story, the subject of John Carreyrou's best-selling and definitive book, *Bad Blood*,[7] has some remarkable similarities to the previous story.

Elizabeth Holmes, the founder of blood-testing company Theranos, similarly convinced many apparently sensible people that her invention could perform tasks that, in fact, it could not. High-profile individuals, including Rupert Murdoch and Henry Kissinger, invested in the firm, in the belief that her amazing technology could perform hundreds of blood tests on just a couple of drops taken from a finger prick. This obviously appealed to many who

What Am I Really Buying?

are not keen on having blood samples taken, a fact that Holmes stressed when promoting her firm.

But while her equipment was perhaps not as bogus as the bomb detectors, it was revealed that only a handful of the 200 tests on offer were carried out on the machines she had invented, and even then the results were unreliable. The other tests were carried out on standard devices, but not always in proven ways. When one of the Theranos laboratories was inspected in Newark, California, in November 2015, the inspectors concluded that 'the deficient practices of the laboratory pose immediate jeopardy to patient health and safety'.[8]

Buying failure comes into this because the pharmacy chain Walgreens spent $140 million with Theranos over seven years, hosting around forty blood-testing centres in their stores.[9] They got very little benefit from that and recovered some $30 million after a lawsuit and settlement following the eventual disclosure of the issues. Amazingly, as *Bad Blood* reports, Walgreens's own laboratory consultant, Kevin Hunter, had seen early on that something wasn't right with Theranos. But the executive in charge of the programme at Walgreens said that the firm should pursue the pilot because of the risk that CVS, their big competitor, would beat them to a Theranos deal.

Again, buyers *wanted* to believe that something was real, even in the face of mounting evidence that it wasn't. This relates back to comments around *believing the supplier* – those earlier examples weren't demonstrating fraudulent behaviour, but the principle is similar. It is easy for a naive or gullible buyer to be sucked into believing what the supplier wants them to believe.

Suppliers will take advantage of this tendency – whether it is the relatively innocent 'Yes, we can install this new IT system in six months' or the more dangerous 'This equipment will find hidden bombs'. And FOMO – the fear of missing out to the competition – is something else suppliers will use, and that can lead to bad decisions. It's not just physical goods, either. The top consulting firm selling its latest 'strategy toolkit' will mention that the potential client's biggest rival is also very interested.

Crazy Horses

Theranos may well have impacted negatively on some final consumers, while in the inferior-oil example, the outcomes might not have been too serious, although lower-grade oil might conceivably affect the longevity of equipment. But on other occasions a fraud related to the substitution of products can have obvious and unpleasant consequences.

In 2013 many people in Europe discovered that tasty beefburgers, cottage pies and other frozen or chilled ready meals, which they had enthusiastically consumed, had an unexpected ingredient – horsemeat. Many of the samples eventually tested were also found to contain pork, which was not good news for Muslim and Jewish consumers.

As well as boosting the popularity of vegetarianism and veganism, this episode caused financial problems for many firms. It raised big questions about understanding the food-supply chain, its vulnerability and corruption within it. Horse-meat itself is not intrinsically dangerous to humans, but this scandal opened up the risk that other harmful ingredients could get into food products – including drugs for horses, which are not suitable for human consumption.

UK-based retailer Tesco suffered; their somewhat ironically named 'Everyday Value Beef Burgers' were found, when tested, to have 29 per cent horse DNA. The source of the meat was Silvercrest Foods in Ireland. Most other samples in the initial testing had much lower percentages – less than 1 per cent – but the scandal soon spread. Food manufacturer Findus announced that in a sample of eighteen beef lasagne products that it tested, eleven contained between 60 and 100 per cent horse-meat. The source was a French firm called Comigel, which made the product in Luxembourg.

Comigel blamed French meat supplier Spanghero. Ultimately, it looked like the meat had originated in Romanian slaughterhouses.

It left there correctly labelled as 'horse', according to investigations, and was supplied to a Cyprus-based firm, Draap Trading Ltd (Draap backwards spells Paard, which is Dutch for horse). Draap was owned by a Dutchman, Jan Fasen, who had previously been convicted for horse-meat fraud in 2007. It was then sold to Spanghero, which insisted that it was labelled as beef when it arrived at their plant. Some processing took place, and then Spanghero sold it on to Comigel.

Fasen only came to trial in January 2019, so the reverberations continue from the scandal. But if nothing else, the case indicated how complex and multi-layered the supply chain for 'fresh' meat is. And clearly the more parties involved, the more scope there is for fraud of various types, including the relatively simple type seen here.

This scam must have required falsification of documents and sticking different labels on batches of meat, but it was not difficult to execute. But surely someone – aside from the perpetrators – must have realized there was something not quite right with the 'beef' they were processing? But perhaps the whole manufacturing process is so industrialized that these days real people don't get to handle the product.

In any case, this was another example of the need for organizations to understand exactly what they're buying. 'Provenance' is becoming a major issue in many markets, not just because of fraud in the food-supply chain, but also in areas as diverse as conflict minerals, timber and military hardware, where questions of social responsibility or adherence to legislation come into play, too. Knowing exactly where what you buy came from, what it is and who is supplying it seem to be basic issues, but establishing the facts is often harder than expected.

Whereas finding out that burgers were made of horse-meat caused real revulsion among some consumers, sometimes the fraud is less emotionally harmful. Having said that, if you really care about your kiwi fruit, prepare to be shocked. In April 2019

French authorities said they had stopped a long-running scam to label and sell Italian kiwis to supermarkets and other retailers as more pricey French fruit.¹⁰

'Kiwis imported from Italy were "Frenchified" during transport so that they could be sold at a higher price,' Virginie Beaumeunier, of the country's DGCCRF anti-fraud agency, said.¹¹

It is unclear quite how you might 'Frenchify' a kiwi fruit, but the fraud was serious, involving some 5,000 tonnes (4,920 tons) of produce over a three-year period. The agency said seven companies were facing charges and made some €6 million ($6.8 million) in illicit profits. This became known as KiwiGate, inevitably, but there was a serious point to it all, as Italian fruit is treated with pesticides that are not allowed in France. Anti-spoiling fungicides are used after harvesting to help shelf-life in Italy, which reduces the production cost compared to that of French growers.

Again, this is not a simple fraud for the buyer in the supermarket chain to uncover. But understanding the provenance of everything you buy, through the whole supply chain, should be the ultimate aim.

Life and Death

These issues are truly global; and away from fruit, there have been cases that are literally matters of life and death. In China, from 2013 to 2016, a series of vaccine scandals killed at least twenty-one children.¹² In 2017 regulators announced that pharma firms Changsheng and Wuhan Institute of Biological Products had sold more than 652,000 ineffective DPT (diphtheria, pertussis and tetanus) vaccines. In 2018 a whistle-blower from Changsheng claimed that the company fabricated production records of rabies vaccines, and China is now strengthening legalization around vaccine products and policies.

'Conflict minerals' or materials is another example where fraud has disguised the sources of what is being bought. These are natural resources extracted or obtained in war zones and sold to help fund and perpetuate the conflict. There is evidence that this process can extend the conflict; the worst recent example being the eastern provinces of the Democratic Republic of the Congo (DRC), where various armies, rebel groups and others from outside the country have profited from mining, while contributing to violence and exploitation. Inevitably this has fraud and corruption at its heart, as the provenance of what is being bought and sold is disguised; often fraud also denies income to governments or other rightful owners of the resources.

Diamond mining and trading ('blood diamonds') is a major example of this problem,[13] and even petroleum can be a conflict resource: ISIS used oil revenues from territories it controlled to fund terrorist and military activities. So there have been international efforts to reduce this trade, such as the Dodd–Frank Wall Street Reform and Consumer Protection Act in the US, which required manufacturers to audit their supply chains and report on the use of conflict minerals.

It is around these high-value items that blockchain technologies may prove important in terms of tracking and monitoring provenance. De Beers Group, the world's largest diamond exploration, mining and trading business, is working on Tracr,[14] a blockchain-enabled solution that will provide a single tamper-proof and permanent digital record for every diamond registered on the platform. If the project is successful, it will be possible to ensure that all the diamonds registered on the blockchain platform are conflict-free and natural. Presumably, buyers will be able to monitor their progress through the supply chain and from owner to owner.

However, it is harder to see how this could be applied to meat, for instance. Where products are not individually identified, but are broken up, disaggregated or combined into different batches, it is more difficult to use blockchain principles. But smart people are working on similar ideas across many products, so the issue around substituted, fake and counterfeit goods is one area where there may be less fraud in the future.

Out on Bail

When the focus switches to services, matters can also be complex. In July 2019 outsourced services provider Serco was fined £23 million by the UK's Serious Fraud Office over its behaviour as a supplier to the UK Ministry of Justice.[15] Serco provided electronic-tagging services for offenders who were not in prison but still under 'observation', but admitted three offices of fraud and two of false accounting between 2010 and 2013. That related to understating profits from its contracts, and the agreement meant that Serco wouldn't face criminal charges.

Serco had already paid £70 million as a settlement to the ministry after the firm, along with fellow services provider G4S, faced allegations that they were charging for tagging people who were either dead, in jail or had left the country. The whole episode has the flavour of a detective story in its own right. A senior executive at the ministry told me that when they got suspicious that the data being provided by suppliers didn't match what was really happening, they phoned some of the offenders charged for by the supplier to check.

'What's that strange background noise?' he asked one.

'That's the warders locking the cells,' the prisoner replied, at which point the executive realized something wasn't quite right (you don't wear tags if you're actually in prison!).

So this was a different type of alleged 'not getting what you paid for' fraud, where the supplier was accused of misleading the buyer about exactly what it was doing in terms of the volume of work provided. The wording of the 2019 settlement also suggests that financial information, which presumably fed into the payments made to Serco, was also manipulated, which is perhaps a somewhat different type of action – but deceitful, nonetheless.

The case highlights the difficulty of knowing exactly what has been provided in the services sector; it's similar to the issue of keeping track of days worked by consultants on major programmes, as mentioned earlier. Organizations face real challenges to ensure they get what they pay for, not just when buying physical goods, but also when services are involved.

Lost in Space

Back to physical products, and one further example that hit the headlines in May 2019. A report from the US National Aeronautics and Space Administration (NASA to its friends) claimed that the Taurus T8 and T9 rocket failures in 2009 and 2011 failed because a single component (a 'single frangible joint') supplied by SAPA Profiles Inc. (SPI) had not performed as expected.

The aluminium joints are designed to be broken apart by a small explosive, to split the nose cone and allow the rocket to shed weight in order to reach orbit. NASA said an internal investigation had found that the joints supplied by SPI to Orbital Sciences, the contractor that built the rockets, had not successfully split in either mission.

But this was not a simple mistake, according to the report. The investigators found that the supplier 'altered extrusion material property test results from failing to passing, and had falsely provided material property certifications',[16] which stated that the materials met specification requirements – which they didn't.

According to the Department of Justice, for nearly twenty years critical tests on the aluminium sold were falsified – 'tests that their customers, including the US government, depended on to ensure the reliability of the aluminum they purchased . . . Corporate and personal greed perpetuated this fraud against the government and other private customers.'[17]

Some elements of the fraud were very basic. A plant manager led a scheme to make hundreds of handwritten alterations to test results. Later, results were altered within the firm's computer systems to provide false certifications. While SPI paid restitution to NASA and others, and at least one executive ended up in jail, the firm disputed some of the claims made by NASA.

But the firm has been punished in terms of government business, being excluded from bidding for contracts since 2015. 'To protect the government supply chain, NASA both suspended SPI from government contracting and proposed SPI for debarment government-wide.'

Again, the case shows how difficult it can be to understand exactly what we are buying, particularly in the case of complex components and raw materials. Vigilance, testing, site inspections, audits ... all have their place, but if a supplier is determined to commit fraud of this type, it can be difficult to spot. However, in other cases the fraud is painfully simple – in concept at least – as the perpetrators simply use the organization's money as if it was their own. But even here, it can be tricky to detect exactly what is going on, as the next chapter will show.

16

Spending Someone Else's Money

A fraud case that was startling in its simplicity hit a National Health Service organization in the UK in 2014. As the *Liverpool Echo* reported, Madeleine Webster from Liverpool was sent to prison, along with her husband, for spending £160,000 of her employer's money on printer cartridges, then reselling them on eBay.[1]

Once the authorities looked into the case, they found that before Webster took over ordering ink cartridges for her department, the spend on this product was £2,500 a year. After it became her responsibility, the bill went up to more than £10,854 in 2008, then £27,000, £45,000 and finally £60,000 in the following years.

Colleagues noticed cartridges lying around for printers they didn't have in the office, and her boss even saw Webster and her husband removing items via the fire escape one night when she was working late. But until Webster accidentally sent an email to a colleague that linked to her eBay account, which showed what she was offering for sale, it didn't click that something improper was going on.

Which leads to the obvious question: how on earth did no one spot the spend on cartridges going from £2,500 a year to £60,000? Is there so much money washing around the NHS that this gets lost in the rounding? And who signed off the budget that *allowed* £60,000 to be spent on printer cartridges in a year? Or, if it wasn't budgeted, who authorized the orders?

The case certainly suggests that controls were very lax – not only in terms of the perpetrator being able to order goods, receive them and presumably authorize payment, but also in after-the-event budget checks. Surely a manager was responsible for the overall departmental budget and noticed the growth in spending on these items? Evidently not, and I'd suggest somebody deserved to be fired.

Thrift Shop

This type of buying-related fraud is driven generally by internal players, whether staff or contractors, and is often pretty basic in terms of its approach, as in the case above. It occurs when people simply buy something using the organization's money that is *not* used for the benefit of that organization. The culprit uses the purchase for their own purposes, or even resells for personal gain, as with Ms Webster. I'm not talking about taking the odd pencil home, or snaffling the biscuits left on the plate at the end of the meeting; the cases here are more serious transgressions.

It does take us into some rather grey areas, however. In earlier chapters I looked at some *failure* examples that weren't fraudulent – but would, in some eyes, be seen as a waste of the organization's money. The portrait of Chris Christie costing $85,000, for instance. And it would be good to know if firms that sponsor Formula 1 motor-racing teams actually do a full business case and cost/benefit analysis, or whether the real reason is that the CEO loves hanging around with racing drivers and fast cars! However, while you or I might think this is wasteful, if the spending decision is corporate and involves multiple parties, it is unlikely to be truly fraudulent, and cases of fraud here are very much focused on personal gain.

Who Let the Dogs Out?

Just because someone works for a charity and appears to be a 'good person' doesn't mean they are necessarily immune from temptation when it comes to stealing from their own organization.

Spending Someone Else's Money

An article in the *Miami Herald* featured an alleged fraud carried out by the head of purchasing for Miami-Dade's Animal Services Department.[2] That is a public agency which runs the county's animal shelter in Doral, along with control and enforcement work. Police called it 'a cagey scam' (think of stray dogs being looked after by the agency) and should themselves have been censured for dodgy puns. The case started with the classic kickback fraud, receiving bribes in return for giving business to a particular supplier, then graduated into a 'spending someone else's money' variant, with an invoicing element, too.

Michael Garateix was charged in April 2019 with four criminal counts, including theft of more than $100,000. Investigators said he and his more junior colleague initially demanded a kickback of 10 per cent from a local supplier of janitorial products. If the firm's owners refused, then according to the arrest warrant, the buyers threatened that they wouldn't be used again as a supplier.

From there, the alleged perpetrators moved on to inventing transactions. The report suggests that they persuaded the firm that had paid kickbacks to collaborate by issuing fake invoices for goods that the department never received, then pocketed the money paid out for the apparent purchases. The two suspects are also accused of using passwords from other colleagues to bypass procurement controls designed to catch fake orders – so clearly the organization did have some controls in place, even if they didn't work fully here.

The police allege that the scheme started in 2016 and that fake orders worth some $140,000 were processed, but the scheme was discovered when other colleagues started to notice unexplained budget overspends. So the organization deserves credit that at least there was a budgeting process and anomalies were noticed, unlike the NHS organization that opened this chapter. The other basic defence against such fraud is to ensure there are multiple approvals and sign-off for significant spends, but here, with two people involved, and perhaps the use of passwords from other staff, there was a level of sophistication from the alleged fraudsters. These activities aren't always easy to spot; but audits are another recommended tool that can help protect against this type of fraud.

Strapped for Cash?

In recent years many organizations have started to use 'purchasing cards' or 'corporate cards' to allow staff to pay for (generally low-value) company purchases. The process is similar to that which we use for personal credit cards, except that the monthly bill is picked up by the organization, rather than the individual card-holder.

This has generally been a useful buying tool, as long as the right safeguards are put in place. It makes life easier for users to buy low-value, irregularly required items they need, and gives the organization a robust audit trail, in terms of what has been bought. In fact it is harder to pick up fraudulent spend when organizations are using 'conventional' methods of ordering and payment for items. If someone can place an order on the phone, sign the invoice, then authorize payment through accounts payable, then spend is usually less visible than that made through the card route.

But some people can't resist the temptation to use the card as if it was their own. Sometimes they try and disguise the spend – so a lap-dancing club becomes a 'restaurant', used for a client meeting. Or a car-repair bill for their partner's car is presented as if it was for their own company vehicle.

Now the availability of spend data has led to newspapers looking to make a good story out of the potential for fraud. In some cases real issues have been discovered. A 2018/19 investigation by *The Post and Courier* newspaper in South Carolina discovered that Charleston County had issued 520 employees, out of the total staff head-count of 1,830, with purchasing cards,[3] – which seems rather a lot.[4] Records obtained by the newspaper showed that one employee used a card to buy 'faux birch logs' and 'owl jars' on Amazon (no, me neither . . .). They were delivered to her house, which was an accident, according to the county. And there was more:

> A department head saw his purchase card spending privileges suspended for a year after he used his card for some of his wife's travel expenses. And an employee responsible for p-card oversight in her department allegedly racked up nearly $25,000 in questionable expenses.

The newspaper has an admirable record of finding such cases, including a publicly funded solicitor who used a card for 'pricey meals, Christmas parties, luxury Uber rides and other perks'.

On the other side of the Atlantic, in 2019, a government employee at the UK's Foreign Office appeared at Southwark Crown Court in London. She was accused of blowing nearly £20,000 on government credit cards in a month-long gambling binge. Laura Perry was alleged to have made almost 250 transactions over thirty days with an online casino, using Foreign Office purchasing cards. She is also alleged to have used a card for a personal restaurant meal. She had been given the cards to book travel tickets, pay for accommodation and make payments for other costs incurred by government and visiting dignitaries.

She claimed that she had accidentally mixed up the card with her own – which can be done – but ultimately pleaded guilty to stealing £2,223.[5] However, she was cleared on the £20,000 accusation relating to the gambling, claiming it was her ex-boyfriend who used the card for that purpose.

In fact it sounded as if the organization had done all the right things – Perry had signed a form stating that cards weren't to be used for personal expenditure, each transaction was logged, and there were checks on spending once it hit a certain level, which was presumably how her fraud was discovered. Again the fact that she was stopped quickly shows that card fraud will be picked up, if the proper processes are in place. Yet unfortunately some people seem unable to resist temptation, even if the chances of discovery are high.

Straight Down the Middle

While there have been horror stories about inappropriate use of company purchasing cards, it is easy to jump to false conclusions. Some years ago I talked to a logistics manager based in the UK Ministry of Defence's head office. He told me he had not long returned from Afghanistan, where he was working as a logistician in a big military camp there. (He couldn't wait to get back there, interestingly – he enjoyed it much more than being stuck behind a desk in Whitehall, he told me.)

We talked about the need for buying processes to be flexible and for buyers and logistics people to be able to react quickly in military situations. The use of the purchasing card came up, and he explained that there had been a bit of an internal furore when Finance had looked at expenditure on the card in use at the camp. One invoice related to expenditure on a range of golf equipment. That looked very strange, possibly fraudulent.

But it wasn't. He explained that opportunities for rest and relaxation were limited for the troops in Afghanistan. Not many friendly bars, and you couldn't just go off for a run through the hills or take a trip to the beach. So someone had the bright idea of buying some golf equipment and rigging up practice nets. Even non-golfers were getting into it, with more expert players offering lessons. The golf kit showed up on the card bill and looked odd, but most people would agree it was actually an appropriate and intelligent use of public money.

As a corporate executive, and on behalf of the firm, I've bought retirement presents, flowers for staff to celebrate a wedding or birth, strange items to be used on corporate away-days, booze and many items that would have looked odd on that card bill. But all were justified and for the organization's benefit, not mine. Another case saw a government body chastised for spending money at a horse-racing venue. But that was explained as the fees for a legitimate business meeting, booked in the hospitality suite on a day when no racing was taking place. Today horse tracks and football grounds often have good meeting facilities and can be cheaper than equivalent hotels, so again this might well have been good buying rather than fraud or failure.

Swimming Pools

If a junior clerk, misusing a purchasing card that should be used to book travel for her bosses, is at one end of the food chain, at the other end we have the corporate top-dogs, the politicians, chairs and CEOs who can start to get a little confused about how they spend the organization's money, and about the line between corporate and personal benefit.

The CEO of an education charity, Philip Bujak, was sentenced to six years in jail in 2018 at Southwark Crown Court in London for swindling some £180,000 out of the organization.[6] Using a company credit card, false invoices and other routes, he got the charity to fund his honeymoon, family events at hotels and the restoration of artworks. One bill for a 'charity conference' was really for his mother's eightieth birthday party.

For some top people the lines between work and personal life seem to get genuinely blurred in their own heads. They put so much time and effort into work, it might seem reasonable to them that the organization should fund some personal costs, too. It is certainly a problem when entrepreneurs build a business, then it is floated and becomes more widely owned, perhaps even publicly quoted. The entrepreneur still feels as if he or she is dealing with their own money.

That same blurring of lines applies to politicians, too. One of the best-known cases in recent years concerned the private home of South African ex-President Jacob Zuma. While this was arguably not 'fraudulent', the use of some R246 million ($18 million, £13 million or €15 million) to improve the property at his Nkandla homestead was controversial, to say the least. The improvements were presented as necessary for security reasons, but this did not always convince the public. A 'firepool' to hold water, in case of fire, looked suspiciously like a normal swimming pool, for example.

A report into the matter by the South African Public Protector found that Zuma unduly benefited from these improvements.[7] Zuma initially resisted the findings, but the Constitutional Court subsequently found that he and the National Assembly had failed to uphold the country's constitution because he refused to comply with the Public Protector's report. Zuma finally apologized for using public money to fund his residence and in April 2016 he was asked to resign by prominent public figures, as well as paying back some of the money. It's this blurring again – what is private benefit and what is truly related to the organization?

When I Kissed the Teacher

But sometimes there is little doubt, when the perpetrator steps way over the line in terms of inappropriate use of their organization's money. In 2019 James Stewart was banned from teaching for life in the UK, after being convicted of fraud and jailed for four years back in 2017.

Stewart was head teacher at Sawtry Village Academy near Huntingdon, responsible for more than 1,000 young people aged eleven to eighteen years old. But in the last few years of his thirty-year reign he spent the school's money to convert his office into what *The Sun* newspaper called a 'sex lair',[8] which provided the setting for drunken romps with his secretary.

He bought sex toys and alcoholic drink, was often drunk during working hours, and the two of them disappeared into his 'office', which was in fact more like an apartment, for hours on end. Staff spoke of hearing sounds of a 'sexual nature, rhythmic moaning and laughing'. Stewart was also seen at a horse-racing event when he was supposedly at a 'meeting'. He pleaded guilty to five counts of fraud, including making false expense claims and unwarranted direct-debit payments to himself, totalling more than £100,000.

The Sun reported that the sex den 'was only uncovered when a roofing contractor spotted a large purple vibrator through the office skylight'. But any amusement in reading about this tends to dissipate when we realize that the fraud had serious consequences for others, as well as for Stewart and his family. The school slipped from a rating of 'good', awarded by the regulatory body in 2007, to 'inadequate' in 2011 and was placed into special measures in 2014, and it seems likely that the lack of leadership and focus from Stewart played a role in that decline.

Money was channelled away from where it should have been spent – on building maintenance, for example – to the detriment of pupils. The Cambridgeshire Live website, reporting the trial, said that a fire inspection found the school was unsafe for students, and there were redundancies and high staff turnover.[9] While Stewart was remorseful and repaid the money, lasting damage was done to the school's reputation.

As in many cases throughout the fraud chapters here, the problems would not have happened, had the processes been appropriate. Stewart was a strong character, and both his deputy and the chair of the school governors allowed him to continue to behave badly. He could be a bully, too, staff said – just the sort of strong man, in a powerful position, who can get away with this sort of behaviour. But if the controls on spending had been sensible, with detailed checks and sign-offs, he might never have been tempted to step so far out of line.

The next category of fraud concerns the invoicing and payment end of the process. Given that this is where money actually exchanges hands, it is not surprising that this is a major focus for fraudsters. It is also an area where, in many cases, the villains don't even need an accomplice inside the organization to carry out their plans; and it is an area where technology (such as online banking) has opened up new opportunities for crime.

17

What Am I Paying For?

In a high-profile, long-running fraud, John Maylam, potato buyer for Sainsbury's, the UK's second-largest supermarket chain, pleaded guilty in court in 2016 to demanding bribes from produce supplier Greenvale AP, in return for awarding them valuable contracts. The bribes were funded by Sainsbury's paying Greenvale slightly over the odds – for lots and lots of potatoes.

One technique used by the conspirators was to overcharge the firm, adding £1 to the market price of a crate of produce and charging Sainsbury's the higher price. That translated to just a few pence per kilo on the final retail price, but with the volumes being transacted, it added up to a lot of fraudulently obtained money. The supplier also supplied smaller packs for the same price, and there were 'illogical prices' for new packs.

Greenvale secured a £40 million contract and opened an account called 'The Fund' into which they paid £8.7 million of Sainsbury's money, giving £4.9 million to Maylam and his associates and keeping the rest. Maylam also allegedly ran up a £200,000 bill at Claridge's hotel in London and enjoyed a £350,000 trip to Monaco with the cash given to him over two years by John Baxter, the Greenvale account manager.

Maylam admitted corruption when he appeared before the court, and Baxter pleaded guilty to making the payments. But it's interesting that, according to the report, when arrests were first made in 2008, the fraud wasn't discovered by Sainsbury's themselves.[1] It was auditors from the parent company of Greenvale who spotted unusual payments and blew the whistle. Following an internal investigation by the holding company for Greenvale AP, the payments made to Maylam were brought to the attention of the supermarket's management.

A scam of this nature is one of the most difficult types of buying-related fraud to uncover, which may be why it took the police four years to put evidence together. Collusion between buyer and seller is very common; and where suppliers are paying bribes, they are likely to recover that 'investment' in some manner, by mechanisms such as over-invoicing in terms of pricing, quantity or both. This is such a common type of fraud that it undoubtedly takes place in every industry, sector and country in the world. All it takes is two dishonest people – one inside the organization with some authority and one outside – plus some weak processes, to make it happen.

Another example of this collusion between buyer and supplier around invoicing, but in this case invoices that didn't even relate to actual goods or services supplied, has been seen in South Africa. In November 2019 Victor Tshabalala, the director of Meagra Transportation, told a court in Johannesburg Specialized Commercial Crimes Court that he worked with a former financial manager at the state-owned electricity firm Eskom to generate fraudulent invoices.[2]

Bernard Moraka is alleged to have created fifty-three fraudulent invoices with a value of some R35 million ($2.4 million) for transportation of coal by Tshabalala's company – work that was not actually done. Moraka, who resigned in October 2018, has indicated that he will plead not guilty to all charges,[3] while Tshabalala has admitted that he was involved, but has offered to disclose more about the case and to pay back money that he received.

Whatever happens here, Eskom has apparently suffered severely from fraud, with Public Enterprises Minister Pravin Gordhan making the not-very-precise statement in 2019 that 'somewhere between R20bn ($1.4 billion) and more were lost' to fraud by the business.

I've Got the Power

It is perhaps unsurprising that so much fraud takes place around the invoicing and payment processes. After all, that's where money actually changes hands. Sometimes organizations end up paying for nothing, in effect, but in the Sainsbury's example the product was delivered to the buyer, and was real, tangible and what they ordered. But the problem lay in *how much* they were paying for it, and why.

While it is not always easy to stop this type of fraud, there are specific steps that can be taken to guard against it, and here are some suggestions.

1. Don't keep decision-making buyers in the same job for too long. As well as almost inevitable complacency, opportunities for fraud grow, as people on each side of the negotiating table get to know each other better.
2. Don't keep power and knowledge concentrated. Why didn't someone else spot that the prices being paid for potatoes were above the market norm? An assistant, a peer, a boss – someone who could say, 'Hang on, why are we buying from Greenvale when they're so expensive?'
3. Price benchmarking is a powerful tool to ensure competitiveness and guard against fraud. Sainsbury's should have had some sort of process in place to check their pricing against competitors, given the competitive nature of the supermarket world.
4. Finally, and most importantly, transparent and competitive buying processes, with a clear audit trail, make this sort of fraud much more difficult. If there had been a more open process – perhaps even an electronic auction, where the various bids were recorded – it would have been difficult for Maylam to show favour without an objective reason; and hard for the supplier to 'load' their prices.

Toy Soldiers

But sometimes this type of fraud based on invoices and payments doesn't need an external party at all. Anyone in the organization who is able to authorize payments, supposedly for items that have been bought, without robust checks on those payments, may be tempted to route money to their own bank accounts for their own purposes.

One example close to home for me (in terms of geography) was a quiet and fairly lowly accounts-payable manager for the now-defunct, previously US-headquartered retailer Toys 'R' Us. But this happened in their UK head office, in Maidenhead, Berkshire, rather than in the US.

Paul Hopes seemed a typical middle-aged accountant to colleagues, living in a semi-detached house near Reading and driving an old Vauxhall car. But actually he lived a double life and was stealing millions from the firm, spending money on sports cars, prostitutes and even an estate in Nigeria for his secret mistresses! He was ordered to repay £3.6 million when he was finally caught, as well as being jailed in 2010 for seven years. (His jail term will increase if he doesn't pay the money back.)

His fraud was simple. He created a fictitious toy manufacturer, a 'supplier' to the firm, and then made regular payments of £300,000 a month over more than two years to that account, which of course he controlled. When this was reported in the press, one reader's comment was amusing: 'so he spent £2.4 million on call girls and sports cars – and wasted the rest'![4] But it's not really funny; this was shareholders' money, and sympathy is due to his wife and family, who knew nothing about it and did not benefit in any way.

Less sympathy is owed to senior management, who must take some responsibility here, in the sense that controls and processes in Toys 'R' Us were clearly weak. Checks on setting up new suppliers should have been in place, and no one should be able to authorize payments of that size without some checks and balances.

Wild Horses (and fake invoices)

'Fake invoice' fraud by insiders, like the previous example, happens in the private sector, in government and even in the charity sector. And it can be the most unlikely people – as in this case, where the former head of counter-fraud at Oxfam, the charity that fights poverty globally, was jailed after stealing more than £64,000 from the organization.

Edward McKenzie-Green, thirty-four, defrauded the organization while investigating fellow charity workers in earthquake-hit Haiti.[5] He filed fake invoices from bogus companies, making £64,612 in nine months, before resigning because of unrelated disciplinary proceedings. The scheme was discovered after an internal inquiry was launched to investigate allegations that he had behaved unprofessionally while leading a team in Haiti in 2011.

McKenzie-Green agreed to resign, was given a £29,000 'golden handshake', but then investigators unearthed seventeen fraudulent invoices from two companies under his control. An audit of his own counter-fraud department revealed payments to 'Loss Prevention Associates' and 'Solutions de Recherche Intelligence' in 2011. Investigators contacted the supposed head of one company, Keith Prowse, for an explanation of invoices for 'intelligence investigation', 'surveillance equipment' and 'Haiti Confidential'. But there was no Mr Prowse – that was, in fact, McKenzie-Green. (The 'real' Keith Prowse founded a very successful corporate hospitality firm in the UK.)

McKenzie-Green got two years in jail, and Judge Wendy Joseph, QC, told him, 'You have taken from those who desperately need it substantial sums of money. Worse, you have undermined the public confidence in a charitable institution. You were head of a department set up to counter fraud. This was a profound abuse of the trust invested in you.'

But again, why weren't there checks at the time on whether goods and services 'bought' were real? To be fair to Oxfam, checking that, in the environment of disaster relief operations, is trickier than in an office in Frankfurt or London, but even so, this suggests somewhat lax processes. However, Oxfam did comment that 'The actions of Edward McKenzie-Green were brought to light by Oxfam's own robust counter-fraud measures.' The organization's annual report also states, 'We do not tolerate fraud or corruption, and have invested substantially in reducing them to an absolute minimum through an approach that tackles every level of risk across all our programmes.'

Some frauds generate amusing headlines, but actually the detailed stories can be rather sad. In another case of fake invoices, an NHS manager stole more than £200,000 to pay for (among other things) horse semen, needed by her stud-farm business.[6] But she went to jail for two years once the fraud was discovered. Louise Tomkins, forty-eight, from West Sussex, fraudulently signed off payments running to just over £200,000, which went into her own bank accounts rather than to genuine suppliers.

The judge said that she was 'a woman of very great ability and up to this point of very high character. The difficulty and sadness of cases such as this is only people of high ability could get themselves in a position where they can defraud people and the NHS of the amount of money you took.'

She worked as interim director of operations at Ealing Hospital NHS Trust in 2008, managing a budget of £57 million. Her fraud was discovered when a successor found 'unusual invoices' for medical photography services. Closer scrutiny by the NHS Counter Fraud Authority and Metropolitan Police found other fake invoices and, when interviewed by police, Tomkins admitted creating invoices for non-existent goods and services, which were helping to pay the costs attached to running her stud farm.

Invoices for a £3,000 titanium skull cap, psychological assessment of patients and conference facilities were fake and in fact funded rent payments, work on her stables, land for grazing, horse saddles and payment for the insemination of four mares. But she had a very good reputation, and her barrister said, 'Nobody ever suspected that she would become a woman that would find herself in the dock and about to go to prison for breach of trust.'

Tomkins had money problems, mental-health issues, her marriage had broken up and she wept in court. It is very sad, but the fact remains that she stole money that should have been used to look after patients. And she'd even acted as a fraud adviser for hospital charitable funds, at the same time as she was falsifying invoices for her own benefit. One of the key lessons again comes through here – you shouldn't trust anyone simply because they are senior and apparently respectable.

But yet again, how was it so easy for her to sign off invoices when she was also the budget holder and buyer, placing orders with supplying firms? Did no one check that goods and services for which the organization was paying were actually real, and received? Or that the apparent suppliers were genuine firms? Apparently not, so some blame has to be allocated to hospitals that had weak processes in place, allowing Tomkins to commit the fraud – inexcusable though it was.

Cook of the House

The whole area of over-invoicing is a fruitful one for fraudsters and contains many different variants. Some are simple – charging for twenty days of a consultant's or interim worker's time when they only did fifteen, for instance. Others are a lot more complex and come into play with contracts that require statements of the costs and purchases made by the supplier.

I came across an interesting example in the catering industry. Some contracts operated by large catering firms work on variants of the 'time and materials' approach. So a firm operating canteens and restaurants for a large business might do so on the basis that the client pays for the actual cost of the catering staff used on-site, the cost of food used, plus a margin to cover overheads and profits. Income from the staff buying food then goes back to the client, but the catering firm has some sort of incentivization to sell more (perhaps a percentage of revenue as a 'bonus').

In this case, the catering firm employed many chefs and other workers on a temporary contract basis (very normal for this industry), engaging them through recruitment and temporary staff agencies. One of those agencies told me that a large global catering firm set an 'official' rate for chefs that was well above the true market rate. So they might pay, let's say, £20 an hour to the agency rather than the £15 that was the true market rate.

However, the catering firm then demanded a significant rebate at the end of every quarter from the agency – at least the £5 per hour excess up-front payment, in our example. What was the point of that? Simple, really. The final client was shown the invoice for the chef's time, at £20 an hour, and paid the catering firm on that basis, under the 'time and materials' charge. The catering firm then pocketed the hefty rebate from the recruitment agency, which was invisible to the end client, as well as receiving their fee from the client.

This is certainly borderline fraud – maybe it is not illegal, maybe it is not even breaking the contract (although a good buyer would put something about transparency in the terms and conditions), but it is certainly unethical at best. However, many clients would have no idea whether the going rate for a grill chef in Charlotte, Cologne or Cheltenham is fifteen or twenty dollars, pounds or euros. So the caterer gets away with it.

Easy Money

In 2019 the US authorities caught up with a Lithuanian citizen, Evaldas Rimasauskas, who almost pulled off one of the biggest and most audacious buying-related frauds of recent years.[7] He was arrested in Lithuania in March 2017, extradited to the United States and, in a Manhattan court in March 2019, pleaded guilty to wire fraud.

Rimasauskas managed to extract more than $100 million from a number of firms, including a 'multinational technology company' and a 'multinational online social media company'. The US court prosecutors did not name the companies, but a Lithuanian court order in 2017 identified Google and Facebook as the victims.

He incorporated a business in Latvia with the same name as an 'Asian-based computer hardware manufacturer' (also not named in the US hearing, although Taiwan-based Quanta Computer Inc. confirmed, after Rimasauskas's arrest, that it was the genuine supplier firm involved). He then used email accounts that looked as if they came from staff at this manufacturer and contacted staff in victim firms, instructing them to send payments they owed to the genuine supplier to his own bank accounts.

That money was quickly transferred to bank accounts in locations throughout the world, including Latvia, Cyprus, Slovakia, Lithuania, Hungary and Hong Kong. Rimasauskas also provided faked documents such as contracts, invoices and letters that backed up his fraud and kept the banks happy, in terms of where all this money was coming from. Happily, he was caught and will spend many years in a US prison. He has also agreed to return some $50 million, according to the court filings.

This is just one of many examples of 'business email compromise' that has developed in recent years. When linked to invoicing and payments, it is also known as 'invoice misdirection' fraud.

One of the highest-profile invoice frauds of recent years saw the London Olympics hit by an initially successful ruse. The UK Olympic Delivery Authority (ODA), which ran the event, and the construction firm Skanska were the victims of the simple but elegant invoice-misdirection fraud. Ansumana Kamara contacted the ODA at a time when a large invoice payment was due to Skanska, which carried out major construction work, including the transformation of a former industrial site into the Queen Elizabeth Olympic Park.

He managed to obtain £2.3 million by pretending to be Skanska's finance director and writing to the ODA with a change of

What Am I Paying For?

account details, just as a major transfer was due. Of course the new details were of his own bank account, and the money appears to have been paid. Very simple really, although it seems possible that some inside information was passed on – perhaps the fact that a payment was due, and maybe the invoice amount or the names of the usual Skanska contacts – in order to give the letter some credibility. Questions will also have been asked about who authorized the payment within the ODA, given the obvious suspicions that would have arisen.

The ODA said that 95 per cent of the money had been 'recovered', which is good news, but that does imply that the payment was indeed initially made. It was always going to be discovered, of course – a short time later one assumes that Skanska called the ODA and said, 'Hey, where's our money?' Or the equivalent in Swedish.

At which point, after some desperate tracking down of payment details, all would become clear. It seems that the fraudsters relied on speed to cover their tracks before the fraud was discovered, and they had worked out a fascinating money-laundering route, whereby the money would initially go off to bank accounts in Nigeria, but would then be used to 'buy a parade of shops in Wolverhampton'.

If you were the unexpected recipient of a couple of million, what would you do with it? Of course you would buy a parade of shops in Wolverhampton! Each to their own, as they say. But the fraudsters were caught: Kamara was sentenced to three and a half years in prison, Abayomi Olowo received a sentence of four and a half years, and Ayodele Odukoya got three years and nine months.[8]

While the money was recovered in this case, in many other cases of this nature it disappears for ever. So how can organizations make sure they are less vulnerable than the ODA? The starting point is to have accurate supplier data, and robust technology and processes around seemingly mundane issues, such as recording supplier bank details. Obviously it doesn't say much for the ODA's processes in this area, although they stated that 'Our payments system was reviewed and strengthened immediately after the incident to further limit the risk of fraud.'

It's also interesting to speculate that if the ODA and Skanska (or Facebook and Quanta in the earlier example) had communicated quickly, a rapid check would have discovered the fraud before money changed hands. That's a potential benefit from the trend towards more collaborative 'social media'-type technology that links buyers and sellers – tools that have come into the business world recently. Some try to mimic consumer social-media tools; so a quick communication from ODA, via a collaborative platform, to their finance contact in Skanska, saying, 'Hi, Sue, can you confirm you've changed the bank details', would have stopped this fraud in its tracks.

This is still a global problem, though, and every type of organization is vulnerable. Lazio football club was reportedly conned into paying €2 million into the wrong bank account after the club bought defender Stefan de Vrij from Dutch club Feyenoord.[9] Lazio were paying in instalments and received an email, supposedly from Feyenoord, asking for the next and final tranche to be paid into a different bank account.

Dublin Zoo was another organization targeted (not for an invoice relating to a footballer, it should be said), although reports suggest that it managed to recover most of the €500,000 that they handed over in error to fraudsters.[10]

In many cases, this type of fraud is based on a *genuine* requirement to pay a supplier, but the payment is redirected to the criminals. However, there is another variant where the fraudsters don't even need an invoice to start the process.

Payin' the Cost to Be the Boss

This scam has become known as 'CEO fraud' and hinges on the blatant impersonation of a senior executive. Someone in the finance department receives an email (or even a phone call) purporting to come from the Chief Executive. If it is an email, it looks very convincing of course. It explains that the CEO is out of the country or working from home, but urgently needs some money sent to a particular firm and bank account for some 'secret consulting work', a sensitive acquisition or similar. The junior accounts clerk is basically bullied and threatened into making the payment, which of course is fake.

According to Action Fraud, the UK's national reporting centre for fraud and cybercrime, the largest reported amount of money ever transferred by an employee in this fashion to a fraudster was £18.5 million.[11] The company, a global brand in the healthcare industry, has preserved its anonymity, not surprisingly! But it seems that a man impersonating a senior executive phoned a financial controller in the firm's Scottish office. The controller genuinely believed the man to be the senior member of staff and even exchanged several calls with him, as well as emails. In the end, funds were transferred to accounts in Hong Kong, China and Tunisia and disappeared, never to be seen again.

Again, controls on payments and understanding exactly where the money is going is vital. This case highlights the importance for staff to have the appropriate training, too – and it's interesting that it wasn't a junior clerk here, but a senior, professional finance person who was duped. So be careful.

The cases in this chapter highlight that fraud and corruption occur in every sector and every country. But government entities are particularly vulnerable, for a number of reasons, not least the sheer amount of money that flows through them. The temptation for politicians or officials to dip into that pot sometimes becomes too much, as we'll see in the next chapter.

18

Politics and Fraud

The Sochi Winter Olympics in 2014 were the most expensive to stage in Olympic history, and have been dogged by rumours of fraudulent activity, particularly in terms of the construction costs. Capital spend connected with the event seemed to be way out of line with the cost of similar projects elsewhere, a sign (say critics) that money was siphoned off to fund corrupt activities. The event cost an amazing $51 billion in total – five times as much as the previous Winter Olympics in Vancouver.

Russian authorities have claimed that some of the analysis is unfair, in that much of the cost apparently associated with the Olympics was actually used to build infrastructure in the region, rather than for specific Olympic facilities such as new stadia. But even President Putin had some doubts, firing the vice-president of the Russian Olympic Committee, Akhmed Bilalov, in 2013. He had been in charge of building the RusSki Gorki ski-jumping centre, where spending had risen to $267 million, more than six times the original budget. Bilalov left the country and was accused in his absence of embezzlement.[1]

The most detailed insight into Sochi was provided by the report produced by the Russian Anti-Corruption Foundation.[2] It looked in detail at the cost for individual projects and benchmarked that against similar work elsewhere – considering equivalent sports stadia, for instance. It found considerable variation, but a general theme was that the Sochi infrastructure cost several times more than expected.

Politics and Fraud

The other circumstantial evidence of buying failure (at best) and fraud (at worst) comes from the way contracts for Sochi were handed out. There was little in the way of transparent processes for determining which firms won major contracts, and experience did not seem to matter too much. A construction company owned by Siberian politicians, with no experience of sports arenas, built the hockey arena and bobsled course – at $260 million over market price.

Many of the lucky businesses were also connected to those at the top of Russian politics. For instance, Arkady Rotenberg, the ex-judo partner of President Putin, won Sochi contracts worth some $7.4 billion. In some cases businesses agreed to provide financing for projects, but ultimately failed to do so and the government stepped in. There was also considerable use of agents and middlemen. In the context of defence contracts and offsets (more on that later), they can be used to disguise the final destination of payments made by the buyer, to fund bribes and backhanders, and to divert fraudulently money that should be spent on proper contracted work.

Everybody Wants to Rule the World

While many of the frauds covered in other chapters occurred in or around public-sector organizations, some are similar to those seen in the private sector. Any organization can suffer from staff falsifying invoices to pay themselves with the organization's money, for instance, or see a low-level buyer tipping off a favoured supplier about other firms' bids, in return for a few thousand dollars.

But the link between politics, politicians and buying-related frauds is so significant that it deserves a chapter to itself (frankly, it could be a book). As already described, the scale of the frauds can be enormous, compared to those executed by the greedy manager or disgruntled accounts clerk.

The Petrobas and Odebrecht case study (page 246) is an example of this crossover into the political world. This fraud was based around buying (public or quasi-public bodies buying goods and services from firms in return for bribes), but the proceeds of the fraud then fed the funding of political parties and a circular corruption process. But there are many other cases globally where politics has been, in some sense, at the heart of proven or alleged fraudulent activity.

However, it is important to be careful. Political opponents tend to cry 'corruption' when they don't agree with a decision. When a British firm loses out to a French competitor in a government tendering process, the London newspapers and opposition politicians will grumble and hint of dodgy decision-making or worse. In many countries, accusing opposing politicians of corruption is standard practice to influence voters. While many countries do have real problems, not every accusation is true.

I Need a Dollar

While the media and taxpayers will rightly shake their heads at corrupt behaviour by politicians and officials, there is almost always another party to this sort of fraud – the firm that pays the bribes. In the developed world it is easy to be rather snobbish about the way these scandals unfold in African nations and other developing countries. But often it is 'Western' firms that are the paymasters fuelling this corruption.

One high-profile example was the case against the giant Canadian engineering and construction firm SNC-Lavalin and two of its subsidiaries, as discussed in Chapter 14. It stemmed from the company's dealings in Libya between 2001 and 2011, when a senior executive established close ties with Saadi Gaddafi, son of the country's dictatorial leader, Muammar Gaddafi.

However, this case and many others involving Western firms and their activities in Libya have not yet come to court, in part because of the turmoil caused by ISIS in recent years, as well as the difficulty in obtaining hard evidence since the Gaddafi regime collapsed. But investigations have featured firms from countries including Croatia, the US, Greece, the Netherlands and China, which had dealings in Libya that might have been lubricated by corruption. Firms from the 'developed' world have a lot to answer for when it comes to global corruption.

The scandal that engulfed engineering firm Siemens back in 2008 remains one of the most significant bribery cases in modern times. It seemed particularly shocking at the time because the firm had a reputation for strong technology, high-quality products and responsible behaviour. But in 2008 the firm paid the US Securities and Exchange Commission $350 million to settle bribery charges, which involved government officials and politicians in various South American countries, Israel, Vietnam, China and Russia.[3] As the SEC filing stated:

> This pattern of bribery by Siemens was unprecedented in scale and geographic reach. The corruption alleged in the SEC's complaint involved more than $1.4 billion in bribes to government officials in Asia, Africa, Europe, the Middle East, and the Americas.

As the Siemens case demonstrated, it is not only the developing world that is the location for these crimes. For instance, the notorious Gürtel case grew into one of the biggest political scandals ever seen in Spain, eventually bringing down the government.[4] It started with firms controlled by the main protagonist, Francisco Correa, winning contracts in local districts and towns by bribing officials and mayors, but grew into a national-level corruption scandal. As in other cases here, it wasn't just a matter of the envelopes of cash handed over as bribes, but also perks such as sex parties for politicians involved in the scheme.

Correa was eventually sentenced to almost fifty-two years in prison on multiple counts of bribery, money-laundering, tax fraud and misappropriation of public funds. A number of People's Party city councillors and advisers also went to jail, and the party itself was convicted as a direct beneficiary of the fraud. It had maintained a parallel accounting system to collect the money from kickbacks, which was used to fund its activities, the court heard.

These cases are global in both the origin and the destination of the corrupt activities. As China increases its global ambitions, with huge programmes such as the Belt and Road initiative, there are growing suspicions that the country is exploiting corrupt regimes to further its own purposes. For example, a *Wall Street Journal* investigation has alleged Chinese involvement in the Malaysian government fund at the centre of a billion-dollar corruption scandal.[5] As we see the US and European governments trying to use legislation more aggressively to dissuade firms from offering bribes in other countries, it would be disappointing, from an anti-corruption perspective, if China decided to step into that gap to promote its own interests.

Where's the Money?

Once politics and business become entwined, cases can be very complex and hard to untangle. In July 2019 Kenya's finance minister Henry Rotich and other officials were arrested on corruption and fraud charges, related to the multimillion-dollar Arror and Kimwarer dam projects in Kenya's Rift Valley region.[6] This was just the latest in a string of alleged frauds in the country that President Uhuru Kenyatta has pledged to address, and follows a 2018 report from the Auditor General that showed the government couldn't account for more than $400 million of public funds on the project.

The Kenyan *Star* website reported that the government accounts reviewed by the auditor 'showed a possible loss of over Sh5.6 billion, part of it being advance payments for no work done, variations in costs of tender, and goods not put to use after purchase'.[7] Too often, purchases were made without competition, and certain firms were given preferential treatment in the contracting process.

Politics and Fraud

This case led to the first arrest of a sitting Kenyan minister. Director of Public Prosecutions Noordin Haji ordered the arrest of Rotich and twenty-seven other top officials on charges of fraud, abuse of office and financial misconduct. The chief prosecutor highlighted a contract award to Italian firm CMC di Ravenna in a manner that he said flouted proper procurement procedures, and despite financial problems, which had led to it failing to complete three other mega-dam projects.

There were also issues over cost control. The project was supposed to cost a total of $450 million, but the finance ministry had increased this amount by $164 million 'without regard to performance or works,' said Mr Haji. And $180 million has already been paid out, with little construction to show for that investment, he claimed. Mr Rotich has previously denied any wrongdoing, as has CMC di Ravenna.

According to Reuters, advance payments of nineteen billion shillings were made, including eleven billion shillings in unnecessary debt insurance.[8] The prosecutors claim that was shared out in accounts belonging to the conspirators and their agents. But CMC has responded saying these payments referred to conditions of the bank financing and that the firm was not involved in those negotiations at all. The firm also disputes the claim that no work has been done on the dams.

The country's Auditor General has also claimed that as much as 50 per cent of the government's $5-billion-a-year spend is lost to fraud and corruption.[9] The arrest of Rotich, who has denied doing anything wrong, is part of a wider anti-corruption drive aimed at drastically cutting that waste.

Whatever the outcome, this case shows the complexities of large capital projects with government funding and international suppliers, banks and other parties involved. That leads to opportunities for fraud and corruption, but political machinations can come into play as well and make it difficult to know exactly what has happened and where the money has actually gone. This case, like others of its type, may take some time to be resolved.

Independent Women

It's probably not a good sign when a government's own anti-corruption head gives up, citing a 'hostile environment', which presumably wasn't coming just from the private sector and suppliers to government. That's what happened in Romania in 2019 when Anca Jurma, head of the National Anti-Corruption Directorate (DNA), stood down.[10] She had only been in the role for months, after the previous head was dismissed by the government in July 2018.

The former DNA chief, Laura Codruta Kövesi, is challenging her own dismissal at the European Court of Human Rights. Most observers felt that the DNA made major strides under her leadership – she managed to prosecute dozens of mayors, five MPs, two ex-ministers and a former Prime Minister in 2014 alone. She was appointed as the first European chief prosecutor by the EU in 2019, in a recognition of her ability. But in Romania her capability wasn't appreciated by the political and business classes, who claimed the DNA was too powerful and was overstepping its authority.

Wherever organizations – from the smallest to the largest, and including government bodies – are seen to be unwilling to take basic anti-corruption steps, often that reluctance is based not solely on a lack of knowledge or time to take action. It usually suggests that someone, probably at a senior level, wants to make it easier for fraud and corruption to flourish.

At a local level, I've worked with organizations whose top management didn't even want to put in place a clear 'conflict of interest' policy. That would mean staff having to disclose any interest they (or close family/friends) have in another business that might be a supplier or a customer of the organization for which they work. But there's usually a reason for that hesitancy. Where you see organizations that won't support anti-corruption activities, you might draw obvious conclusions.

A Promise

Corruption and fraud cases can have serious consequences for the politicians involved and their political parties, and can even swing the entire political momentum in a country. That can happen even if the corruption didn't actually take place, but was only proposed or discussed.

The Austrian populist right-wing FPÖ party increased its share of the vote to 25 per cent in the 2017 national elections and became a partner in a coalition government with the ÖVP. But in May 2019 the German newspapers *Süddeutsche Zeitung* and *Der Spiegel* published evidence from hidden camera video-recordings made in Ibiza in 2017.[11]

They showed the then leader of the FPÖ party and Vice-Chancellor of Austria, Heinz-Christian Strache, and a party colleague, Johann Gudenus, apparently talking business with an elegant woman introduced as Alyona Makarova, the purported niece of Igor Makarov, a Russian oligarch close to President Vladimir Putin. The woman was a decoy, involved in what was a mysterious entrapment operation – and it still isn't clear who was behind it.[12] In the meeting, as well as explaining that the sources of her wealth weren't exactly legal, she claimed to be in the process of buying a controlling stake in *Kronen Zeitung*, Austria's biggest-selling tabloid newspaper.

Makarova told the men that she would be prepared to use the paper to promote the FPÖ in elections in Austria later that year. Strache expressed approval for that potential media takeover, but went further, suggesting that to pay her back for her political support, government contracts could be directed her way. There was even a discussion around inflating the prices charged within the contracts.

Once the video became public, Strache apologized, saying that his comments were fuelled by alcohol and designed to impress the 'attractive female host', but the consequences have been dramatic.[13] Strache and Gudenus resigned and the governing coalition fell apart, leading to further elections, in which the vote for the FPÖ was down to just 16 per cent.

It is by no means certain that the politicians could actually have delivered on Strache's boastful promises. There are procurement rules in place in Austria, as in the rest of the European Union, and while corruption does happen, it isn't that easy to simply direct business to any firm, and it would have been even harder if that was a Russian-owned entity, I'd suggest. But simply raising the possibility of a buying fraud in this case contributed to major and national political events.

Leaving on a Jet Plane

The excellent Balkan Insight website ran a lengthy article in May 2019 all about a mysterious private jet that ended up in Serbian government hands, in a manner that no one seems able to understand.[14] The twin-engine Embraer Legacy 600 was built in 2007, and until July 2018 it was owned by Itaubank leasing and used by EMS S/A, the largest pharmaceutical company in Brazil. Eight months before that, in late 2017, EMS won a tender to buy Serbia's state-owned drug-maker Galenika, for €16 million.

But in July 2018 the plane was flown from Brazil to Belgrade, and according to the Balkan Investigative Reporting Network (BIRN), papers show the Serbian government was registered as the new owner in place of EMS or Itaubank. There is no evidence that any sort of formal procurement process was undertaken for the plane, and no record of payments being made by any Serbian government body to EMS or Itaubank. BIRN has asked various Freedom of Information questions, without getting to the bottom of the matter.

So was the plane a 'sweetener' for the company sale? It appears that no one has benefited personally from this, but if there are suspicions of something odd going on, it might suggest other issues around the Galenika deal. EMS paid €16 million for a 93.7 per cent stake, agreed to repay €25 million in debts, invest €5 million over the following two years and maintain a workforce of 900 'for an indefinite period of time'. Was that a fair price for a company 'with 177 registered medicines, 43 hectares of land, four factories, subsidiaries in two neighbouring countries and access to the European Union market', as the Balkan Insight report asked?

Let the Sunshine In

The issue once again in that case is transparency. It may be that nothing fraudulent occurred, and it may also be that Serbia got a fair price for the sale of the firm. But when there is limited visibility of what exactly is happening when money (and planes) change hands in transactions between governments and private firms, then there will always be doubt.

Sunlight is said to be the best disinfectant; if the public can see clearly what public bodies and politicians are doing, there is far less likelihood of fraud and corruption. That means being open and clear about buying processes from the beginning to the end of the process, from setting specifications and deciding what will be bought, all the way through to effective contract management and control over payments and invoicing. Publishing data about the competitive process (such as how many firms bid), and about contracts awarded, is another positive aspect of transparency.

Corruption in public-sector spending is one of the most widespread and corrosive issues globally today, in terms of its economic, social and political effects. Yet it is one that can be addressed – remedies and measures to reduce it are available and understood. It is just the willpower to do so that is sometimes lacking. In the next chapter I'll look in more detail at how organizations, both public and private-sector, can take tangible and practical steps to make fraud and corruption less likely. But first, another worthy award-winning case study, featuring one of the largest, most international and politically charged fraud cases ever.

Bad Buying Award

MOST POLITICALLY SIGNIFICANT AND GEOGRAPHICALLY SPREAD FRAUD – PETROBAS AND ODEBRECHT

If you doubt that buying-related fraud and corruption can have wider effects than simply the organization's money going where it shouldn't, then the case of Petrobas of Brazil – a huge government-owned oil company – and construction firm Odebrecht will surely persuade you otherwise.

This extraordinary case, initially known as 'Operation Car Wash', because some early evidence came from tax investigations into a Brasília car wash, revealed a huge web of corruption. It has led to more than a thousand warrants for search and seizure, prison sentences and plea bargaining from perpetrators. Petrobas is still trying to recover what is thought to be billions of dollars in money extracted one way or another from the business. Later the related scandal around Odebrecht spread to Peru, Mexico and further away, leading to arrests and even the death of a leading politician.

Shareholders of both firms suffered, and Odebrecht filed for bankruptcy in June 2019. Clearly more tax should have been paid to the authorities – and how much money, which could have been used to address poverty in South America, disappeared into the pockets (and Swiss bank accounts) of the corrupt elite?

When the law-enforcement agencies first came across this case, it looked like another standard fraud. Suppliers of goods and services to Petrobas, including construction work, paid bribes to win contracts and were overcharging the firm to fund those payments. One of the principal firms paying bribes was Odebrecht, the biggest contracting and engineering firm in Latin America. In June 2015 Brazilian authorities arrested the former CEO, Marcelo Odebrecht, and in March 2016 he was sentenced to nineteen years in jail, for paying more than $30 million in bribes to Petrobas executives, in exchange for contracts and influence.

But it turned out that the web of corruption spread more widely than just a few crooked Petrobas executives and supplier accomplices.[15] Suppliers were paying between 1 and 5 per cent of the contract value, across a range of services and construction contracts, into a 'fund', which was then used for various purposes. The contract value and pricing were inflated to finance this, and everyone personally involved in the deals seems to have benefited, receiving cash, luxury cars, expensive artworks, Rolex watches, $3,000 bottles of wine, yachts and even helicopters. Huge sums were placed in Swiss bank accounts or laundered via overseas property deals or other companies. Cash was moved around via elderly 'mules' who flew around the country with wads of cash strapped to their bodies!

However, executives also channelled payments to politicians who had appointed them in the first place (Petrobas being a government-owned company). This was the additional dimension to this affair, making it unusual, as payments funded election campaigns for politicians as well as lining individuals' pockets, with the aim of ensuring that the governing coalition stayed in power. A circle of corruption, in effect, because the government then placed senior executives into key positions in Petrobas in order to run the scheme.

The scandal and investigation have implicated dozens of high-level figures in the Brazilian Workers' Party, including former President Luiz Inácio Lula da Silva and his successor, Dilma Rousseff. Corruption allegations also dogged the successor to

Rousseff, Michel Temer, who stood down as President at the end of 2018.

The sums of money involved here are huge. *The Guardian* reckoned this could be the biggest fraud scandal in history.[16] In January 2018 the firm agreed to pay $2.95 billion to settle a shareholder lawsuit in the United States, indicative of the billons that appear to have been extracted throughout the affair. The company itself says it is the victim and has recovered more than $400 million in restitution. Petrobras has stated that it will 'continue to pursue all available legal remedies from culpable companies and individuals'.

Dozens of foreign corporate suppliers (of engineering equipment, power lines, drilling rigs, and so forth) also face regulatory and shareholder enquiries about the bribes they paid to secure contracts with Petrobras. Among them was Rolls-Royce, which agreed to pay Petrobas $25 million in 2017 to settle the claim.[17] Fraud investigations are now also focused on the construction of six out of the twelve stadiums used in the 2014 World Cup and 2016 Olympics.

The scandal spread to other countries, too. Allegedly Odebrecht won contracts by bribing or paying 'campaign contributions' to politicians in many other countries, including Mexico,[18] Peru[19] and Colombia.[20] Tragically Peru's former President, Alan García, shot himself in April 2019 when police tried to arrest him in connection with the scandal.[21] That does not mean he was guilty of course, but no fewer than four former Peruvian leaders are implicated in the scandal.

What lessons can be learned from this? There is no doubt that certain buying policies, approaches and governance could at least have made this more difficult for the fraudsters. It would be interesting to know exactly how many people were involved in the decisions made about the awarding of contracts to the conspiring suppliers. While it is always possible that *everyone* on the buy-side can be corrupt, the more people play a role in the decision-making, the more chance there is that someone won't go along with it.

Transparency in the bidding process, and in how winning suppliers are selected, can also make corruption harder to execute. Did Petrobas and other organizations involved have policies in place around competition, the number of suppliers who should be bidding for work, how the selection process would be managed, and so on? How much of the process was in the public domain? Again, the more suppliers that can be brought into the competitive process, the more open that process is; and the more different firms that win work, the less chance there is for serious fraud and corruption to take hold.

Arguably the biggest lesson, though, is that sometimes fraud can be so widespread, so pervasive, that putting an end to it is no easy matter. Just as in the case of 'Fat Leonard' (see page 196), it seems as if whole institutions were corrupted, including much of Petrobas's senior management and much of the political establishment in Brazil and elsewhere.

19

Preventing Fraud

As I've described, fraud comes in many forms. But there are a relatively small number of key principles that, if followed, make fraud more difficult, although it is impossible to eliminate it altogether. Some of the principles also make it more likely that fraud will be detected, even if it hasn't been cut off at source.

As well as the direct risks, the ability of the organization and the justice system to take action against fraudsters may be compromised if strong anti-fraud policies are not implemented. If policies aren't clear or don't exist, or there is inconsistency in how bad behaviour is handled, then it can be difficult to act against perpetrators. For example, in the National Health Service case mentioned in Chapter 14, the UK's NHS Counter Fraud Authority report stated that 'the Crown Prosecution Service chose not to prosecute. Part of the reason for this were the systemic failings within the health body to control costs and a culture of work being given to favoured contractors with little work being tendered.'[1]

The prosecutors did not believe that successful criminal charges were likely because the rules in general were so lax in that particular organization. So there are multiple reasons for organizations to act against buying-related fraud, or at least make it as difficult as possible for anyone who fancies trying.

Fraud relies on extracting money from an organization, either in return for nothing or in return for less value than the money justifies. So basic counter-fraud principles comprise *controlling the flow of money out of the organization* and ensuring that *full value is provided* by suppliers in return for that money. Here are seven key principles for avoiding buying-related fraud and corruption:

1. It must be clear who is entitled to spend the organization's money and how much they can spend.
2. Any expenditure committed must be authorized properly.
3. All entities ('suppliers' or similar) to which money is paid must be genuine – that is, verified and authorized.
4. Goods or services bought must be checked (for quality, quantity, performance, etc.) to ensure that what was received is aligned with what was expected and contracted for, and with payment made.
5. Supplier selection and pricing of purchased goods and services must be transparent – there must be reasons to believe this is the 'best' supplier and a 'fair' price!
6. Opportunities for collusion with fraudulent aims between suppliers, between suppliers and buyers or between multiple people internally, must be minimized.
7. Perpetrators of fraud (particularly internal staff) must know that proportionate and strong action will be taken against them if they transgress.

If these are the key principles, then there is also an overarching need to put the right detailed *policies and processes* in place to counter potential fraud. These also need to be captured, codified and communicated, so there is no excuse for anyone within the organization not to follow the rules.

The processes should cover how buying is carried out; for instance, how a purchasing card should be used, or the flow of requisitions or orders for goods that a member of staff needs to buy, including approval routing. Policies are the 'rules' – so everything from the need to declare a conflict of interest when choosing a supplier ('But my wife's firm is clearly the best'), to when a supplier can win a contract without the need for a competitive process, to sign-off limits on invoices. But let's go back to those principles and look at each in more detail.

1. **It must be clear who is entitled to spend money in the organization**

    This seems so fundamental, yet organizations often don't follow this advice. Only designated people should be allowed to commit the organization's money to any third party. That doesn't mean they will carry out that buying well or properly, but by restricting the number and making the rules clear, you can at least reduce the size of the overall field of potential fraudsters. Clarity also makes it harder for the exposed fraudster to say, 'I didn't know I wasn't allowed to do it like this.' Too often, when suspected fraud is discovered, it can be shrugged off with that excuse, because policies aren't clear.

    Some firms restrict the number of authorized spenders very vigorously; others may have thousands of authorized staff. The key point is clarity, whatever the chosen policy. That should include how much they can spend, how they must do it and when – which might fit with budgets and financial years, for instance. In some cases it might also relate to what they are buying, too. So, a factory manager might be allowed to buy raw materials, but not management consulting services or complex software (which might need to be routed via IT professionals).

    It is also important to make clear the distinction between a *budget holder* – who has accountability for expenditure generally – and a *buying authority*: the right to agree contracts with external parties. They may be one and the same person, but that is not necessarily so. Many organizations might give a senior manager a huge budget, but stipulate that buying authority in some areas rests with a procurement team, or with an IT head for technical purchases.

2. Any expenditure committed must be authorized properly

The first precaution against buying fraud should be the check on what is being committed to the third party. What are we going to receive, what is the contract and what will we pay to the supplier? Does the commitment that we're making look appropriate? Are the goods or services the sort of things we would expect to see the organization ordering (as opposed to inappropriate purchases)? Does the supplier look genuine? Is the financial commitment within the individual's authority levels, as mentioned in the previous principle?

The buyer (the person placing the order or awarding the contract) could, and should, answer these questions themselves and be satisfied that this is a good use of the organization's money. But for obvious reasons, to avoid temptation and potential conflict of interest, it is sensible in most cases – and certainly for any significant commitment – for someone else to look at what is proposed. So those questions should also be considered by another party, the *authorizer* of the expenditure. That will often be the buyer or budget holder's boss, or it might be finance, procurement or other specialists.[2]

Employees at any level, including senior managers, should not be able to authorize their own purchases, except for those of very low value. The reason is obvious. Although most people are honest, some aren't, and the temptation of being given what is in effect a blank company cheque book is something you shouldn't put in front of anyone.

So having another person authorizing a request to buy is sensible, but I often see senior managers allowed to place orders without any checks. Yet the evidence suggests that it is often middle-level and senior executives who commit fraud, which is simply a case of those people having the *opportunity* to do so. Look at the Fat Leonard case (see page 196) and observe that the corruption went right to the top of the US Navy, while it is also senior individuals who have been arrested with regard to the recent fraud at Patisserie Valerie;[3] and looking back at Enron, Robert Maxwell and many other cases, it is clear that status is no predictor of honesty. It is not an insult to ask a director to have his or her expenditure authorized – it is good practice at any level of the organization.

3. **All entities to which money is paid must be verified and authorized**

The previous principle gives a basic sense-check of the order. But how do we know that order isn't going to a fake or dummy company, perhaps even one controlled by the order placer (the fraudster) or their associates? That 'supplier' may still supply the goods and services required, or something approximating to them, with the fraud being the quality or quantity of what is provided. Or they may supply nothing, relying on no one other than the fraudster realizing that nothing has actually been received. Or perhaps the time-lag before the discrepancy is noticed is long enough for the fraudster to disappear safely, before anyone asks where those 5,000 laptops that have been paid for have got to.

It is vital to check that the entity you're dealing with is genuine. Are they a registered company with a trading history? Do they have a track record? Who are the directors? That also means checking ownership against the organization's own employees. This is an area where appropriate supplier information-management processes, systems and tools absolutely come into their own (experts talk about the importance of accurate and complete 'vendor master data'). You need to understand who your suppliers are, and identify any that aren't genuine.

4. **Goods or services purchased must be checked to ensure that what was received (quality, quantity, etc.) is aligned with the contract and payment**

Even if the requirement is genuine, and the supplier is genuine, fraud can take place where the nature of what is supplied – quality, quantity, specification – is not as contracted for. So a lower value product may be substituted; horse-meat instead of beef, as an example, or counterfeit components and spare parts.

In the case of services, it may be the *quantity* that is supplied rather than quality that lies at the heart of the fraud. A supplier invoices for a certain number of hours worked by a contractor, when the actual number is lower. Or it might be a payment for services delivered on a cost-per-item basis (anything from a courier service to equipment repairs), where the number of events is exaggerated. So there should be a process in place to validate and sign off what is being claimed and invoiced. Again, certainly for higher-value purchases, it should not be only the order-placer who authorizes payment.

Fraud of this nature is hard to pick up at the best of times, and where there is collusion between suppliers and internal staff, it becomes even harder. Perhaps the most difficult buying fraud of all to detect would be a budget holder and a professional services supplier in collusion, with the budget holder signing off that twenty days of consulting or contingent labour services were provided, when really it was only fifteen. This is a real challenge in terms of detection; involving more than just the budget holder in the checking process will help and provides some protection, but this is not an easy risk to address.

5. **Supplier selection and pricing of purchased goods and services must be transparent**
   Even if the goods or services received are as per the contract, how do you know the *price* paid was appropriate and competitive? Even if it was, are you sure the supplier was selected in a fair and transparent manner, rather than because of a bribe or a connection to the buyer? This, like the previous example, is hard to detect. Perhaps the manager paying £1,000 a day for a contractor should only be paying the market price of £800. The senior manager may be taking the £200 'excess' back as a bribe for awarding the contractor the work. Transparency and competition, when it comes to choosing the supplier and arriving at a competitive price, can at least reduce the risk of this particular fraud.

   Such an approach might have stopped the Sainsbury's potato fraud previously described, and may well have shown that the prices paid for potatoes were out of line. It might also have introduced alternative suppliers into the picture. Competition is always an important driver of fair pricing, and benchmarking is another option to determine true market pricing and to help identify any problem areas.

6. **Opportunities for collusion between suppliers, and between suppliers and buyers, must be minimized**
   Many frauds rely on collusion between buyer (or budget holder) and seller, so reducing the opportunity of this reduces the chances of fraud. Organizations should ensure there is always more than one person involved with any major purchase and in signing-off work with suppliers. Moving staff regularly is another option, so there is less time for the relationship, and perhaps the fraudulent plans, to mature. Some organizations have a policy that no one in a decision-making buying role will

stay for more than three years in that same job role, for this very reason.

It is not just professional buyers (procurement staff) to whom this applies. Indeed, it can be stakeholders such as budget holders or service users who by the nature of what is being bought find themselves getting too close to suppliers. I once discovered that my firm's major IT equipment supplier was sponsoring our internal IT budget holder's expensive car-racing hobby!

It may be very innocent, but when a marketing or IT manager makes it clear they don't want professional procurement or finance colleagues involved in 'their' relationship with a key supplier, that can be a warning sign that it isn't totally innocent. Organizations should look at discouraging closeness that goes beyond the need to work well with a supplier to get a job done. This should influence the organization's policy on hospitality, gifts and entertainment, which should be clear and should err on the side of caution.

7. **Perpetrators of fraud (particularly internal staff) must know that action will be taken against them if they transgress, and that action will be proportionate and strong**

Many fraud examples relate to government bodies, because such cases often get into the public domain. On the other hand, private-sector firms don't wish to see their dirty, fraud-related linen washed in public or in the courts. The danger is that potential internal fraudsters feel that the consequences, if they are exposed, might not be too serious, as the firm will want to avoid publicity.

I saw different approaches to both buying failure and outright fraud during my management career. In one case my firm was happy to see the staff involved leave, and no further action was taken. But in another case, at NatWest Bank, the internal perpetrator was prosecuted successfully, and in fact there was remarkably little publicity at the time.

Organizations need to have a robust set of policies in place and make it clear that fraudsters will face strong action. They also need to do everything they can to encourage whistle-blowers. One sad aspect of cases such as the 'Fat Leonard' fraud is that there were whistle-blowers, but they were either ignored or the people with whom they raised the issue were themselves part of the corrupt network. Having whistle-blowing routes outside the normal line-management chain is vital.

A Better Future

If organizations followed these principles, fraud would happen far less often. The good news is that technology is introducing new tools in the fight against fraud and corruption, although it has also introduced new opportunities for villains. On the positive side, organizations can run more transparent processes for selecting suppliers, such as online auctions. They can capture details of what firms have bid, and integrated systems can carry the information about the prices the suppliers offered directly into the systems that monitor their invoices.

In the future, artificial-intelligence-based tools will look at the history of orders and invoices to seek patterns that might suggest fake transactions. If invoices are always submitted on a Friday, when they might get less attention from the managers authorizing them, could that be a sign of something fishy going on? Blockchain technology also has some interesting possibilities in terms of establishing provenance and the history of goods that are bought through the entire end-to-end supply chain.

However, it would be naive to think technology will ever totally eliminate fraud and corruption. Even reducing the corruption seen in political circles might be tough, particularly as the trend to more populist politicians – who don't always play by the rules – seems to be blossoming. Democracy itself is under threat in some parts of the world, but even in that political sphere, more regulation that makes large firms hesitate before offering bribes to politicians or government officials is having some impact, as seen in major prosecutions in the UK, US and elsewhere.

If you care about these issues, then pushing organizations to take the right steps to avoid corruption and fraud in the buying sphere is something we can all do. Cutting this sort of waste will benefit every honest citizen, every country and every business.

PART 3

How to Avoid the F*ck-ups

20

Ten Principles for Good Buying

There are thousands of books, ranging from very readable to highly specialist, academic and sometimes impenetrable, that aim to educate organizations and individuals about buying better. Some favourites are listed in the Bibliography (page 281). I can't compete with this library in a few pages, but here are 'Ten principles for good buying', with key suggestions identified under each heading. Any organization that takes on board these points, along with the anti-fraud principles outlined in the previous chapter, will go a long way towards avoiding the waste and failures highlighted in earlier case studies.

KEY PRINCIPLE I
Your suppliers contribute to the creation of competitive advantage and the achievement of organizational goals.

Having worked in this field for most of my working life, my goal is to see all organizations taking a 'strategic' view of spending money with third parties. That means an approach to buying starting with the *strategic aims of the organization*. Those aims should translate into buying strategies, processes and activities, because buying well at its heart means finding, engaging, contracting and working with third-party organizations that can best help your own organization achieve its goals.

Firms spend on average close to 70 per cent of their revenues with suppliers. If you don't get that right, how can you really succeed? Similarly, in the public or charity sectors, virtually every organization depends for its success on buying mission-critical goods, services and support. So success requires you to get the most out of your suppliers.

What is 'good buying' in a strategic organizational context? It is *not* simply spending as little as possible on everything you buy. Any organization that sees cutting costs as its primary aim will not last long. That route ends up with quality issues (see the Schlitz case study on page 56), no product development, no marketing budget, no investment in equipment, systems or staff training – and, eventually, no sustainable business.

In the private sector we look at *competitive advantage* as the foundation of long-term success. There are a number of ways that organizations seek to create advantage, and theorists such as Michael Porter have suggested strategies such as *differentiation, cost leadership, innovation* or *customer intimacy* as routes to success. Understanding how the organization is positioned, its goals and sources of advantage leads on to thinking about how buying can best be 'strategic'.

For example, if a firm pursues an approach based on launching innovative new products, then suppliers, contracts and buying have to support that. The focus needs to be on identifying suppliers that can help innovation and work quickly (speed to market will be key), and with whom buyers can develop collaborative relationships. That's not to say you should be happy to be ripped off and pay more than you need to for *anything* you buy. But screwing the last couple of per cent out of a deal is probably not critical, if a deep relationship with a supplier can help launch a brilliantly innovative product in record time.

On the other hand, a buyer working for a retailer whose pitch is 'Everything for $1' (or £1 or €1) will rightly have a very different focus. However, even here it is not just the lowest price that matters. The product has to sell (preferably very well) at that price point, and must be safe and legal, of course. But the focus of buyers in such firms will be very different from those in Apple, BMW or the BBC.

Competitive advantage does not apply in the same way to government (public-sector) organizations, which exist to deliver policies decided upon by politicians and citizens. The specific policy goal might be defending the country, providing health and education services or running the local cemetery. But suppliers contribute hugely to every major public-sector activity, as we saw during the 2020 pandemic, when the supply of personal protective equipment, ventilators and other medical items was central to the response. So this strategic linkage is just as important here.

KEY PRINCIPLE 2

For everything you buy, consider how that item or spend category contributes towards strategic goals, and conduct your buying appropriately.

I've come up with nine different ways in which buying can contribute to organizational goals, deliver value and, ultimately, provide a competitive advantage. Most of these, shown in the table below, are fairly obvious, but a couple require explanation. *Gain access to scarce resources* might cover how the organization buys rare minerals or other materials and might even look to capture a supply market, so that competitors struggle to buy that material. Or, in the marketing industry, it might mean getting hold of the very best freelance designers and artists to work on a campaign – again, denying your competitors access to that 'raw material'.

Motivating staff is not something many people consider, in terms of buying. Yet it can be very demotivating to be forced into using lousy systems in our jobs, or to work with equipment that constantly breaks down, or have our firm book us into the worst hotel in town. At the more positive extreme, leading firms compete to offer staff great facilities, amazing catering, fussball and ping-pong tables . . . so sometimes the purpose of the buying concerns that motivational element.

Even in a single organization, different products or services that are bought ('spend categories') will fulfil different strategic roles,

> ### NINE 'GOOD BUYING' OBJECTIVES
>
> 1. Reduce costs, in terms of what is bought from third parties.
> 2. Contribute to top-line revenue growth through supplier input and action.
> 3. Optimize assets, cash and working capital (balance-sheet management).
> 4. Capture external market/supplier innovation to benefit the organization.
> 5. Support internal organizational efficiency and effectiveness.
> 6. Manage the (supplier-related) risk profile.
> 7. Promote monopoly positioning – gain access to scarce resources.
> 8. Help to motivate staff.
> 9. Contribute to wider 'sustainability' initiatives (which benefit organizations' reputations).

Table 1: Ways to provide a competitive advantage

and there is also variation *across* organizations. To illustrate that, consider an upmarket whisky firm. Their product packaging is a strategically important spend category: both the bottle itself and the box it is placed within. How their new brand looks on the display stand in duty-free shops or supermarkets is vital to its success. Novel, attractive and innovative packaging can contribute very directly to revenue and market share growth.

This doesn't mean the firm should be prepared to pay more than a fair market price for their elaborate engraved bottle and fancy outer case, but cheap packaging is not the buyer's prime motive. The real issue is finding suppliers who can respond to the need for quality and innovation and who will offer their best ideas proactively, rather than taking them to other competitors.

But for the whisky firm, their core IT systems – perhaps an ERP (Enterprise Resource Planning) system, CRM (Customer Relationship Management), and so on – are important, but aren't going to directly drive revenue. Technology will contribute towards the smooth running of the firm, however, and might even affect staff motivation. Cost minimization will also be more significant here in terms of priorities, I'd suggest.

Now consider an online clothing retailer. Their packaging doesn't really matter too much, in terms of innovation and revenue. It must meet basic quality thresholds, of course, so that clothes don't get damaged in transit, but if buyers can find ways of shaving a few pence or cents off the cost for every delivery, that will make a very useful contribution to profitability.

On the other hand, for this firm technology is potentially a source of real competitive advantage. From their website, to the slickness of the online catalogue search and ordering capability, how they manage stockholding, customer information, data analytics to tailor promotions to individual customers, perhaps even a highly automated warehouse, tech is key. (Amazon's growth is heavily attributable to their ability to extract advantage in these and similar areas.) Saving a few per cent on the technology spend is immaterial compared to what it can do strategically for the business.

But there is commonality, too. Both our whisky producer and our clothing retailer want to know that key suppliers of packaging *and* technology are reputable, and aren't using slave labour in their factories or polluting rivers. Reputational risk looks similar for both spend areas and both firms.

KEY PRINCIPLE 3

Good data allows you to understand what you're buying and is central to good performance.

Some of these principles, including this one, apply across every organization and type of purchase and, as in most areas of business life today, data is vital. To get buying into shape it is essential to know where money flows out of the organization, who receives it, what it is spent on, who those suppliers are (with deeper insight about the more important ones), which budget holders are spending what . . . and the list goes on. But it is not enough simply to have data. It must be accurate, up-to-date, easily accessible and available quickly to those in the organization who need it.

The old 'garbage in, garbage out' mantra is valid here, too. Analysis will be difficult if data has been entered into buying systems in such a way that thirty-eight different suppliers are listed on a spend report, when actually they are all variants of 'IBM'. Growth in technology solutions that help with the accuracy of 'supplier master data' is helping to address this now, and even once the true IBM spend is identified, you need to know what was bought (equipment, software, consulting services, training) and who bought it.

Once data quality is appropriate, there is more that can be done to support better buying. Looking at where there are too many suppliers of the same item – or, indeed, not enough suppliers – examining price variations or trends, identifying where users aren't buying from contracts that you want them to, even fraud detection . . . all this and more can be driven through analysis of spending, contract and supplier data. You can also look at data around suppliers' performance and their status in terms of issues such as corporate social responsibility, or you might want to identify smaller or minority-owned suppliers for corporate reporting purposes.

If your organization hasn't got decent systems in place, enabling you to look at this sort of data quickly and translate it into 'actionable intelligence' for better buying, then resolving that should be a high priority on any improvement agenda.

KEY PRINCIPLE 4
Define clear and differentiated processes for buying, enabled by technology and balancing ease of use with control and rigour.

For many years executives tended to talk about 'business processes' in the buying context. Today it's increasingly common to hear discussion around 'target operating models', or the TOM. The meaning is the same in the buying context: the TOM is the set of process descriptions that explain how people in the organization should go about buying goods and services. It should cover all eventualities, from day-to-day ordering of stationery or basic equipment through to buying the most complex programmes, IT systems or construction projects.

In most cases an organization needs a number of variants within the overall model, so that the buying process is defined differently, depending on the size and risk of the purchase. The aim should be to make buying low-value, low-risk items as simple as possible, with minimal and sensible controls, while complex purchases require a more rigorous approach.

In my first job in charge of corporate buying, as an organization's procurement director, I was told by the CEO that he didn't want to listen to my strategic ideas until I solved a basic problem. Why was the task of his PA getting him a new diary such a 'pain in the backside'? He made a fair point – the bureaucracy and time taken (some six weeks, he reckoned) to acquire something that also ended up costing more than he would have paid in his local shop was ridiculous.

Successful organizations look at both the *transactional/operational* buying processes and the *strategic* processes. So how my CEO (or his PA) orders routine items, gets them delivered and pays for them is what buying professionals call the 'purchase to pay' process. How the organization buys goods or services in complex spend areas will be termed 'strategic sourcing' or 'category management' processes, which may well be led by specialist professional staff (see Key Principle 10.) But the main point is that all processes need to be carefully thought out, well defined and communicated, to ensure understanding throughout the organization.

When I carry out organizational reviews of buying as a consultant, a revealing question to ask senior managers is simply, 'Do you know how you are supposed to buy things?' It is amazing how often the answer is 'Not really'. That might mean lack of clarity in the systems they should be using to place an order; confusion about when they are supposed to consult a subject-matter expert or buying professional; or how many quotes they should get for a fairly small purchase. But this lack of clarity and understanding adds up to a much higher chance of buying failure or, indeed, of fraud.

KEY PRINCIPLE 5

Put clear buying policies in place, communicate, ensure understanding and take action where staff don't respect the policy.

This principle supports both day-to-day good buying performance and acts as a key element of the anti-fraud effort. You need clear policy statements so that staff know exactly what they should and shouldn't do, along with the willingness to take action against transgressors. If buying behaviour is poor and you don't have clear policies and controls in place, then even if real fraud or corruption is detected, it will be harder to take action. The poor management of buying will allow culprits to say, 'Well, everyone here does that' or even 'I didn't know there was anything wrong with giving work to my brother's firm.'

Good buying governance means setting the right buying policies – the rules for those spending money with suppliers. Defining the right operational processes for buying (Key Principle 4) sits alongside policy definition: the two are inextricably linked. That's because if staff don't understand the processes and policies around spending money, how can they choose the best firms to work with, negotiate good contracts and manage suppliers well? Lack of clarity inevitably leads to buying failure.

These are fundamental organizational governance issues and should be high on the priority list when boards exercise their due diligence and compliance role. In my own experience as a non-executive director, capital investment sometimes gets covered, but often huge non-capital projects or contracts simply aren't considered at board level, let alone a discussion of supply-chain risk or fraud prevention. Indeed, few directors – executive or non-executive – would get far if you asked them to name the ten biggest suppliers to their organization, never mind identifying where the risks of major buying-related failure lie. Yet it's clear how much can go wrong through buying problems, up to and including total organizational failure.

KEY PRINCIPLE 6

Define what you want carefully (the specification) and always seek value and innovation from suppliers.

This is a vitally important part of the overall process, but it is not something that is easy to codify completely. However, the principles are clear. You must define carefully what you want to buy or run the risk of the supplier providing something that is not fit for purpose – completely or at least in part. Or it might be over-specified, so that it works, but is more expensive or risky than is needed.

Equally, if specifications are too detailed, or tie down the supplier with brand or technical details, then you lose the chance of the supplier applying some initiative and perhaps providing interesting ideas about how they can fulfil your needs. This relates also to the strategic aims discussed in Key Principle 2. When you actively seek innovation from suppliers, it is particularly important to give them the chance to propose different ideas; an overly prescriptive specification will kill this.

But tread carefully where your requirement and needs are likely to change over time, possibly rapidly. The right specification today may not be appropriate in several years' or even months' time. Retaining some flexibility for change, and building that into the contract, is important in many spend areas.

KEY PRINCIPLE 7

Understand your key suppliers.

Every organization needs to understand their suppliers, for many reasons – from the basic anti-fraud details (are they who they say they are?), to recognizing their advanced capabilities so that you can make the most of their potential strategic contribution.

New opportunities and risks have emerged around the supply base, too. No doubt a buyer somewhere is now being accused of discrimination by a supplier (or potential supplier), on the grounds that the owner of that supplier is transgender – that is just the latest topic on the long list of wider corporate social responsibility (CSR) issues that impinge on the business agenda. 'Policy' in a buying context used to mean internal rules, but now it encompasses more of an outward-facing nature, with organizations expected to be aware of, and consider, many of these wider issues in day-to-day operations.

CSR topics are generally categorized as in the table below under social, environmental and economic headings. For some time larger organizations have generally been aware of their responsibility in areas such as carbon and pollution reduction. Human-rights issues, such as modern slavery, have come to the fore more recently, along with a drive to persuade firms to help the less fortunate in society – perhaps by employing ex-offenders or the disabled. Discrimination on sexual, religious, racial or other grounds is obviously to be avoided. And issues around climate change, the use of plastics, deforestation and water shortages all fit into a greater awareness of the fragility of life on our planet.

Organizations must consider their own internal operations, policies and approaches, but in many cases they are also expected to look at their suppliers and use their influence to address what is going on in the supply chain. Indeed, most organizations can have more positive impact by persuading *suppliers* to behave in certain ways than they can through purely internal initiatives.

This requires policies that face inwards and outwards. How will we make sure modern slavery, forced labour or similar is not happening in our own business (of course), but also is not present among our suppliers in distant countries? How can our suppliers reduce carbon use in their production and in the shipping of what we buy from them? Are we sure we aren't prejudiced against

Ten Principles for Good Buying

SOCIAL	ENVIRONMENTAL	ECONOMIC
Buying from/supporting charities, social enterprises	Sustainable consumption of natural resources (timber, water, etc.)	Creating 'decent' jobs and employment
Support to local communities	Energy conservation	Fair employment (living wage, employment contracts, etc.)
Workplace diversity and inclusion – race, sex, religion, disability, etc.	Recycling and obsolescence including the 'circular economy' and plastics	Employment of disadvantaged or displaced people (natural disasters, political conflicts, etc.)
Diversity in the supply base	Waste reduction and landfill	Supporting smaller firms (suppliers)
Providing apprenticeships	Carbon footprint reduction	Supporting-minority owned firms
Rehabilitation of ex-offenders	Pollution (emissions, discharges and toxic waste)	Promoting innovation in the supply base
Modern slavery and human rights (including 'conflict minerals')	Farming issues, food safety and provenance	Fair treatment of suppliers (e.g. formal and fair contracts, prompt payment)

Health-and-safety issues	Eco-friendly products and services	Paying taxes (no evasion)
Wage discrimination	Local issues – noise, litter, traffic	Fraud and corruption – awareness and action
Data protection (GDPR)	Wildlife conservation	Sanctions programmes

Table 2: Corporate Social Responsibility/'Procurement with Purpose' issues

minority-owned suppliers – indeed, can we help disadvantaged groups become more successful (and better suppliers to us) by supporting certain firms?

And this is no longer just a 'nice to have'. The right positioning and action around supplier-based CSR issues – or 'procurement with purpose', as it is becoming known – can not only avoid some of the buying failures discussed earlier, but can actually contribute positively towards competitive advantage and business success.

KEY PRINCIPLE 8

Competition (and the threat of competition) are always good.

Many failures arise because of lack of competition. That may be because of market structure (monopolies or cartels), supplier dependence or because buyers fail to seek competition and thus weaken their negotiating position, increasing their chance of a bad deal.

Real competition, particularly if it is open and transparent, can also be effective protection against fraud. Globally, a government giving large contracts to suppliers *without* competition is often a warning sign of corruption. Closer to home, what many feel are ridiculously high returns are often made in industries where there is limited competition or buyers tend to award contracts without competitive processes – for example, in the way firms appoint investment banks to work on acquisitions or run a bond issue.

But doesn't competition contradict ideas around close, long-term relationships with key suppliers, enabling them to provide the competitive advantage that buyers seek? There is some truth in that, so let's add a caveat to this enthusiasm for competition. The *threat* of it can be just as good as actual competitive processes. In other words, you don't have to ask every supplier to rebid for their work every year, to keep them motivated. But having the *possibility* to substitute a supplier undoubtedly keeps any firm on its toes, even if you then choose to work with them closely and collaboratively.

And do note that suppliers will try and persuade you against competitive processes or even tough negotiations! 'But we're strategic partners,' they will say, looking hurt. Two of my friends in the industry, ex-procurement directors, are both making good money now on the 'dark side', training sales teams how to respond to buyers. One of their favourite topics is how to exploit and get the most out of any customer who says they want 'a strategic partnership'!

KEY PRINCIPLE 9

The job isn't done when the contract is agreed.

Contract and risk management were covered through Chapters 9, 10 and 11, so I won't labour those points again here. But this principle emphasizes how important it is to manage the contract and the supplier *after* the initial buying exercise has been carried out. Too many organizations think that the job is done at this point; however, for all but the most basic contracts, success is absolutely dependent on what happens once the supplier is actually delivering the goods or services in question.

Contract and supplier management requires appropriate systems and tools to keep records and to support proactive performance, risk and value management. It also inevitably involves many staff within the organization. They must have the right skills, both technical and behavioural, in order to carry out their duties effectively.

They must be able to understand incentivization and what motivates suppliers, know how to negotiate, be structured in their approach, but also possess good interpersonal skills. This isn't an easy job, particularly when suppliers are performing critical tasks for the organization.

It isn't enough to leave 'buying' to just a few people in the organization, no matter how good and talented they may be; or to consider that good buying performance stops the day a contract is signed. That thought also leads on to the final principle.

KEY PRINCIPLE 10
Everyone who plays a role in the buying process must be appropriately knowledgeable and skilled, to get the most out of your suppliers.

I've worked for large parts of my career in departments that had titles such as Purchasing, Buying or Procurement (this last being the favoured term now in much of the English-speaking world). Sourcing is another word used; and the British public sector has confusingly decided that 'commercial' is a good word for 'buying' or 'procurement', despite the fact that most private-sector businesses consider 'commercial' to mean the sales and marketing side of the business.

I have deliberately not talked much about 'procurement' in this book, for two reasons. First, many people and organizations see it as a specialist team or function that has little to do with day-to-day business objectives, activities and success, and therefore executives tend to view these buying-related activities as not particularly important to them personally.

And second, the presence of a 'procurement department' has led some organizations into a false sense of security. One of the main sources of failure that I've described comes from organizations not understanding the need for broad competence in buying, across many different people, functions and teams, and from the beginning of the process (specifying what is to be bought) to the end (managing the contract and supplier right through to contract end or renewal). Having a department with that 'procurement' label can mean top management is tempted to sit back, in the naive belief that everything is fine and money is being spent well.

In fact avoiding buying failure and getting the most value from suppliers involves *everyone* who holds a budget, helps to decide what is bought or from whom, or manages suppliers once they are working for the organization.

From the technologist who specifies the new IT system, to the accounts clerk who checks invoice payments; from the CEO who gives consulting contracts to her friends, to the regional manager who fails to manage a difficult services supplier in his region – a large organization will have thousands of staff involved in what I've called the buying process. Indeed, every time someone in your organization talks to someone in a supplier organization, the conversation is potentially part of the negotiation process – and, sometimes, it can be a critical part.

That need for wider capability doesn't mean organizations should not put a procurement department in place. Larger organizations in particular will benefit from having key relevant capabilities (such as deep negotiation and technical contracting skills) centred in a specific department. Someone needs to define the buying policies, processes and systems discussed earlier. And if you want to look at major strategic suppliers across all their interactions with your organization, which might touch upon hundreds of staff, then someone has to coordinate this and take an organization-wide approach to key supplier management. But don't think that having a capable buying or procurement department is enough to guarantee success; more is needed.

Culture Shock

It's also important to understand the culture of your organization, both in terms of what to buy and in terms of ensuring appropriate best practice. Sometimes failure is caused when buyers go against the culture of the organization rather than doing something intrinsically 'wrong'. Some organizations love experimentation, innovation and new ideas. Others don't. Some are very centralized, with a controlling management style. Others, even if they are huge, run more as a loose confederation of quasi-independent bodies. Family-owned businesses often have a different feel from publicly quoted firms – the family may have a longer horizon for investments, but might also be tighter with costs in some areas.

I worked for the Mars Group early in my career, and the culture of that family-owned business certainly coloured our approaches to buying. The firm took very long-term views and didn't worry about one bad quarter, or even one bad year, in terms of performance. But assets were sweated hard, because funding for growth and capital investment had been largely internally generated. One of the 'Five Principles of Mars'[1] was, and still is, *mutuality*, which meant that we aimed to treat suppliers in a certain way (fairly), and the firm's obsession with product quality (another one of the Principles) led to great care around buying ingredients.

Adult Education

How do you make sure staff don't unwittingly promote bad buying? The right approach runs through policy and processes, from having the right technology in place to support good buying, to the training and education that staff need to develop the correct capability. That needs to be appropriate for all those different participants in the process, from those who develop specifications to specialist buyers, from contract managers to the organizational leaders who will have top-level contact with strategic suppliers. Everyone should understand their role and how they can execute it effectively.

There are three relevant skill types here. *Technical management skills* can be quite specific to buying, such as market analysis or developing incentivization models that encourage suppliers to work in the desired way. Some are very important to buyers, but are also critical elsewhere – such as negotiation skills, which are required in a range of roles. Then there are those technical skills that arguably every executive should possess, such as project management or basic financial understanding.

The second set of skills is *behavioural*. There has been increased focus on these skills in business in recent years, as bosses realized that executives need to be persuasive, tenacious, creative, good listeners, empathetic, have good judgement, show initiative, and so on, in order to succeed. This is certainly true of anyone who is interfacing regularly or at a high level with suppliers. Technical skills themselves are not enough.

The final skill type is more focused on buying and relates to knowledge around *specific spend categories*. So whether they are a 'procurement professional' or someone sitting in IT, operations or elsewhere in the organization, if someone is buying marketing services, or major construction projects, or ERP software, or automotive components, then they need to know a bit (or more than a bit) about that area. The depth of understanding required depends on how important the spend area is to the organization, and different spend areas also have various levels of complexity. You can quickly pick up enough about the office stationery market to the point where you can be an effective buyer. Becoming a real expert on the energy market, or the market for complex outsourced IT services, or aerospace engine components, can take many years.

Future Proof

It seems appropriate to finish with a focus on people, because success for any organization comes down largely to that in the end, and improvement almost always requires people to change their behaviour. Politicians need to drive for more open, competitive and transparent processes in public-sector buying, driven by commercially skilled public servants. Senior business leaders should be driving improvement from the top – from ensuring the basics of process, policy and systems are in place, to building skills across a wide range of staff and understanding how suppliers contribute strategically (and knowing who their top ten suppliers are wouldn't be a bad start).

Will this change in the future? I've touched on innovations such as blockchain, which might alter the way the provenance of what is bought and sold can be tracked. Advanced data analytics are already helping organizations analyse what is being spent, and artificial intelligence might help us match our needs seamlessly with the most appropriate suppliers in the market. For more routine buying, automated marketplaces may well commoditize what is being bought, with pretty much standard pricing and specifications across the supply market.

But even if technology helps point you towards an interesting new start-up firm as a potential supplier, I believe you will still want to go and meet the firm and their people. You'll want to negotiate, discuss creatively how you can best work together to mutual benefit, and how they can provide you with goods or services that can help your organization develop or protect competitive advantage.

And if something goes wrong once a supplier is working with you, I strongly suspect you are going to want to talk to a real live person about how they are going to put that right. If my mission-critical system crashes, I want to speak *now* to a human at the software firm that provides it, not simply record the fault with a 'chatbot'. So although technology will be increasingly central and useful in the buying world, the difference between bad buying and good buying will continue to rest largely with humans, not machines.

Good buyers lead to good buying, and bad buyers lead to bad buying.

Bibliography

If this book has stimulated your interest in buying, procurement and supply-chain management, here are some books that are potentially interesting and useful to a general business reader. I have avoided 'textbook'-type publications, many of which are excellent, but are aimed at deep specialists. There are also various reports and papers mentioned in the source notes throughout the book, many of which are well worth reading in their totality.

GENERAL BUSINESS AND BUYING

Business Success by Andrew Cox: An original and stimulating view of supply-chain power in a wider business context from Professor Cox.

The CPO by Christian Schuh, Michael F. Strohmer, Stephen Easton, Armin Scharlach and Peter Scharbert: A decent attempt to create a novel from a case study concerning a procurement change programme.

The Prince by Machiavelli: Dealing with people, managing change and power, from the sixteenth century and still relevant today.

Thinking, Fast and Slow by Daniel Kahneman: Behavioural psychology and economics, including insight into negotiation and risk management.

BOOKS BY BUYING PRACTITIONERS

The Lean Supply Chain: Managing the Challenge at Tesco by Barry Evans and Robert Mason: Another company case-study, interesting even if Tesco's issues dented their pedestal somewhat.

Procurement Mojo by Sigi Osagie: A very readable and engaging mix of personal experience and theory from a successful practitioner.

Strategic Procurement by Caroline Booth: A strong focus on the business aspects of buying, from a top procurement director and consultant.

Bibliography

Strategic Sourcing and Category Management by Magnus Carlsson: The subtitle 'Lessons learned at IKEA' indicates the content, and the IKEA approach is fascinating.

Winning Selling . . . to Impress the Buyer! by Tim Ussher: Aimed at salespeople, so that they can understand how senior buyers (like the author) think.

NEGOTIATION

Getting to Yes by Roger Fisher, William Ury and Bruce Patton: Still the best book ever about business negotiation.

The Negotiation Book by Steve Gates: From the founder of the Gap Partnership, a detailed guide packed with good ideas and wisdom.

Negotiation for Purchasing Professionals by Jonathan O'Brien: O'Brien is a prolific author in the procurement space; all his books are educational and very useful.

OTHER SPECIALIST TOPICS

Buying Professional Services by Fiona Czerniawska and Peter Smith: How to spend money wisely with consultants, lawyers, auditors . . .

Contract Management – Core Business Competence by Peter Sammons: An exhaustive guide to an often-overlooked topic.

Financing the End-to-End Supply Chain by Simon Templar, Erik Hofmann and Charles Findlay: The definitive guide to supply-chain finance, a growing area of importance.

Plundering the Public Sector by David Craig and Richard Brooks: A readable and shocking exposé of how consultants 'rip off' the UK public sector (allegedly).

A Practical Guide to Public Procurement by Abby Semple: Everything you need to know about European public procurement – clear and useful.

A Short Guide to Procurement Risk by Dr Richard Russill: Russill has written many procurement-related books, from textbooks to a novel – all worth reading.

Notes

INTRODUCTION

1 Joint third globally in the Transparency International Corruption Perceptions Index: https://www.transparency.org/cpi2018
2 https://www.ago.gov.sg/docs/default-source/report/103c3319-e3df-4300-8ce0-e0b831e0c898.pdf
3 Not the most reliable reference, I know – a discussion on Quora: https://www.quora.com/How-does-the-sum-of-all-the-worlds-public-company-revenues-compare-to-the-total-of-private-company-revenues
4 Unpublished data – it should be public by early 2020.
5 Study from consulting firm Proxima: 'Corporate Virtualization – A global study of cost externalization and its implications on profitability'; https://bit.ly/2YxJ6N8

1. GETTING THE SPECIFICATION RIGHT

1 https://www.washingtontimes.com/news/2018/oct/23/chuck-grassley-demands-answers-air-forces-1280-cof/
2 https://www.theregister.co.uk/2019/11/27/irish_government_printer/
3 https://www.theguardian.com/world/2019/nov/26/irish-parliament-red-faced-printer-too-big-doors
4 https://www.forbes.com/sites/duenablomstrom1/2018/11/30/nobody-gets-fired-for-buying-ibm-but-they-should/#913f6c348fc3
5 https://citywire.co.uk/wealth_manager/share-prices-and-performance/share-factsheet.aspx?InstrumentID=483
6 https://sverigesradio.se/sida/artikel.aspx?programid=2054&artikel=5776337

Notes

7 https://www.nao.org.uk/wp-content/uploads/2019/05/Progress-delivering-the-Emergency-Services-Network.pdf
8 https://www.centreforpublicimpact.org/case-study/air-forces-expeditionary-combat-support-system-ecss/
9 https://spectrum.ieee.org/riskfactor/computing/it/bipartisan-senate-condemns-us-air-force-ecss-program-managements-incompetence
10 https://www.nist.gov/nist-time-capsule/nist-beneath-waves/nist-reveals-how-tiny-rivets-doomed-titanic-vessel
11 https://www.theguardian.com/uk-news/2017/jan/27/bath-tipper-truck-crash-haulage-boss-and-mechanic-jailed
12 That's a poor analogy really, as camels are brilliantly adapted to their environments, but the point is valid.

2. UNDERSTANDING THE MARKET

1 Market: 'an area or arena in which commercial dealings are conducted'
2 https://www.ghanaweb.com/GhanaHomePage/NewsArchive/Procurement-of-GHC-23M-anti-snake-serum-The-twist-and-turns-750102#
3 http://webcache.googleusercontent.com/search?q=cache:XQITe57L FRgJ:archive.spectator.co.uk/article/15th-september-1973/26/skinflints-city-diary+&cd=1&hl=en&ct=clnk&gl=uk&client=firefox-b-d
4 https://www.instituteforgovernment.org.uk/blog/ministry-justice-was-wrong-outsource-probation
5 https://www.nao.org.uk/report/transforming-rehabilitation-progress-review/
6 https://feweek.co.uk/2019/02/15/major-training-firm-with-contracts-across-government-goes-into-administration/
7 https://en.wikipedia.org/wiki/Pacific_Solution
8 https://www.amnesty.org.uk/press-releases/manus-island-refugees-moved-one-hellish-situation-another
9 https://www.independent.co.uk/news/world/australasia/manus-island-refugees-suicide-attempts-australia-election-papua-new-guinea-a8927051.html

Notes

10 https://www.straitstimes.com/asia/australianz/australia-under-pressure-to-explain-lucrative-png-refugee-contracts-award-to
11 https://www.b2bmarketing.net/en-gb/resources/blog/half-money-i-spend-advertising-wasted-trouble-i-dont-know-which-half
12 https://cdn2.hubspot.net/hubfs/2364596/Reports_and_Documents/Ad%20Fraud%20Reports/Pixalate%20-%20Q3%202018%20Ad%20Fraud%20Report.pdf?utm_campaign=Ad%20Fraud%20Benchmark%20Reports&utm_source=Post-download%20FU%20email
13 https://www.refinery29.com/en-gb/2018/11/217422/instagram-purge-fake-followers-accounts-remove
14 https://www.thetimes.co.uk/article/fake-social-media-bots-con-200m-out-of-brands-gwd57s7xl

3. CHOOSING SUPPLIERS

1 https://www.theguardian.com/business/2019/mar/01/charity-t-shirts-made-at-exploitative-bangladeshi-factory
2 https://www.fairwear.org/news/over-1000-women-workers-interviewed-for-research-on-gender-based-violence-in-vietnam/
3 https://www.theguardian.com/technology/2017/jun/18/foxconn-life-death-forbidden-city-longhua-suicide-apple-iphone-brian-merchant-one-device-extract
4 China's suicide rate is around eight people per 100,000, similar to the UK's and considerably lower than in the US.
5 https://www.theguardian.com/technology/2017/jun/18/foxconn-life-death-forbidden-city-longhua-suicide-apple-iphone-brian-merchant-one-device-extract
6 https://www.fairlabor.org/sites/default/files/documents/reports/foxconn_investigation_report.pdf
7 http://electronicswatch.org/en
8 https://www.theguardian.com/global-development/2017/oct/06/laptop-firms-accused-of-labour-abuses-against-chinese-students-sony-hp-acer
9 http://electronicswatch.org/cal-comp-statement_2556108.pdf

10 https://www.theguardian.com/politics/2019/oct/07/jennifer-arcuri-five-unanswered-questions-tv-interview-boris-johnson
11 https://www.thetimes.co.uk/edition/news/big-brands-fund-terror-knnxfgb98
12 https://www.thedrum.com/news/2017/01/30/pg-review-all-agency-contracts-2017-four-step-plan-bring-transparency-media-supply
13 https://www.thetimes.co.uk/edition/business/revealed-how-tsbs-spanish-owner-sabadell-sparked-it-meltdown-xjt678hvq
14 https://www.bbc.co.uk/news/business-50471919
15 https://www.tsb.co.uk/news-releases/slaughter-and-may/

4. DON'T GET TOO DEPENDENT

1 https://www.welt.de/wirtschaft/article157743075/Lieferstopp-zwingt-VW-Golf-Produktion-in-Stillstand.html
2 https://uk.reuters.com/article/volkswagen-cartrim-dispute-IDUKFWN1B00MM
3 https://uk.reuters.com/article/us-volkswagen-suppliers-IDUKKCN1150OG
4 https://europe.autonews.com/article/20160819/ANE/160819826/vw-to-halt-golf-production-as-supplier-dispute-hits-4-german-plants
5 https://www.washingtonpost.com/news/the-switch/wp/2018/05/09/ford-suspends-f-150-production-after-supplier-fire/
6 https://www.ft.com/content/50c272c4-dce9-11e3-ba13-00144feabdc0
7 https://pfeyeblog.wordpress.com/2015/05/05/beyond-britain-how-do-other-countries-use-pfi/
8 https://www.telegraph.co.uk/news/politics/8973557/Hospitals-being-charged-extortionate-sums-by-PFI-sums-to-carry-out-basic-DIY-jobs.html
9 https://www.bbc.co.uk/news/uk-england-merseyside-47046477
10 http://constructionblog.practicallaw.com/learning-from-carillion-under-bidding-in-the-age-of-austerity/
11 https://www.ft.com/content/a4dd80be-f9f1-11e7-a492-2c9be7f3120a
12 https://www.theregister.co.uk/2018/11/20/uk_government_suppliers_living_wills/

Notes

13 Adam Smith, *An Inquiry into the Nature and Causes of the Wealth of Nations*, 1776, Book I, Chapter X
14 This would not have been allowed in countries such as Germany, which still has strict laws relating to 'beer purity'.
15 https://www.youtube.com/watch?v=hC8mqPLHDVU

5. HOW TO NEGOTIATE

1 https://www.safc.com/history/the-roker-roar/charlie-hurley
2 https://www.smtmagazine.co.uk/school-photocopier-rip-offs-just-the-tip-of-the-iceberg/
3 https://www.computerweekly.com/news/2240234478/How-to-deal-with-the-next-software-sales-renewal-call-from-the-big-four-suppliers
4 https://japantoday.com/category/national/exclusive-as-north-korea-expands-arsenal-japan%27s-missile-defence-shield-faces-unforeseen-costs-sources
5 http://spendmatters.com/uk/mod-paying-5000-day-plus-vat-consultants-evidence/
6 https://www.penguinrandomhouse.com/books/324551/getting-to-yes-by-roger-fisher-and-william-ury/9780143118756/
7 https://www.procurious.com/procurement-news/how-to-deal-with-monopoly-suppliers
8 https://www.bbc.co.uk/news/world-africa-47574463
9 https://www.nao.org.uk/wp-content/uploads/2018/01/The-Ministry-of-Defences-arrangement-with-Annington-Property-Limited.pdf
10 https://www.penguin.co.uk/books/56314/thinking--fast-and-slow/9780141033570.html

6. UNDERSTANDING INCENTIVES

1 https://webcache.googleusercontent.com/search?q=cache:cZ8FgJAkbe4J:https://www.bloomberg.com/graphics/infographics/most-efficient-health-care-around-the-world.html+&cd=14&hl=en&ct=clnk&gl=uk&client=firefox-b-d
2 https://www.ft.com/content/e92dbf94-d9a2-11e9-8f9b-77216ebe1f17

Notes

3 https://jamanetwork.com/journals/jama/fullarticle/2752664?guest AccessKey=bf8f9802-be69-4224-a67f-42bf2c53e027&utm_source= For_The_Media&utm_medium=referral&utm_campaign=ftm_ links&utm_content=tfl&utm_term=100719
4 https://www.npr.org/sections/health-shots/2019/09/27/765230011/ u-s-justice-department-charges-35-people-in-fraudulent-genetic-testing-scheme
5 https://www.nytimes.com/2016/03/10/upshot/medicare-tries-an-experiment-to-fight-perverse-incentives.html
6 https://www.birminghammail.co.uk/news/local-news/call-centre-not-blame-frustration-186673#ixzz1xZjnDlCG
7 https://blogs.lse.ac.uk/latamcaribbean/2019/10/17/good-incentives-bad-timing-crop-substitution-coca-cultivation-and-aerial-spraying-in-colombia/
8 https://hbr.org/2004/11/aligning-incentives-in-supply-chains
9 https://www.theguardian.com/business/2018/jan/15/the-four-contracts-that-finished-carillion-public-private-partnership
10 By Cees J. Gelderman, Janjaap Semeijn and Sjerp De Vries; https:// www.mdpi.com/2412-3811/4/3/41/htm
11 https://www.bbc.co.uk/blogs/thereporters/robertpeston/2009/05/ bt_selfinflicted_wounds.html
12 https://www.theguardian.com/business/2009/may/24/executive-pay-bonuses-royaldutchshell

7. HOW *NOT* TO BE STUPID

1 https://eu.northjersey.com/story/news/new-jersey/governor/2018/ 04/19/chris-christie-nj-official-portrait/485434002/
2 https://www.independent.co.uk/news/education/education-news/ university-of-bath-glynis-breakwell-vice-chancellor-resignation-fat-cat-pay-row-salary-a8081291.html
3 https://www.independent.co.uk/news/education/bath-university-painting-glynis-breakwell-vice-chancellor-a8845331.html
4 https://www.autonews.com/executives/ghosns-image-boosting-campaign-undermined-party-video

Notes

5 (you probably think this airport is about you)
6 https://euobserver.com/justice/126948
7 https://www.rollingstone.com/politics/politics-features/pentagon-budget-mystery-807276/
8 https://www.nao.org.uk/wp-content/uploads/2019/02/The-award-of-contracts-for-additional-freight-capacity-on-ferry-services.pdf
9 https://www.thetimes.co.uk/article/brexit-no-deal-ferry-boss-ben-sharp-of-seaborne-freight-left-tax-debt-when-old-firm-failed-kwwlohvnq
10 https://www.deccanherald.com/national/national-politics/what-all-can-the-court-overlook-in-the-rafale-matter-723110.html
11 https://www.politico.eu/article/germany-biggest-enemy-threadbare-army-bundeswehr/
12 https://www.casemine.com/judgement/uk/5a8ff72e60d03e7f57ea932c
13 https://www.bbc.co.uk/news/business-41573951
14 https://www.dailymail.co.uk/news/article-2164686/How-U-S-Army-spent-5BILLION-failed-pixel-camouflage--wanted-look-cooler-Marines.html
15 https://www.warhistoryonline.com/instant-articles/army-camouflage-that-cost-more.html
16 https://www.nao.org.uk/report/review-of-the-final-benefits-statement-for-programmes-previously-managed-under-the-national-programme-for-it-in-the-nhs/
17 https://www.theguardian.com/business/2010/mar/21/nhs-national-program-problems
18 https://itknowledgeexchange.techtarget.com/public-sector/open_letter_from_79_doctors_nu/
19 https://www.theguardian.com/government-computing-network/2012/feb/10/csc-nhs-npfit-writedown
20 https://webcache.googleusercontent.com/search?q=cache:QgAQ2Ui6buQJ:https://www.hsj.co.uk/technology-and-innovation/exclusive-government-agrees-multimillion-settlement-over-it-contract-row/7023325.article+&cd=1&hl=en&ct=clnk&gl=uk&client=firefox-b-d

8. TRUST NO ONE

1. https://www.slaughterandmay.com/media/926429/bskyb%20v%20eds_feb_2010.pdf
2. https://www.justice.gov/opa/pr/japanese-fiber-manufacturer-pay-66-million-alleged-false-claims-related-defective-bullet
3. https://www.theguardian.com/education/2008/aug/15/sats.schools
4. http://news.bbc.co.uk/1/hi/education/7562835.stm
5. https://assets.publishing.service.gov.uk/government/uploads/system/uploads/attachment_data/file/229092/0062.pdf
6. https://www.lemonde.fr/culture/article/2019/10/21/philharmonie-de-paris-jean-nouvel-contre-attaque_6016278_3246.html
7. https://www.thetimes.co.uk/article/architects-plan-for-philharmonie-strikes-535m-wrong-note-in-paris-tcdchrxk6
8. https://publications.parliament.uk/pa/cm201314/cmselect/cmpubacc/471/471.pdf
9. https://www.theguardian.com/business/2013/dec/13/serco-lose-contract-gp-services-nhs-outsourcing
10. https://www.panmacmillan.com/authors/jon-ronson/the-men-who-stare-at-goats/9780330375481
11. https://nationalinterest.org/blog/buzz/its-official-us-navy%E2%80%99s-littoral-combat-ship-complete-failure-58837
12. http://news.bbc.co.uk/1/hi/8561286.stm
13. https://www.itv.com/news/2020-01-20/hs2-could-cost-106-billion-says-review/
14. NAO link: https://www.nao.org.uk/wp-content/uploads/2020/01/High-Speed-Two-A-progress-update.pdf

9. COPING WITH CHANGE

1. https://www.bbc.co.uk/news/world-us-canada-24613022
2. https://gcn.com/Articles/2013/10/22/healthcaregov-woes.aspx?Page=2
3. https://regmedia.co.uk/2019/04/23/hertz-accenture-website.pdf
4. https://www.slideshare.net/jmramireza/the-foxmeyer-drugs-bankruptcy-was-it-a-failure-of-erp-2332065

5 https://aisel.aisnet.org/cgi/viewcontent.cgi?article=1437&context=amcis1999
6 https://www.theguardian.com/uk-news/2018/feb/21/hundreds-of-kfc-shops-closed-as-storage-depot-awaits-registration
7 https://www.thesun.co.uk/money/5610538/kfc-branches-stores-shut-chicken-delivery-reopen/
8 https://www.kapsch.net/ktc/ir/Announcements/adhoc-announcements/ktc_20190619_adhoc
9 https://www.bbc.co.uk/news/business-47967766
10 https://www.wired.co.uk/article/crossrail-delay-2019

10. WHAT'S THE RISK?

1 https://economictimes.indiatimes.com/maruti-suzuki-temporarily-suspends-production-due-to-fire-at-subros-manesar-facilities/articleshow/52503400.cms
2 https://www.pedestrian.tv/news/a-potato-shortage-is-causing-mcdonalds-in-japan-to-ration-fries/
3 https://www.independent.co.uk/news/world/asia/chips-run-out-for-kfc-in-japan-due-to-potato-shortage-that-has-also-affected-mcdonalds-9998378.html
4 https://usa.oceana.org/press-releases/oceana-finds-seafood-fraud-persists
5 https://www.bbc.co.uk/news/10565838
6 https://www.thetimes.co.uk/article/more-patients-die-in-listeria-outbreak-from-hospital-sandwiches-jxddpkqpd
7 'Reputation Risk – A Rising C-Suite Imperative'; https://www.mmc.com/content/dam/mmc-web/Files/Reputation-Risk-Final-web.pdf
8 A 'stock-price drop' is defined as a drop in the company stock price greater than 20 per cent within a ten-day period relative to changes in the industry average.
9 https://riskandinsurance.com/putting-a-price-on-reputational-damage/
10 http://electronicswatch.org/en
11 https://www.autonews.com/article/20180907/OEM01/180909837/toyota-halts-production-in-japan-after-deadly-quake

12 For example, from the German software firm riskmethods: https://www.riskmethods.net/
13 https://www.autonews.com/article/20091130/OEM01/911309997/customers-cry-foul-over-bankruptcy-tactic
14 https://www.thetimes.co.uk/article/what-a-bore-rome-buries-its-drills-bptmzkb2t
15 https://www.wantedinrome.com/news/rome-metro-roma-metropolitane-to-go-into-liquidation.html
16 https://roma.repubblica.it/cronaca/2019/10/01/news/roma_metro_c_al_capolinea_ne_soldi_ne_progetti_sepolte_le_due_talpe-237387780/?refresh_ce
17 https://www.dw.com/en/berlin-airports-pr-man-sacked-for-honesty/a-19181604
18 https://www.tagesspiegel.de/berlin/hauptstadtflughafen-ber-eroeffnung-im-oktober-2020-kannste-vergessen/24061576.html
19 https://www.thelocal.de/20161013/berlin-airport-employee-jailed-for-taking-huge-bribe

11. THE JOYS OF CONTRACT MANAGEMENT

1 https://www.birminghammail.co.uk/news/midlands-news/revealed-how-bollards-led-bill-16064141
2 https://www.constructionnews.co.uk/news/knowledge-news/amey-fined-48m-for-birmingham-bollard-dispute-05-04-2019/
3 https://www.ft.com/content/dd2392ba-98d8-11e9-9573-ee5cbb98ed36
4 https://www.nao.org.uk/wp-content/uploads/2016/12/Good_practice_contract_management_framework.pdf
5 http://www.ijimt.org/vol8/717-MP0022.pdf
6 H. S. Heon et al., 'Analyzing schedule delay of mega project: Lessons learned from Korea train express', *IEEE Transactions on Engineering Management*, 2009, vol. 56, no. 2, pp.243–56
7 D. H. Pickrell, *Urban Rail Transit Projects: Forecast Versus Actual Ridership and Cost*, Department of Transportation, Washington DC, 1990, p.164
8 https://www.mckinsey.com/business-functions/digital-mckinsey/our-insights/delivering-large-scale-it-projects-on-time-on-budget-and-on-value

9 http://www.healthpayrollinquiry.qld.gov.au/__data/assets/pdf_file/0014/207203/Queensland-Health-Payroll-System-Commission-of-Inquiry-Report-31-July-2013.pdf
10 https://www.nao.org.uk/wp-content/uploads/2011/07/10121272es.pdf

12. THE FUNDAMENTALS OF FRAUD

1 https://www.sfo.gov.uk/2018/12/19/five-convictions-in-sfos-alstom-investigation-into-bribery-and-corruption-to-secure-e325-million-of-contracts/
2 https://um.dk/en/news/newsdisplaypage/?newsid=7868548c-37b8-4d9e-9393-e99e6e545e7c
3 https://www.hbs.edu/faculty/Publication%20Files/self-serving+justifications_132c2494-d4a5-4292-94ee-26249cb46b6e.pdf
4 https://www.bbc.co.uk/news/uk-scotland-edinburgh-east-fife-47816701
5 https://www.transparency.org/news/feature/where_are_africas_billions

13. WHO AM I REALLY BUYING FROM?

1 https://www.theguardian.com/business/2008/jun/12/corporate-fraud.ukcrime
2 http://europa.eu/rapid/press-releASE_IP-09-137_en.htm?locale=en
3 Adam Smith, *An Inquiry into the Nature and Causes of the Wealth of Nations*, 1776, Book I, Chapter X
4 https://www.oxfordscholarship.com/view/10.1093/acprof:oso/9780199551484.001.0001/acprof-9780199551484-chapter-4
5 http://www.oecd.org/germany/33841373.pdf
6 https://www.internationallawoffice.com/Newsletters/Shipping-Transport/Germany/Arnecke-Sibeth-Dabelstein/Impact-of-truck-cartel-on-transport-sector#ori
7 http://europa.eu/rapid/press-releASE_IP-16-2582_en.htm
8 https://www.telegraph.co.uk/finance/newsbysector/construction-andproperty/2788332/Construction-cartel-may-have-cost-taxpayer-300-million.html

9   https://www.gov.uk/cma-cases/construction-industry-in-england-bid-rigging
10 https://www.gov.uk/government/news/construction-suppliers-accused-of-colluding-to-keep-prices-up
11 https://www.theguardian.com/law/2011/feb/10/mabey-johnson-directors-guilty-kickbacks-saddam-hussein
12 https://www.gov.uk/government/case-studies/water-tanks-cartel-case-study
13 https://www.telegraphindia.com/business/cci-lens-on-beer-cartel-anheuser-busch-inbev/cid/1714499
14 https://therealdeal.com/2017/04/13/contractor-hid-ties-to-mob-to-win-1wtc-contract/
15 https://af.reuters.com/article/worldNews/idAFBRE89817A20121009
16 https://businesstech.co.za/news/government/192632/why-bee-fronting-is-a-bad-idea-and-wont-be-tolerated-any-longer/
17 https://www.justice.gov/usao-ma/pr/owner-sham-veteran-owned-company-sentenced-100-million-fraud
18 For more detail on Sochi, see Chapter 18.

14. FIXING THE SUPPLIER SELECTION

1   https://gulfnews.com/uae/crime/uaes-federal-audit-body-detects-dh60-million-fraud-1.2295484
2   https://businesstech.co.za/news/general/116470/60-of-tenders-are-corrupt-in-some-way-deputy-public-protector/
3   https://www.unian.info/society/10457937-new-study-warns-of-worrying-levels-concerning-openness-of-foreign-ngos-in-ukraine.html
4   https://assets.publishing.service.gov.uk/government/uploads/system/uploads/attachment_data/file/783688/Mr_Thomas_Marshall_15858_-_SoS_decision.pdf
5   https://www.bbc.co.uk/news/world-us-canada-47408239
6   https://globalnews.ca/news/5768603/snc-lavalin-trudeau-supreme-court-judges/
7   https://business.financialpost.com/news/rcmp-charges-snc-lavalin-with-fraud-and-corruption-linked-to-libyan-projects

8 https://en.crimerussia.com/corruption/transparency-told-about-billion-worth-state-contracts-concluded-without-tender/
9 https://uk.reuters.com/article/us-india-agustawestland/india-scraps-agustawestland-helicopter-deal-agrees-to-arbitration-IDUKBREA02oLU20140103
10 https://cfa.nhs.uk/resources/downloads/guidance/NHSCFA%20Pre-contract%20procurement%20fraud%20guidance%20-%20v1.0%20July%202018.pdf
11 https://www.ibtimes.com/who-fat-leonard-navy-corruption-scandal-sees-malaysian-defense-contractor-plead-1786076
12 https://www.washingtonpost.com/investigations/navy-repeatedly-dismissed-evidence-that-fat-leonard-was-cheating-the-7th-fleet/2016/12/27/0afb2738-c5ab-11e6-85b5-76616a33048d_story.html?utm_term=.6ed83e62fa18
13 https://www.nytimes.com/2017/05/17/business/us-admiral-in-fat-leonard-navy-scandal-sentenced-rear-adm-robert-gilbeau.html
14 https://www.rollingstone.com/politics/politics-news/fat-leonards-crimes-on-the-high-seas-197055/
15 https://www.justice.gov/opa/pr/us-navy-admiral-sentenced-prison-lying-federal-investigators-about-his-relationship-foreign
16 https://www.justice.gov/opa/pr/active-duty-us-navy-commander-pleads-guilty-conspiring-foreign-defense-contractor-defraud-us
17 https://www.navytimes.com/news/your-navy/2019/02/09/navy-captain-who-moonlighted-as-fat-leonards-pr-man-is-headed-to-prison/
18 https://www.forbes.com/sites/charlestiefer/2018/04/03/naval-contender-to-head-the-military-chiefs-was-tainted-by-ethics-scandal/#2dfoc1736990

15. WHAT AM I REALLY BUYING?

1 https://www.bloomberg.com/news/articles/2015-03-13/glencore-told-to-pay-40-million-over-romania-oil-fraud-by-court
2 https://www.casemine.com/judgement/uk/5a8ff7cc60d03e7f57eb22dd
3 A blockchain is a way of recording data, using an open, distributed ledger that can record transactions between parties efficiently in a

manner that is verifiable and permanent. It is (virtually) impossible to corrupt or hack into a blockchain.
4 An interesting phenomenon, but not a connection to the spirit world; https://www.bbc.com/future/article/20130729-what-makes-the-ouija-board-move
5 https://www.theguardian.com/politics/2014/jan/26/fraudster-paid-government-promote-fake-bomb-detectors
6 https://www.bbc.co.uk/news/uk-29459896
7 https://www.nature.com/articles/d41586-018-05149-2
8 https://www.inc.com/business-insider/theranos-violated-clinical-laboratory-regulations.html
9 https://www.cnbc.com/2017/06/21/theranos-walgreens-reportedly-reach-a-deal-to-settle-suit-for-under-30-million.html
10 https://www.france24.com/en/20190325-sweet-scheme-france-uncovers-massive-italian-kiwi-fraud
11 How exactly do you 'Frenchify' a kiwi fruit – get it to smoke a Gauloises in a sexy manner? Dress it in a beret and stripy top? Ensure it takes a long lunch break, with wine, of course?
12 https://www.asiapacific.ca/blog/vaccine-scandals-china-why-do-they-keep-happening-over-and
13 https://www.brilliantearth.com/conflict-diamond-trade/
14 https://www.tracr.com/
15 https://www.theguardian.com/business/2019/jul/03/serco-fined-229m-over-electronic-tagging-scandal
16 https://www.nasa.gov/sites/default/files/atoms/files/oco_glory_public_summary_update_-_for_the_web_-_04302019.pdf
17 https://www.justice.gov/opa/pr/aluminum-extrusion-manufacturer-agrees-pay-over-46-million-defrauding-customers-including

16. SPENDING SOMEONE ELSE'S MONEY

1 https://www.liverpoolecho.co.uk/news/liverpool-news/nhs-manager-ripped-over-160000-6758080
2 https://www.miamiherald.com/news/local/community/miami-dade/article228969369.html

3. https://www.postandcourier.com/news/charleston-county-doles-out-credit-cards-to-roughly-half-its/article_dc2bc540-0241-11e9-bde1-6b474d68abd8.html
4. https://www.postandcourier.com/news/charleston-county-will-cut-number-of-employees-who-can-charge/article_c781f226-18e0-11e9-817b-3737d094379e.html
5. https://www.standard.co.uk/news/crime/foreign-office-executive-who-stole-2k-public-funds-with-government-credit-card-may-face-jail-a4092626.html
6. https://www.nurseryworld.co.uk/nursery-world/news/1165017/ex-montessori-charity-boss-jailed-for-fraud
7. https://bit.ly/2VosXMU
8. https://www.thesun.co.uk/news/9196084/headteacher-sex-lair-banned-from-teaching/
9. https://www.cambridge-news.co.uk/news/cambridge-news/live-headteacher-sawtry-stewart-court-13724890

17. WHAT AM I PAYING FOR?

1. http://www.telegraph.co.uk/news/uknews/1581837/Sainsburys-potato-buyer-arrested-in-3m-bribe-case.html
2. https://www.iol.co.za/the-star/news/tycoon-admits-to-using-fake-invoices-to-swindle-eskom-of-r35m-36813370
3. https://www.iol.co.za/the-star/news/ex-financial-manager-accused-of-swindling-eskom-of-r35m-36197594
4. https://www.dailymail.co.uk/news/article-1290821/Sex-mad-Toys-R-Us-boss-Paul-Hopes-stole-millions-pay-girls-sports-cars.html
5. https://www.telegraph.co.uk/news/uknews/crime/10858267/Oxfam-anti-fraud-boss-jailed-for-stealing-65000-from-the-charity.html
6. https://www.telegraph.co.uk/news/uknews/crime/7071820/Ambassadors-daughter-stole-200k-to-buy-horses.html
7. https://www.justice.gov/usao-sdny/pr/lithuanian-man-pleads-guilty-wire-fraud-theft-over-100-million-fraudulent-business
8. https://www.bbc.co.uk/news/uk-england-17829021
9. https://www.thelocal.it/20180328/serie-a-club-lazio-email-scam

Notes

10 https://www.irishtimes.com/news/crime-and-law/dublin-zoo-victim-of-500-000-internet-based-fraud-by-organised-gang-1.3410481
11 https://www.actionfraud.police.uk/alert/action-fraud-warning-after-serious-rise-in-ceo-fraud

18. POLITICS AND FRAUD

1 https://www.latimes.com/sports/more/la-sp-sochi-corruption-20140203-story.html
2 http://sochi.fbk.info/en/report/
3 https://www.sec.gov/news/press/2008/2008-294.htm
4 https://www.theguardian.com/news/2019/mar/01/spain-watergate-corruption-scandal-politics-gurtel-case
5 https://www.wsj.com/articles/how-china-flexes-its-political-muscle-to-expand-power-overseas-11546890449
6 https://www.telegraph.co.uk/news/2019/07/22/kenyan-finance-minister-arrested-corruption-charges-mega-dam/
7 https://www.the-star.co.ke/news/2019-09-15-how-procurement-officers-swindle-billions-through-tenders/
8 https://www.reuters.com/article/kenya-corruption/italys-cmc-di-ravenna-denies-any-wrongdoing-in-kenya-dams-scandal-IDUSL8N24P21Y
9 https://www.bbc.co.uk/news/world-africa-49081916
10 https://euanticorruption.com/2019/01/13/romanias-anti-corruption-chief-resigns/
11 https://projekte.sueddeutsche.de/artikel/politik/caught-in-the-trap-e675751/
12 Strache called the video 'a honey trap stage-managed by intelligence agencies'; https://www.theguardian.com/world/2019/may/20/austrian-government-collapses-after-far-fight-minister-fired
13 https://www.theguardian.com/world/2019/may/18/austrian-government-in-crisis-over-secret-strache-footage
14 https://balkaninsight.com/2019/05/21/red-flags-raised-over-serbias-procurement-of-official-jet/

Notes

15 https://www.nytimes.com/2018/01/03/business/dealbook/brazil-petrobras-corruption-scandal.html
16 https://www.theguardian.com/world/2017/jun/01/brazil-operation-car-wash-is-this-the-biggest-corruption-scandal-in-history
17 https://www.offshoreenergytoday.com/rolls-royce-pays-25-million-to-petrobras-in-bribery-settlement/
18 https://www.wsj.com/articles/odebrecht-testimony-in-brazil-points-to-pemex-bribe-in-mexico-1494025050
19 https://www.theguardian.com/world/2017/feb/10/peru-ex-president-alejandro-toledo-faces-arrest-on-bribery-charges
20 http://www.xinhuanet.com//english/2017-06/01/c_136329639.htm
21 https://www.theguardian.com/world/2019/apr/17/peru-alan-garcia-ex-president-shoots-himself-critical-condition

19. PREVENTING FRAUD

1 https://cfa.nhs.uk/resources/downloads/guidance/NHSCFA%20Pre-contract%20procurement%20fraud%20guidance%20-%20v1.0%20July%202018.pdf
2 This is often termed 'separation of duties' – having different people involved in terms of order placing, authorization and signing off on payment.
3 https://www.accountancyage.com/2019/10/16/toppled-from-the-top-the-risk-of-fraud/

20. TEN PRINCIPLES FOR GOOD BUYING

1 https://www.mars.com/about/five-principles

Index

3D printing 4

Accenture 103, 121–2
Action Fraud 235
'act of God' 132, 135–6, 137–8
advertising 16, 26–7, 38–9, 41, 58,
    189–90 *see also* marketing
Aegis Ashore 63
AgustaWestland 191–2
airlines
    'group deals' with hotels around
        major airports 50–51, 67
    premium pricing 6
    see also individual airline name
algorithms 38
Alix Partners 65
Alstom Group 163–4
Amann, Horst 145
Ambani, Anil 95
Amey plc 148–9
Amundson, Troy 201
analyse (supplier testing) 109, 112, 117, 126
Analyse, Reference, Test (supplier testing advice) 109–10, 112, 117, 126
Andersen Consulting 122–3
Anheuser-Busch 56, 57, 176
Annington Property 69
Apple 6, 31, 264
Arcuri, Jennifer 33, 34
Arias, Jesus Marcelo Ramirez 122
Arklow Shipping 93
arrogance 13, 87, 88, 89, 98, 100, 102, 157

Arror dam project, Rift Valley, Kenya 240
artificial intelligence (AI) 7, 177, 258, 280
Australian government refugee 'transition centres' 24–5
Austrian government collapse (2019) 243–4
authorization of payments/orders 87
    fraud and 183, 186, 215–16, 218, 227, 233
    fraud prevention and 251, 252, 253–4, 255, 258
awards, Bad Buying *see* Bad Buying Awards

Bad Buying Awards 55, 56–88
    buying failure, existential (Schlitz Beer) 56–8
    collaborative fraud (Fat Leonard and the US Navy) 196–202, 249, 254, 258
    international project, worst (Berlin Brandenburg Airport) 144–7
    politically significant and geographically spread fraud, most (Petrobas and Odebrecht) 246–9
    technology failure, most impressive (NHS National IT Programme) 100–103
balance-sheet management (optimizing assets, cash and working capital) 266

301

Index

Balkan Insight 244, 245
Balkan Investigative Reporting Network (BIRN) 244
ballistic-missile interceptor stations 63
Banco Sabadell 39, 40
Bangladesh, garment manufacturers in 29–30, 134
bankruptcy xv, 80, 124, 132, 139–40, 145, 178, 246
Barron, Sir Donald 17
Bartels, Hans-Peter 96
BATNA ('best alternative to a negotiated agreement') (fall-back position or backstop) 65–6, 70, 72
Baverstock Academy, Birmingham 185
Baxter, John 224–5
B-BBEE Commission, South Africa 181
Beaumeunier, Virginie 210
behavioural skills 279
believe, need to 114–16, 206, 207
Beliveau, John 199
Belt and Road initiative 240
Berlin Brandenburg Airport 144–7
Berlin Tegel Airport 144
bid evaluation/bidders *see* evaluation process
Bidvest 124
Bilalov, Akhmed 236
Birmingham 115, 148–9, 185
Birmingham Council 148–9
biscuit flour 66
'black swan' events 155
Blair, Tony 101
blockchain-based technology 204, 211, 258, 280
Bloomberg 203
    healthcare efficiency study (2018) 74
BMW 6, 37–8, 264
Bolton, Gary 205
Bombardier 14

Bonanno crime family 179
bonuses 54, 83, 84, 231
bots 26, 27
brand specification 5–6, 7
Brazilian Workers' Party 247
Breakwell, Dame Glynis 86–7
Breslau, Jeff 201
Brewin Dolphin 8
brewers
    buying failure and 56–8
    price-fixing and 176–7
Brexit 91–3
bribery 35, 61, 98, 170, 184, 188, 256, 259
    AgustaWestland and 191–2
    Alstom Group and 163–4
    Berlin Brandenburg Airport and 147
    Fat Leonard and 164, 165, 193–4, 196–202, 249, 254, 258
    invoice fraud and 224, 225
    kickback fraud and 39, 61, 170, 217, 240
    'magic wand' detectors and 205, 206
    offsets and 95
    politics and 237, 238, 239, 240, 246–9
    sexual 193–4, 196–202
    SNC-Lavalin affair and 186–7, 238–9
    social media and 135
    strengthening of laws against 169, 187
Bribery Act (UK) 169, 187
Bridgestone 171
British Airways 6
British Army 190, 205
BSkyB 104–5
BT (British Telecom) 82–3
budget holder (accountable for expenditure generally) 32, 61, 83, 230, 252, 253, 255, 256, 257, 268
Budweiser 57

Index

buffer stocks 78
Bujak, Philip 221
bulletproof vests, US law-
  enforcement purchase of 106–7
Bundeswehr (German armed-forces
  organization) 96
bureaucracy 90–91, 188, 269
Busch InBev 176–7
buyer savings 82–4
buying authority (right to agree
  contracts with external parties)
  252
buying power 20, 45–6, 64
buying spend (spend on buying by
  organizations worldwide) xiv

Cabinet Office 49–50
Capita 52, 76, 79
call centres 76, 79
carbon emissions 32, 272, 273
careful, being 116–17
Care International 30
Carillion 48–50, 55, 80
Carlsberg 176
Carreyrou, John: *Bad Blood* 206, 207
cartels 171–8, 182, 274
category management processes 269
catering 23, 24, 27, 47, 48, 53, 196,
  230–31, 265
Car Trim 44
Cayman Islands 178
Centers for Medicaid and Medicare
  Services (CMS) 118–19
CEO fraud 234–5
CFO (Chief Financial Officer) 66,
  106, 165
CFP consortium 97–8
change management 118–29, 132, 158
  changing your mind, difficulty of
  129

completion dates 127–9
contract 119, 129
internal organizational 120–21
legal position, clarifying 126–7
specification 118–19
supplier 124–6
technology industry pace of
  change 121–4
Changsheng 210
Channel Tunnel 154
Charleston County, South Carolina
  218–19
Chesterman, Hon. Richard 155
Chillgarde 57
China 18–19, 30–31, 133–4, 137, 205, 210,
  235, 239, 240
  Belt and Road initiative 240
Christie, Chris 86, 87, 216
Cisco 78–9
climate change 30, 272, 273
CMC di Ravenna 241
cocoa 17–18, 26, 111
coca cultivation 76–7
collaborative fraud 196–202, 249, 254,
  258
collusion between suppliers/suppliers
  and buyers 186, 190, 225, 251, 255,
  256–7 *see also* cartels *and*
  price-fixing
Colombian farmers 76–7
Combat 18 38
Comigel 208–9
commercial models 23, 61
Community Rehabilitation
  Companies (CRCs) 23
Compaq 6
competition 20, 23, 25, 34, 35, 41, 278–9
  brand specification, restricting
  competition through 7
  cartels and *see* cartels

303

Index

competition – *cont'd*
  choosing suppliers and 29, 30, 34, 35, 39, 41
  competition (and the threat of competition) is always good (Key Principle 8) 274–5
  competitive advantage, your suppliers contribute to the creation of (Key Principle 1) 263–7, 274, 275, 280
  competitive process 25, 34, 65, 175, 189, 199, 245, 249, 251, 275
  FOMO and 207
  negotiation and 64–6, 67, 71
  political fraud and 240, 245, 249
  preventing fraud and 251–2, 256
  price benchmarking and 226
  risk and 139
  supplier dependence and 44, 46–7, 48, 54, 55
  supplier selection and 188–9, 190, 194, 197, 199, 256
  ways to provide 266
Competition and Markets Authority (CMA) 175, 176
Competition Commission of India (CCI) 176
conflict minerals/materials 209, 210–11, 273
conflict of interest 185, 201, 242, 251, 253
consistency 51–2, 131, 143, 250
construction projects *see individual project name*
'Construction Projects Cost Overrun: What Does the Literature Tell Us?' (Abdulelah Aljohani, Dominic Ahiaga-Dagbui and David Moore) 154
Continental 46, 171

contract management 68, 69, 81, 111, 124, 143, 146, 148–59, 245
  cost overruns 153–8
  importance of 150–53, 151, 275–6
  IT projects 155–8
  length/breakdown of contracts 148–9
  National Audit Office (NAO) 'Good practice contract management framework' 151–2
  neglected art of 150
  ongoing 149–50
contracts, unprofitable 82–3, 113
copyright 42
Córdoba Airport 90
corporate hospitality 194–5, 228
corporate social responsibility (CSR) 29–32, 41, 79, 268, 272–4
Correa, Francisco 239–40
corruption *see* fraud/corruption
Corruption of Foreign Public Officials Act (CFPOA), Canada (1998) 187
counterfeit materials 133, 211, 255
cover pricing 174–5
creative, be 70–71
credit cards, business 165, 218–21, 251
Crimson Construction Corporation 178–9
CRM (Customer Relationship Management) 267
Crossrail 127–9
crude oil 203–4
CSC 103
CTS Eventim 126, 127
Culliname, David 5
culture, organizational 277–8
customer intimacy 264
cyber risk 140

Index

DAF 173
Daimler 173
Dassault Aviation of France 94–5
data analytics 13, 177, 267–8, 280
data protection 140, 274
De Beers Group 211
Deccan Herald 95
decision-making buyers, length of time in office of 226, 256–7
Defence Science and Technology Laboratory 205
defence sector *see individual company and entity name*
deforestation 133, 272
Dell 6, 20
Democratic Republic of the Congo (DRC) 211
Denmark 93, 164
Denso 46
Department for Communities and Local Government 156–7
Department for Transport (DfT) 91–2, 93
Department of Health 100–101, 103
Department of Justice, US 86, 106–7, 200–201, 213
Department of Trade and Industry 205
dependence, supplier *see* supplier dependence
Der Spiegel 243
DGCCRF (French anti-fraud agency) 210
DHL 124–5
diamond mining/trading 211
diesel scandal 44
differentiation 264
discrimination 272, 273, 274
dis-economies of scale 50–51, *51*
Dodd-Frank Wall Street Reform and Consumer Protection Act, US 211

Draap Trading Ltd 209
Dublin Zoo 234
Dunlop Oil & Marine 171

EADS Defence & Security 156, 157, 158
Ealing Hospital NHS Trust 229
economic or financial risk 132–3
economies of scale 50–51, *51*, 67
EDS 104–5
Electronics Watch 31–2, 135
electronic-tagging services 212–13
email compromise 231, 232, 234–5
Emergency Services Network (ESN) programme, The 9
Emirates 6
EMS S/A 244–5
end-of-contract clauses 126
Energy Solutions Ltd 97–8
engineering, procurement and construction (EPC) services 62, 186
Enron 254
enterprise resource planning (ERP) 62, 122, 267, 279
Ericsson 137, 138, 142
ES Automobilguss GmbH 44
Eskom 225
ETS Europe 110–12
European Commission 171–2, 173–4
European Court of Auditors 89–90
European Court of Human Rights 242
European Court of Justice 126, 127
European Union 89–90, 91, 171–2, 179, 242, 244, 245
Eurotunnel 93–4
evaluation process xiii, 13–14, 36–8, 41
fraud and 192–3
three key stages 36–8
see also supplier selection

Index

exchange rates 19
Expeditionary Combat Support System (ECSS), The 9–10
expenses 86, 166, 219

Facebook 232, 234
Fair Labor Association 31
Fair Wear Foundation 30
'false front' arrangements 179–80
FARC 77
Fasen, Jan 209
Fat Leonard (Leonard Francis) 164, 165, 193–4, 196–202, 249, 254, 258
FBI 171, 205
Ferrovial 25
ferry contracts, UK 91–4, 97
Feyenoord 234
Figaro software package 8
Financial Times 46, 49
Findus 208
Finmeccanica 191
FiReControl 156–8
fixed-price contract 79–80
Flughafen Berlin Brandenburg Holding GmbH (FBB) 144–7
FOMO (fear of missing out) 207
Food and Drug Administration (FDA), US 56
Ford 38, 45, 54, 140
Foreign Corrupt Practices Act, US 169, 187
Foreign Office 219
FormTech 140
Formula 1 216
Foxconn 30–31
FoxMeyer 122–4
FPÖ party (Austria) 243–4
Francis, Leonard (Fat Leonard) 165, 193–4, 196–202, 254

fraud/corruption xi, xii, xiii, xiv, 17, 24, 25, 26, 34, 35, 39, 61, 62, 67, 68, 87–8, 94, 95, 98, 104, 112, 133, 135, 142, 146–7, 151, 158–9, 161–259, 263, 268, 270, 271–2, 274
 believe, need to 206, 207
 blockchain-based technology and 204, 211 *see also* blockchain-based technology
 bribery *see* bribery
 cartels 171–8, 182
 CFOs and 165
 comical nature of some low-level frauds 164–5, 204
 component failure 213–14
 conflict minerals/materials 210–11
 credit cards, business/purchasing cards 165, 218–21, 251
 efficacy of products 204–7, 208
 expenses fraud 166, 219
 fundamentals of 163–70
 importance of 168–9
 invoice fraud 223, 224–35, 256–7 *see also* invoice fraud
 misrepresentation 178–82
 motives for 165–8
 personal life and work, blurred lines between 221–3
 politics and 236–49 *see also* politics, fraud and
 preventing *see* fraud prevention/ seven key principles for
 provenance 204, 209, 210, 211
 regulatory imperative and changes in 169
 reputational risk and 168–9
 services sector, difficulty of knowing exactly what has been provided in the 212–13

Index

spending someone else's money/
buying something using the
organization's money that is not
used for the benefit of that
organization 215–23
substitution of products/disguising
sources 203–4, 208–11
supplier selection, fixing the
183–202, 256 *see also* supplier
selection, fixing the
whistle-blowing 176–7, 202, 258
wider effects of 169–70
fraud prevention/seven key principles
for 250–58
1. It must be clear who is entitled
to spend money in the
organization 251, 252
2. Any expenditure committed
must be authorized properly
251, 253–4
3. All entities to which money is
paid must be verified and
authorized 251, 254
4. Goods or services purchased
must be checked to ensure that
what was received is aligned
with the contract and payment
251, 255
5. Supplier selection and pricing of
purchased goods and services
must be transparent 251, 256
6. Opportunities for collusion
between suppliers, and between
suppliers and buyers, must be
minimized 251, 256–7
7. Perpetrators of fraud
(particularly internal staff) must
know that action will be taken
against them if they transgress,
and that action will be
proportionate and strong 251,
257–8
technology, future and 258–9,
280
Freedom of Information 48, 65, 87,
198
friendships/conflicts of interest and
32–4, 41
F*ck-ups, how to avoid 261–80
Fujitsu 53, 103
fungicides 210

Gaddafi, Muammar 188, 238, 239
Gaddafi, Saadi 187, 238
Galenika 244–5
Galloway, Joe 104–5
Garateix, Michael 217
'garbage in, garbage out' 268
García, Alan 248
Gartner Research 62
Gauke, David 23
GDPR (General Data Protection
Regulation) legislation 140,
274
German government motorway toll
scheme 126–7
Getting to Yes (Roger Fisher, William
Ury and Bruce Patton) 65–6
G4S 212
Ghana health ministry (anti-snake-
venom serum buying failure)
16–17
Ghosn, Carlos 88
Gilbeau, Rear Admiral Robert
199–201
Glemsford Primary School 61
Glencore 203
Glenn Defense Marine Asia (GDMA)
196–202
good buying objectives 266

good buying principles, ten 263–77
   Key Principle 1 *Your suppliers contribute to the creation of competitive advantage and the achievement of organizational goals* 263–5
   Key Principle 2 *For everything you buy, consider how that item or spend category contributes towards strategic goals, and conduct your buying appropriately* 265–7
   Key Principle 3 *Good data allows you to understand what you're buying and is central to good performance* 267–8
   Key Principle 4 *Define clear and differentiated processes for buying, enabled by technology and balancing ease of use with control and rigour* 268–70
   Key Principle 5 *Put clear buying policies in place, communicate, ensure understanding and take action where staff don't respect the policy* 270–71
   Key Principle 6 *Define what you want carefully (the specification) and always seek value and innovation from suppliers* 271
   Key Principle 7 *Understand your key suppliers* 271–4
   Key Principle 8 *Competition (and the threat of competition) are always good* 274–5
   Key Principle 9 *The job isn't done when the contract is agreed* 275–6
   Key Principle 10 *Everyone who plays a role in the buying process must be appropriately knowledgeable and skilled, to get the most out of your suppliers* 276–7

Gopher 205
Gorch Fock (German Navy training ship) 96
Gordhan, Pravin 225
Gorski, David 181
government/public procurement
   change management and 118–19, 126–9
   contract management and 148–9, 156–8
   fraud/corruption *see* fraud/corruption
   incentivization and 73, 74–5, 77, 81–2
   market, failure to understand the 16–17, 20, 22–6
   negotiation and 63–5, 67–70
   private finance initiative (PFI) 47–8, 50, 54, 148
   risk management and 142–3
   specification failure and 4, 5, 8, 9–10, 14–15
   stupidity and 87–99
   supplier dependence and 46–50, 52, 53
   supplier selection and 29, 32–3, 37–8, 180, 183–5, 186–7, 189, 191
   trust and 107, 110–11, 112, 113–17
Granger, Richard 102
Grayling, Chris 23, 91
Greenvale AP 224–5
Großmann, Jochen 146–7
'group buying' 59
Guardian, The 29, 80, 82–3, 124–5, 248
Gudenus, Johann 243–4
Gürtel case 239–40

Haji, Noordin 241
Harvard Business Review (*HBR*) 78
Harvard Negotiation Project 65
HealthCare.gov 118–19

Index

healthcare system, US 74–5
Health Service Journal 103
Hertz 121–2
Hewlett Packard (HP) 20, 32, 53, 104
HHI 140
high-speed 2 (HS2) programme 115–16
Hokkaido region earthquake, Japan (2018) 135
Holmes, Elizabeth 206–7
Hopes, Paul 227
horse-meat scandal 208–10, 255
hotel 'group deals' 50–51, 67
Howard, Mark 105
human-rights violations 30–31, 79, 133, 141, 272
Hunter, Kevin 207
Hurley, Charlie 59–60, 70
Hussein, Saddam 175

IBM 6, 21, 53, 155–6, 268
'ideomotor effect', the 204
'illogical' contract awards 178
impatience 88–9
Imtech 145, 147
incentivization 49, 72, 73–85, 106, 123, 231, 276, 279
   appropriate and proportionate 84
   BT 82–3
   Capita call centres 76, 79
   Cisco 78–9
   Colombian coca farmers 76–7
   construction projects 79–82
   costs of getting it wrong 78–9
   fixed-price contracts and 79–80
   government invoicing 75–6
   individual incentivization 82–4
   judging the efficacy of 79–82
   performance-based contracts and 81–2
   perverse incentives 74–7
   staff and 85
   suppliers acting against the interest of their customer 78–9
   'time and materials' agreement and 80–81
   US healthcare system and 74–5
incompetence xi, xiv, 61, 64, 68, 87, 105, 150, 165, 193
Indian government
   AgustaWestland helicopter contract 191–2
   Rafale fighter jets purchase 94–5
information asymmetry 121–2, 154
innovation 8, 11, 16, 23, 25, 36, 46, 50, 51–2, 54, 114, 179, 192, 264, 266–7, 271, 273, 278
   buyers value consistency above 51–2
   capturing external market/supplier 266
   seeking from suppliers 271, 273
insiders/inside information 165–8, 180, 188, 190, 193–4, 197, 201, 228
Instagram 27
'intelligent client' (knowing enough to manage suppliers properly) 122–4, 147
International Association for Contract and Commercial Management (IACCM) 150
Interserve 23
invoice fraud 165, 175, 180, 186, 197, 217, 221, 223, 224–35, 256–7
   CEO fraud and 234–5
   collusion between buyer and seller and 224–7
   insiders, fake invoice fraud by 227–30
   invoice misdirection fraud xiii, 165, 223, 231–4

invoice fraud – *cont'd*
   over-invoicing 225, 230–31
   steps that can be taken to guard against 226
IP protection 42
Irish government Komori printer purchase 5
Islamic State (ISIS) 38, 39, 211, 238–9
IT (information technology) *see individual company or project name*
Iveco 173

Japan 47, 130–31, 135–6
   government ballistic-missile interceptor stations purchase 63–4
Jobs, Steve 31
Johnson, Boris 33, 34, 35, 41
Johnson Controls 46
Joseph Schlitz Brewing Company 56–8, 264
Jurma, Anca 242
just-in-time production techniques 46

Kahneman, Daniel: *Thinking, Fast and Slow* 71
Kamara, Ansumana 232–3
Kapsch 126, 127
Kardashian, Kim 27
Kastoria Airport, Greece 90
Kentucky Fried Chicken (KFC) xiii, 119, 124–5, 130
Kenyan government 67–8, 158
   Arror and Kimwarer dam projects corruption 240–41
   fence construction/inflated prices 67–8
Kenyatta, Uhuru 240
Kernow Clinical Commissioning Group 113
kickback fraud 39, 61, 170, 217, 240

Kimwarer dam project, Rift Valley, Kenya 240
Kissinger, Henry 206
kiwi fruit fraud 209–10
Komori printer 5
Körtgen, Manfred 145
Kövesi, Laura Codruta 242
KPIs (key performance indicators) 81
Kronen Zeitung 243

La Repubblica 142
Lazio, SS 234
Legion Construction Inc. 181
Le Monde 113
Lexus 135
liability 105, 107
Libya 186–7, 238–9
lightning strikes 137–8
Lloyds TSB Group 39
Lockheed Martin 63, 115
Locklear, Admiral Samuel 201–2
Loss Prevention Associates 228
LSE (London School of Economics) 77
Lulu (dog) 104, 105

Mabey Hire Ltd 175
Machiavelli, Niccolò: *The Prince* 129
M&A (mergers and acquisitions) 35, 83
mafia 178–9
'magic wand' detectors 204–6
Makarov, Igor 243
Makarova, Alyona 243
Malaysian government fund (1MDB) 240
Malunga, Kevin 184
MAN 173
management consultants 6, 19–20, 35, 83, 120–21, 136, 152, 154–5
   see also individual company name

Index

mandated contracts/preferred vendors 91
Manuli 172
Manus Island offshore refugee 'transition centres' 24, 25
Marc Rich & Co. 203
marginal pricing 66–7
Marie Curie 38
marine hoses 171–3, 177
market (in which potential suppliers operate), understanding 15, 16–28, 271–2
   advertising 26–7
   assumption there will always be a range of credible suppliers who can execute the contract 22–5
   commodity markets 17–19
   government sector and 16–17, 20, 22–5
   largest firms and 20–21
   market research 27–8
   market size/dynamics 19–20
   supplier preference model 21–2, 22
   understand your key suppliers (Key Principle 7) 271–4
marketing 10, 26–7, 38, 58, 73, 105, 107, 171, 257, 264, 265, 276, 279
Mars Confectionery 11, 42–3, 130, 132
Mars Group 19, 59, 278
Marshall, Thomas 185
Maruti Suzuki India 130
Mauro, Alfredo di 146
Maxwell, Robert 254
Maylam, John 224–5, 226
McCormick, James 205, 206
McDonald's 130, 131, 136
McKenzie-Green, Edward 228–9
McKinsey 6, 83, 154–5

Meagra Transportation 225
Mercedes-Benz 38
Meridian Magnesium Products of America 45
Metropolitan Police 229
M.G.F. (Trench Construction Systems) Ltd 175
Miami-Dade Animal Services Department 217–18
Michaud, Gilles 187
Microsoft 21
milk-powder scandal, China 133–4
minimum wage 31
Ministry of Defence (MOD) 64–5, 68–9, 190, 220
Ministry of Justice 212
misrepresentation 104–6, 178–82
monopoly 42–8, 52, 53–4, 66, 173, 266, 274 *see also* cartels
Moraka, Bernard 225
Morgan, Sir Terry 128–9
Morse, Amyas 156
motivation
   fraud 167, 168, 189
   staff 265–6, 267
   supplier 275, 276
Mott MacDonald 92
MTN Group 181
Müller, Matthias 45
Murdoch, Rupert 206

NASA Taurus T8 and T9 rocket failures 213–14
National Audit Office (NAO), UK xiii, 9, 23, 69, 91–2, 100–101, 115, 116, 156–7, 158
   'Good practice contract management framework' 151–2
national curriculum tests (SATs) 110

Index

National Health Service 47, 100–103, 165, 174, 192–3, 215–16, 217, 229, 250
National Programme for IT (NPfIT) 100–103
National Interest website 115
NATO 94
NatWest Bank 21, 258
NatWest financial-services group 12–13
Naval Criminal Investigative Service (NCIS), US 198–9
Ndrangheta crime syndicate 179
negotiation 3, 12, 33, 34, 59–72, 74, 77, 83, 95, 117, 119, 140, 149, 151, 153, 158, 176, 226, 241, 276, 277, 279, 280
  BATNA ('best alternative to a negotiated agreement') (fall-back position or backstop) 65–6, 70, 72
  commercial models and 61
  competition and 64–6, 274, 275
  creative, be (understand what the other party really wants) 70–71
  failed 60–62
  objective, try to be 71
  personal issues/people involved 59–60, 70
  planning, importance of 72
  poor negotiation and fraud, diagnosing the difference between 61–2
  poor negotiation skills, cost of 68–70
  position you find yourself in relative to the other party 62–4
  priming effects 71
  strength of your position, understanding 64–6
  supplier dependence and 43–4, 45–6, 48, 55
  tactical and aggressive buying and 66–7
  training 71–2
Networks South Africa 181
Newton, Paul 86
New York Times, The 75
Nigeria 47, 169, 227, 233
Nokia 137–8, 139
Nokia Solutions 181
Nouvel, Jean 112–13, 114
Nuclear Decommissioning Authority (NDA) 97–8
nuclear-power stations 97–8

Obamacare (Affordable Care Act, 2010, US) 118–19
objective, try to be 71
Oceana 133
Odebrecht 237–8, 246–9
Odebrecht, Marcelo 247
Odukoya, Ayodele 233
Office of Fair Trading 171, 174
Office of Government Commerce 111
offsets 94–6, 237
oligopoly (limited number of firms dominate a market) 46, 53
Oliver Wyman 134
Olowo, Abayomi 233
Olympic Delivery Authority (ODA) 232–4
Olympics
  (2012) (Summer) 232–4
  (2014) (Winter) 168–9, 181–2, 236–7
  (2016) (Summer) 248
OMV Petrom S.A. 203
One World Trade Center, New York 178–9
online/electronic auctions 67, 258
opening dates/job completion schedules 127–9

Index

Operation Car Wash 237–8, 246–9
operational risk 132–3
optimism bias 143
Orbital Sciences 213
organizational efficiency and
    effectiveness, supporting
    internal 266
Osofsky, Lisa 163–4
outcomes, specification and defining
    7–8, 11
over-optimism 10, 113–14
ÖVP party (Austria) 243–4
Oxfam 228–9

Pabst 58
packaging 11, 56–7, 134, 178, 204, 266, 267
Paladin Holdings 25
palm-oil plantations 133
P&G 39
P&O Ferries 93
Parker ITR 172
patents 42
Patisserie Valerie 254
payroll management 19, 20–1, 155–6
pen, space xiii, 4
Pentagon 90–91
performance-based contracts 81–2
Perry, Laura 219
personal feelings, buying
    negotiations and 59–60, 70
personal life, blurred lines between
    work and 221–3
Pester, Paul 39
Petex 203
Petrobas 185, 237–8, 246–9
petroleum 203–4, 208, 211
pharmaceutical firms 74, 122–4, 210–11, 244
Philharmonie de Paris 112–13, 114

Philips semiconductor plant, New
    Mexico 137–8, 139
Pillinger, Dr Jane 30
Pinnacle 122–3
Pitts, Commander Bobby 201
Pixalate 26
planning, importance of xiv, 5, 64,
    70, 72, 101, 126, 129, 135, 138, 139,
    141–2, 143, 145, 155, 158
plastic use 30, 32, 272, 273
political risk 133
politics, fraud and 236–49
    anti-corruption authorities and
        236, 240, 242–3
    bribery 237, 238, 239–40, 246–7, 248
    governmental collapse 239–40,
        243–4
    government capital projects 240–41
    Petrobas and Odebrecht (most
        politically significant fraud
        award) 237–8, 246–9
    'sweeteners' for sales given to
        governments 244–5
    transparency and 244–5
    Western' firms as paymasters
        for fraud in developing world
        238–9
pollution 73, 79, 137, 267, 272, 273
Porsche 44
Port Authority of New York and New
    Jersey 178–9
Porter, Michael 264
portraits 86–7, 216
Post and Courier newspaper 218–19
potato buying 224–5, 256–7
power and knowledge, concentration
    of 226
'practice contract management
    framework' 151–2
Prevent DEV 44

Index

price
   benchmarking 71, 84, 226, 236, 256
   contract management and 149, 153, 156
   'fair' 43–4, 62, 71, 79, 86, 245, 251
   fixed 79–80
   fraud and *see* fraud/corruption
   low 48–50, 55, 112–13
   marginal pricing 66–7
   market and 16, 17, 18, 19, 21
   negotiation and 59, 61, 62, 65–6, 67, 69–70, 71, 74, 77, 78, 79, 80, 83, 84
   price-fixing 172, 173, 175, 176 *see also* cartels
   specification and 6, 10–11
   supplier dependence and 43, 45, 46, 48, 49, 50, 54
   supplier selection and 29, 34, 36–8, 40, 41, 256
prime vendors 91
priming effects 71
principles
   good buying xiv, 263–77 *see also* good buying principles, ten
   fraud prevention 250–58 *see also* fraud prevention/seven key principles for
Pritchard, Marc 39
private finance initiative (PFI) 47–8, 50, 54, 148
procurement 276–7
Procurious 66
provenance 204, 209
Prowse, Keith 228
public sector procurement *see* government
purchase to pay process 269
purchasing cards/corporate cards 165, 218–21, 251
Purple Futures 23

Pust Siebel 8
Putin, Vladimir 181, 189, 236, 237, 243

Quadro Tracker 204
Qualifications and Curriculum Authority (QCA) 110–12
quality 5–6, 11–12, 31, 36, 37, 40, 67, 76, 80, 110–11, 113–14, 149, 192, 203, 204, 239, 251, 254, 255–6, 264, 266, 267, 278
Quanta 232, 234
quarterly price competition 67
Queensland Health 155–6

Radio Sweden 8
Rafale 94–5
Raggi, Virginia 142
Rana Plaza, Bangladesh 29–30, 134
RBS 22
recycling 273
reference (technique to avoid being duped or misled) 109, 112, 126
Reggio Calabria 179
Reliance 95
Renault 88, 173
reputational risk 24–5, 29, 133–5, 139, 141, 168–9, 267
Reynolds, Nicholas 163
Rijkswaterstaat (RWS) (Dutch public agency) 81
Rimasauskas, Evaldas 231–4
risk management 130–43
   acts of God and 132, 135–6, 137–8
   change management and 132–40
   competition/competitors and 139
   cyber risk 140
   diversity of risk 142–3
   economic or financial risk 132–3
   four ways to handle each risk 141–2
   plans 138–9, 141–2
   political risk 133

Index

reputational risk 24–5, 29, 133–5, 139, 141, 168–9, 267
spreading the risk 136
supply-chain risk/suppliers and 102, 130–31, 136–7, 139–40, 141, 266
technology and 135–6
transferring risk to suppliers 102, 141
Rolling Stone 90–91, 200
Rolls-Royce 248
Romania 203, 208, 242–3
Rome subway Line C 142–3
Rosenberger, JoAnn 62
Rotenberg, Arkady 237
Rotich, Henry 240, 241
Rousseff, Dilma 247–8
Rowntree Mackintosh 17–18, 19
Royal Engineers 205
Royal Liverpool University Hospital 48–9
Russia xiii, 7, 74, 168–9, 181, 189, 236–7, 239, 243, 244
Russian Anti-Corruption Foundation 236
Russian Olympic Committee 236

SABIS 40
Sainsbury's 224–5, 226, 256
sales team/people 8, 11, 20, 43, 59–60, 62, 63, 64, 82–3, 104–6, 126, 179–80, 194, 204, 206, 275, 276
SAPA Profiles Inc. (SPI) 213–14
SAP R/3 ERP system 122
Sawtry Village Academy 222
Scania 173
scarce resources, gaining access to 265, 266
Schauss, Dave 198–9
Schlitz *see* Joseph Schlitz Brewing Company

schools, overcharging of 61, 67
Scott, Judy E. 122
Seaborne Freight 92–3
seafood, mislabelling and fraud around 133
Securities and Exchange Commission, US 239
Serbian government private jet scandal 244–5
Serco 52, 113–14, 212
Serious Fraud Office 163, 175, 212
Service Birmingham 76
Service Disabled Veteran Owned Small Business (SDVOSB) Program 181
sex, offers of 193–4, 196–202, 239
shareholders 3, 8, 73, 82, 88, 138, 165, 169, 227, 246, 248
Sheldon, Tim 205
Shelter 23
Sicilian Mafia 179
sickness pay 31
Siemens 239
Silva, Luiz Inácio Lula da 247
Simpkins, Paul 199
Singapore xiii, 25, 196, 199, 201
Singapore Air 6
Skanska xiii, 232–4
Sky 104–5
Slaughter and May 40
slavery 32, 133, 267, 272, 273
Smith, Adam 54, 172, 176
SNC-Lavalin affair 186–7, 238–9
social media 27, 38–9, 41, 135, 232, 234
software 6, 8, 9, 40, 42, 53, 62–3, 69, 83–4, 102–3, 108, 121, 124, 128–9, 136, 140, 252, 268, 279, 280 *see also individual company and piece of software name*

Index

Solutions de Recherche Intelligence 228
South Africa 181, 184, 225
Spanghero 208–9
specification 3–15, 16, 17, 89, 100, 107, 109, 111, 115, 271
  brand specification 5–6, 7
  changing mid-process 8–10, 14–15
  define what you want carefully (Key Principle 6) 4–5, 271
  fraud prevention and 190–92
  miscommunication/ misunderstanding between internal user of what is being bought and those who will actually do the buying 12–14
  output, outcomes and performance, defining 7–8
  over-specified 3–4, 271
  supplier dependence and 52
  trust and 5–7
  value, true 10–12
Spectator, The 17, 18
spend-analytics systems 177
spend categories 265–7, 279
Spend Matters UK/Europe website 64–5
staff
  fraud and *see* fraud/corruption
  incentivizing 85
  motivation 265–6, 267
  put clear buying policies in place, communicate, ensure understanding and take action where staff don't respect the policy (Key Principle 5) 270–71
State Audit Institution (SAI), Abu Dhabi 183
State of Queensland, Australia 14–15
Stewart, James 222–3
Stone Educational Consultants 185
Strache, Heinz-Christian 243–4
Straits Times, The 24–5
strategic aims/goals of the organization 263–7, 269, 271
stupid, how not to be 86–99
  arrogance and 88
  bidding process, fairness in 97–9
  corruption and 87–8, 99
  cost overruns 96
  drivers of buying failure, three major 87–9
  expenses and 86
  ferry contract, UK government 91–4, 97
  impatience and 88–9
  offsets 94–6
  portraits 86–7
  US military spending and 90–91, 99
  vanity projects 88, 89–90, 98–9
Subros 130
Süddeutsche Zeitung 243
Sunday Times, The 27, 40
Sun, The 125, 222–3
supplier, choosing 28, 29–41
  avoiding suppliers/keeping work within the company 39–40, 41
  bad choice, realization of 34–5
  corporate social responsibility (CSR) and 29–32, 41, 79, 272–4
  competition and 41
  digital advertising and 38–9
  evaluation process 36–9, 41
  friendships/conflicts of interest and 32–4, 41
  questions, posing well-thought-out 41

Index

supplier dependence 42–55, 62, 102, 115, 123, 158, 172, 180
  'buying power' and 45–6
  car manufacturers and 44–6
  competition to mitigate the risk of, using 54, 274
  dis-economies of scale and 50–51, *51*
  government contracts and 46–50, 52, 53
  monopoly and 42–8, 52, 53–4
  oligopoly and 46, 53–4
  risk issues and 131–2, 135, 136–7, 141
  supplier bidding too aggressively and 48–50, 55
  warning signs in terms of how your organization buys, five 50–54, *51* see also warning signs in terms of how your organization buys, five
supplier market, understanding 15, 16–28
  advertising 26–7
  assumption there will always be a range of credible suppliers who can execute the contract 22–5
  commodity markets 17–19
  government sector and 16–17, 20, 22–5
  largest firms and 20–21
  market research 27–8
  market size/dynamics 19–20
  supplier preference model 21–2, 22
  understand your key suppliers (Key Principle 7) 271–4
supplier promises, trust and 104–17
  believe, need to 114–15
  brands and 5–7
  careful, be 116–17
  checking/testing 108–10, 112, 126
  completion dates and 115–16
  exaggeration, salespeople/supplier 104–8, 110–12
  extraordinary suppliers claims 110–12
  low prices to win the deal, bidding 112–16
  walking away from unprofitable contracts 113–14
  yes, natural tendency to say 107–8
supplier selection, fraud and 97–9, 183–202, 256
  bribery 186–7
  competition, avoiding 188–9
  evaluation process 192–5, 238, 240, 244, 250
  extent/widespread nature of 183–5
  inside information 190
  methods for 188–94
  preventing fraud and 251, 256
  restricting the field 189–90
  specifications and requirements designed to favour a particular firm 190–92
sustainability initiatives 266
Sutherland Review 111
'sweatshop' conditions 30–31
Sweden 8, 233

target operating models (TOM) 269
target pricing 81
tax havens 178
technical management skills 278–9
technology *see individual piece of technology and project name*
Temer, Michel 248
tendering process *see* supplier selection
Tesco 208
testing supplier promises (Analyse, Reference, Test) 107, 108–11, 112, 117, 126

Index

Thanet District Council 92
Theranos 206–7, 208
Thessaloniki airport, Greece 90
3D printing 4
'time and materials' approach 80–81, 230–31
Times, The 38, 102–3, 142
Titanic 11–12
Tomkins, Louise 229–30
top-line revenue growth, contribute to through supplier input and action 266
total facilities management contract 53
Toyobo Co. Ltd of Japan 106–7
Toyota 38, 135–6, 140
Toys 'R' Us 227
Tracr (blockchain-enabled solution) 211
training
  good buying 278–9
  negotiation 71–2
transactional/operational buying processes 269
transparency 34, 39, 95, 98, 226, 231, 237, 245, 249, 251, 256, 258, 275
Transparency International (TI) xi, 94, 170, 183–4, 189
Trelleborg 172
truck manufacturers, European 173–4
Trudeau, Justin 186–7
trust 104–117, 178, 188, 228, 230
  believe, need to 114–15
  brands and 5–7
  careful, be 116–17
  checking/testing 108–10, 112, 126
  completion dates and 115–16
  exaggeration, salespeople/supplier 104–8, 110–12
  extraordinary suppliers' claims 110–12
  low prices to win the deal, bidding 112–16
  walking away from unprofitable contracts 113–14
  yes, natural tendency to say 107–8
TSB 39–40, 41
Tshabalala, Victor 225
Twitter 135

UAE (United Arab Emirates) 183–4
UK government
  Carillion liquidation and 49–50
  contract managers and 152–3
  ferry contract with firm that had no boats 91–4, 97
  FiReControl project 156–8
  'magic wand' detectors and 206
  Ministry of Defence (MOD) spending *see* Ministry of Defence (MOD)
  PFI contracts 47–8, 50
  probation-services outsourcing 22–4
  Serco electronic-tagging services for offenders and 212–13
  volume deals with large suppliers 20
Ukraine 184
unfair procurement 92–4
Unilever 27
United Breweries 176
University of Bath 86–7
University of Oxford 154
US Air Force 4, 9–10
US Army 99
US government
  bulletproof vests, law-enforcement purchase of and 106–7

318

healthcare system 74–5, 118–19
military spending 4, 9–10, 90–91, 99, 114–15, 116, 165, 193–4, 196–202, 254
NASA Taurus T8 and T9 rocket failures and 213–14
US Navy
    Fat Leonard corruption case 165, 193–4, 196–202, 254
    programme for littoral ships 114–15, 116

vaccine scandals, China 210–11
vanity projects 88, 89–90, 98–9, 116
Vertuccio, Vincent 178–9
Vietnam 30, 31, 200, 239
Volkswagen (VW) 44–5, 46, 54, 153
Volvo 173
Vp plc 175
Vrij, Stefan de 234

Waitrose 38
Walgreens 207
Wall Street Journal 240
Wanamaker, John 26
warning signs of how your organization buys, five 50–54, *51*
    buyers aggressively aggregate their own spend 50–51, *51*

buyers group together disparate bundles of goods or services 52–3
buyers make tendering and bidding for work difficult 53
buyers set tighter and tighter specifications 52
buyers value consistency above innovation and experimentation 51–2
Washington Post, The 33, 198, 199
Washington Times, The 4
Webster, Madeleine 215, 216
Weed, Keith 27
Wharton Business School 134
whistle-blowing 176–7, 202, 258
Wilson-Raybould, Jody 186–7
Working Links 23
World Cup
    (1996) 60
    (2014) 248
World Reader 29, 30
Wuhan Institute of Biological Products 210

yes, natural tendency to say 107–8
Yokohama Rubber Company 171
YouTube 38–9

Zuma, Jacob 221–2

PENGUIN PARTNERSHIPS

Penguin Partnerships is the Creative Sales and Promotions team at Penguin Random House. We have a long history of working with clients on a wide variety of briefs, specializing in brand promotions, bespoke publishing and retail exclusives, plus corporate, entertainment and media partnerships.

We can respond quickly to briefs and specialize in repurposing books and content for sales promotions, for use as incentives and retail exclusives as well as creating content for new books in collaboration with our partners as part of branded book relationships.

Equally if you'd simply like to buy a bulk quantity of one of our existing books at a special discount, we can help with that too. Our books can make excellent corporate or employee gifts.

Special editions, including personalized covers, excerpts of existing books or books with corporate logos can be created in large quantities for special needs.

We can work within your budget to deliver whatever you want, however you want it.

For more information, please contact
salesenquiries@penguinrandomhouse.co.uk

About the author

Originally from the Isle of Man, I now live in Stafford with my partner, Natalie, and two rabbits, Harry and Willow. I have an undergraduate degree in English Literature with Creative Writing and a Master's in Publishing. My working background is in marketing and working with special needs children, but I've always had an active interest in writing. My favourite authors are Stephen King and C.J. Sansom, but it's J.K. Rowling that allowed me to fall in love with fiction. In my spare time, I enjoy playing tennis and table tennis.

THE FIELD OF SPIES

Seán Cassidy

THE FIELD OF SPIES

Vanguard Press

VANGUARD PAPERBACK

© Copyright 2021
Seán Cassidy

The right of Seán Cassidy to be identified as author of
this work has been asserted by him in accordance with the
Copyright, Designs and Patents Act 1988.

All Rights Reserved

No reproduction, copy or transmission of this publication
may be made without written permission.
No paragraph of this publication may be reproduced,
copied or transmitted save with the written permission of the publisher, or in
accordance with the provisions
of the Copyright Act 1956 (as amended).

Any person who commits any unauthorised act in relation to
this publication may be liable to criminal
prosecution and civil claims for damages.

A CIP catalogue record for this title is
available from the British Library.

ISBN 978 1 784658 92 2

*Vanguard Press is an imprint of
Pegasus Elliot MacKenzie Publishers Ltd.*
www.pegasuspublishers.com

First Published in 2021

**Vanguard Press
Sheraton House Castle Park
Cambridge England**

Printed & Bound in Great Britain

Dedication

For my parents

Chapter One

Mark could not bring himself to feel anything when he found out his father lay critically ill in hospital. He could not figure out whether or not he really cared. He knew he should, and he knew he wanted to care, but how could he? John Formby, his father, had never been a loving parent to him. Ever. He was always critical, always finding fault. This in itself was not abnormal, he knew that. All parents nagged their children in some shape or form, but not all parents were like his father. Out of love, parents give out because they have their kids' best interest at heart.

This was not the case for Mark's father.

As far he was concerned, Mark did not exist, and may as well be a part of the décor in the living room. Mark felt he was an unwanted inconvenience to John, as if fatherhood was a binding contract he had been tricked into. Everything Mark did seemed to irritate him. He did not care for Mark's welfare as far as Mark could ever tell. Some may put it down simply being a bad father, but Mark felt coldness directed at him every time his father looked his way. It was almost as if John was blaming him for forcing his attention, like the very act of acknowledging his son's existence was a chore. They never embraced or played football together in the garden and he had never come to see shows Mark was in at school. They never joked with one another, nor did they converse unless it was necessary and even then, it was forced and stilted. He just didn't seem to care at all. That was what was abnormal.

After a few years of trying, Mark had given up on him; if his father was going to make it impossibly hard to form any sort of positive relationship with his own son, then Mark was going to put in the same effort too. In many respects, John Formby was a stranger to him.

Why should he try to make the effort to care whether John survived or not? For most sons, even thinking of that question would be terribly callous, but Mark felt he had good reason to think it. Caring about your own flesh and blood should be a natural instinct, not a chore. Although he did feel guilty at the thought of not caring – or at least, enough – about whether or not John died.

As he sat in the waiting room, Mark's mind steered back, as it often did, to a conversation he had overheard two years ago which had fundamentally been the focal point of his resentment towards his father.

"You have to put in more effort with him John, he's your son! He won't be a kid forever, and he'll end up resenting you before long. Do you want that? Do you?" Mammy demanded.

"I don't care what he thinks of me."

"You don- you don't care?! How can ya say that?"

Silence.

"But he's your son!" Sylvia Formby was doing her best not to raise her voice too much, but she was shaking in anger.

"Yeah… and I didn't ask for one, did I?" he snarled back.

More silence. Mark had been stunned by such a petty remark.

"I don't care what you want or don't want," Mammy eventually replied, "I let you get away with a lot, and I know you're busy at work but that don't excuse you. Any man can be a father to their kid if they tried; you just haven't made any attempt! I've seen Mark recently not give you the time of day. That's 'cause of you!"

"I don't care. He's just a dumb kid like the rest of 'em. He knows I'm busy so he will be used to it by now," he replied in a tired, casual voice.

"How cold are you?" she whispered, horrified.

Mark had leaned in slightly by the door of their bedroom at this point, and could only just decipher Mammy's words, but he could still hear each one. His heart was pounding, yet he was transfixed; the shock was paralysing him.

More silence from John.

"You really don't love him, do you?"

Mammy uttered this question just as she had the epiphany.

Further silence ensued. She gasped in disbelief.

The silence here was an answer enough—he didn't even try to defend any sense of paternal love for his own kid, and as Mark stood there, he felt as if an anchor had attached itself to his chest and torn his insides away. He could not move but felt like a weight was dragging him into a depth of sadness.

"This can't seriously be because of the affair," Mammy moaned, agonised tears escaping from the corners of her eyes.

Mark's head shot up, and he moved his ear away from the door which was slightly ajar and looked through the slim gap between door and frame. He could just about see John's face look up at Mammy. He was sitting upright against the bedframe. John glared at her.

"What do you think?" he asked without emotion, but his eyes were burning with rage.

Mark's view, although limited, could just about see the edge of Mammy's jaw drop.

"But he is still your kid, John," Mammy insisted. "The DNA test proved that."

John continued to glare. "That's not the point."

"So, you're just going to let him suffer just because I made a stupid mistake? One stupid mistake! This is not his fault!"

"He could just as easily not have been mine."

"So? He is your son! That's surely all there is to it."

"No!" John said, standing up now, "when we tried for a baby, I was misled. The idea of having a kid with you excited me. I loved you, but once you did that... I never did again. And I no longer wanted a kid with you. By then, it was too late, and you were pregnant. We made him when I thought you were faithful, and I wanted a child. If I don't love you, how could I love anything we made? Don't you get it? He's a reminder of your betrayal."

Now it was Mammy's turn to be silent. The hammer had been swung and it was a heavy blow.

"Leave, then!" Mammy eventually shouted at him. "Go on, feckin' leave! It wouldn't make any difference."

To Mark's bewilderment, John laughed. He was laughing in a pleasurable way, although twisted with heavy irony.

"Oh no, no, no. Oh dear Lord, no," he chuckled to himself, "I'd never make it that easy for you."

And then his laughter subsided immediately into an icy cold stare.

Mark shivered. He could not digest what he had heard. He now knew why John was the way he was with him, but he could not believe how unreasonable he was about it. His Dad *should* love him. He was shocked at Mammy, but he wasn't thinking too hard about that at that moment.

This memory resurfaced fairly often. Each time, his heart sank a little deeper. The night of that memory was the last time he shed tears over his father. But he still felt the pain.

Mark relived this painful memory as he sat in the hospital waiting room at St James' Hospital in Dublin, oblivious to the scurrying of many student nurses worriedly scampering about the place. Was there any worse form of betrayal than your own parent not loving you? It was clear that he didn't love Mark, the silence said all that was needed to be said. Neither of his parents to this day knew he was stood outside the door, listening in. It was evident John would not be overly bothered if Mark died, so why should he care if things were the other way around? He knew he sounded melodramatic, but it didn't prevent it from being fact, at least to him.

It was Mammy he really felt sorry for, now distraught as she sat helplessly. Mammy had never once betrayed Mark like John had—she was his only real friend as well as being a mother to him. She didn't want to change him; she didn't ask for much. So, what if she made a mistake? That didn't stop her from being a brilliant mother. She simply loved him and accepted him for who he was. Sure, he was friendless, and she did want him to make friends, but she never pressed him too hard or put him in a position that made him uncomfortable. And that was something Mark would never forget; he would never leave her alone with John or run away just to escape him. The thought had struck him before. He knew that if he ran away, Mammy would break completely. He would never be so selfish. Although he wished she would leave John, he knew that there was nowhere for her to go; she had no family and like him, no friends either.

Mark looked at her with genuine pity as she sat next to him, hunched forwards with hands to her face. He knew she could not help but love John; that much was clear. Even if he was a terrible husband who did not treat her the way she deserved to be treated, Mark knew, or at least learned over time, that love was not a choice.

Mammy had been crying silently in a pain that seemed too much for sound to justify. The tears clung at her eyes like a see-through parasite, but that was not what struck Mark most. No, it was the emptiness and seeming absence of hope which caused Mark to stare helplessly at her, grabbing both her hands in his for comfort. He was never really one affection, but neither was Mammy one for giving up; she was always positive and looking for some form of hope, so seeing her like this was both surreal and disturbing.

His touch seemed to wake some part of her that had temporarily been hidden.

"Thanks," she whispered with a weak smile.

She looked at Mark then with an edge of feeling back in her eyes. It was as if somehow, he had refuelled her spirits slightly.

"That's better," he said, smiling whilst wiping away some tears, vaguely being aware of the role reversal that was going on.

She laughed a mucous laugh as if the same thought had occurred to her; it appeared all her sinuses were open.

"Why can't I cry with style like... I dunno... Meryl Streep?"

Mark smirked. He was not expecting that one.

"Meryl Streep?"

"Yeah, you know. The good actress – betcha she could cry without spurting snot darts at everyone."

She attempted her pretend-annoyed expression. Instead she looked more like a clogged-up volcano about to burst.

Her quirky humour was something which Mark had always found amusing; her unexpected one-liners always made for laughter and it was something which helped Mark cope growing up when John was not around for him. She indirectly allowed him to develop his own humour— not that he had friends to experience it with. But still, sometimes it is important to laugh.

"I dunno, I reckon she'd be impressed to spurt 'snot darts' at people on cue," Mark replied, bemused.

Mammy laughed. A genuine, happy laugh. It made Mark feel a little less helpless.

"You're a good boy, Mark."

"Only 'cause of you."

Another tear trickled down her face.

"I'm glad I have you for a son," she said seriously. She blew her nose before continuing: "You are always there for me. Without you, I don't know what I'd do. I know you'll be a good Da someday. Even if you won't always let them beat ya at Cluedo," she added, winking.

Mark smirked. That was their thing. Mark and Mammy had often played Cluedo together. Mark was so good at the game that he couldn't remember the last time he hadn't guessed the criminal correctly. However, their Cluedo time together was tarnished slightly after a comment Mammy made during one game:

"You're amazing at this game! You're just like your Da!" Mammy exclaimed.

Mark never forgot it. If he was honest, he was surprised his father had ever played a board game with Mammy – or anyone, in fact. It seemed he was a different person before the affair Mammy had. Playing Cluedo never felt the same afterwards nonetheless, and his father was not someone Mark particularly wanted to be like.

Like father, like son, they say. He felt the connection too; something that was inherited, a part of him. He didn't want to be like his father in the way other boys wanted to be like theirs. He did not look up to him in the same way. Mark already looked like him with his gingery-brown hair and green eyes—that was bad enough to deal with. But to be told by *Mammy* that he was just like him? That hurt. But he knew she didn't mean how it sounded.

He had often wondered how Mammy could have ever been attracted to his father.

Mark's recollections were interrupted by a small man walking over towards them. He was in blue overalls and had removed his latex gloves and medical hygiene mask that covered his mouth.

"Mrs Formby?"

"Y-y-yes?" Mammy stuttered, looking up at the doctor, dreading the worst. The upset but bemused facial expression was now replaced with terror.

"My name is Doctor Larmer. I'm the leading surgeon treating your husband. John is in a critical condition, but he is stable. For now. Miraculously, none of the bullets punctured any vital arteries or organs. He is extremely lucky to be alive, in short, but I must warn you, it is a long road to recovery if that proves to be possible—nothing is certain yet. Right now, we cannot be sure whether he will make a full recovery, but that is a long-term objective. We can only focus on the present now, and that is keeping John alive and trying to keep him in a stable condition in order to give him a fighting chance. It's a waiting game I'm afraid. But you should be proud. Your husband is a hero."

The surgeon smiled to try and reassure Mammy, and she had responded with a small smile.

"Thank you, Doctor," she replied, slightly hoarse.

A part of Mark was glad he was not dead. John Formby didn't deserve to die, but it was for Mammy that he was most relieved. Mark was not sure how she would react if he had died... or later did die. She was in quite a state even when she was first told.

The Head Commissioner had come to visit their small, terraced house in Dublin soon after John was hospitalised. Mark had known straight away that something must have happened to him—it was the first time a guard was at the door of their house and he knew the head commissioner's presence meant something serious had happened. He had seen enough TV to know that.

"Good afternoon, Mrs Formby. My name is Commissioner Derek Dalone. I'm the Head Gardaí Commissioner of the Dublin Branch. I work with John, but of course you already knew that." He shuffled awkwardly. "I'm here to talk about John. Can I come inside?"

Mammy looked at him, a little shocked. She wasn't expecting a visit from the commissioner out of the blue.

Understandably. It was never good news when two Guards appeared at your door.

"Erm, yes, of course. Come in," she said, letting him in.

Mark had been standing at the bottom of the stairs. He was curious. He wanted to find out what had landed his father in such a state, so he walked tentatively down the stairs, intimidated by the presence of authority.

"Hello, Mark, how are you?" the commissioner asked him formally.

Mark shrugged timidly but replied that he was fine.

After the three of them sat down in the living room, the commissioner had got to the point. Mark supposed that was a force of habit.

"Mrs Formby, I'm sorry to tell you that John was shot a few hours ago. He was involved in a drug-raid from what our sources tell us. John went in trying to catch a group of drug dealers and producers, and in the process, he was caught sneaking in and was shot at on the spot. For legal reasons, I can't tell you any names or anything like that. He was found unconscious when the ambulance arrived and was presumed dead. He had a weak pulse and the paramedics and doctors managed to stabilise his condition, although he was and remains in a critical condition. We tried to call, but there was a signal problem, I think." His expression was full of sympathy.

Mammy looked alarmed, fear instantly hitting her. Mark had been less shocked; his father was in a dangerous occupation. Instead of worrying for him, Mark could not help thinking just how well-spoken the guards were, especially the commissioner. The news hadn't sunk in.

"Was he alone?" was all Mammy asked in a barely audible voice. It had not properly hit her yet either.

The commissioner shuffled uncomfortably then. It was of course illegal to send in someone alone for a high-risk operation.

"Well, unfortunately, yes. But that's what is most surprising about the whole thing. No one sent the command for him to do so. In fact, no one knew where he was for hours—not until we had heard he had been shot. The situation is very confusing."

Whilst Mammy was desperate to get to the hospital, Mark's mind had been distracted by the most obvious question: why had he gone on his own? Surely John knew that going alone was sheer stupidity.

But John Formby was not a stupid man, not usually. Not with something like this. Something didn't add up. Maybe he had done something illegal? Gone in guns blazing without a warrant or something? His father didn't reveal any details about his work, nor had anyone else. John was strictly private in all areas of his life, but he would not put it past his father to be corrupt. Mark doubted they would ever find out the answers.

But it was this mystery which haunted Mark. And would continue to do so.

Man has a flaw in his character as Mark discovered: they can't be every man society expects them to be. In Mark's mind, there were only two types of men in the world: the business man, where everything in life is your career, and where nothing in the world was more important; or you were a family man, where nothing mattered more to you than your family, and a job was just a means to make ends meet and keep the family afloat.

John Formby was a career-led man. He was always at work, and Mark was always hard work to him. But could anyone blame Mark for being 'hard work'? Most kids could be but there was an added layer of bitterness in his approach to John.

Yet John Formby remained oblivious.

Several weeks passed and John was still an in-patient. In contrast to these previous conversations, Mark had not spoken to his father all through his time in hospital. Mammy had tried to get him to visit, and for her, he did visit for a while. But John was silent. Most of the time Mark was at the hospital, John had either been sleeping or looking out of the window, avoiding Mark's gaze. Besides, nothing had changed in him. Mammy had talked to John, but every answer from him was given in the form of grunts, and she often cried when she was away from the hospital. It was Mammy's tears that made Mark all the more resentful. They weren't drugged-up grunts either, like he was trying to speak but couldn't. Mark could have forgiven him if that were the case.

No. John chose not to talk to them, and it wasn't long until he told both of them to leave. Mark knew he was well able to talk. Mark didn't know how Mammy could deal with that.

Still, time went by and, gradually, John was getting fitter. As soon as Mark was told he was regaining consciousness for longer periods of time, Mark made the decision not to go, not even for her.

She had come into his small, football-poster-filled bedroom and sat next to Mark, putting a hand on his knee, asking him to go to the hospital with her and talk to him.

"Ma, I just can't go."

"Mark, please," she implored desperately.

"No, Ma. I won't watch him treat us like that again. Not even once more!"

"M-M-ark…" her voice trailed off. Tears had formed in her eyes and Mark noticed something in her teary gaze that almost broke his heart.

Mark looked at her with genuine sympathy.

"I'm sorry," he stammered.

She tried explaining to him how frustrating it was for John lying there, unable to do anything in a bed. Mark felt like she always had an excuse for him. She hoped, like any mother, that the incident might make her family stronger. Mark noticed that him refusing to go with her now was some sort of significant moment for Mammy. All these hopes were unravelling, and she saw that nothing of the family union could be salvaged.

Both of them knew it. The pain of knowing they were a crumbling family exploded out of her. It was torn out of a good and honest woman in the form of tears cascading down her face, her hands shaking, trying to hold her face as if she herself felt as if she were physically breaking into pieces.

How could anyone defend a man that reduced his wife to a mental breakdown like this? Affair or not. Mark could barely watch as she wept; her pain was so palpable that he himself felt tears in his eyes, making his vision blurry. She did not deserve this, and it upset him to see her in such a state.

He could only imagine how the red-hot tears felt, like lava, whilst she shook. It was like she was an active volcano filled with a sadness.

All he could do was to try and hold her.

Chapter Two

Many weeks after being admitted to hospital, John was discharged. He had made his decision for them to leave Dublin and head to the country to 'relax and have a fresh start'. It was a bit out of the blue but neither Sylvia Formby nor Mark had any say in the matter. But that was normal.

Just like at the hospital, some local journalists swarmed around their new house, waiting for someone to leave so that they could bombard them with questions. At least at the hospital the staff could shield them from the press; at home, they were vulnerable. It was on the local news for a while, before it became old news. *'The Guard who Survived Five Bullets!'* was the standout headline for the locally famous John Formby. The Gardaí did its best to keep it hushed up but stories like this always got out. Eventually, the journalists and reporters became bored by the lack of statements by the Formby family. The hospital staff were also dismissive, as they were peeved that the interference was a hindrance to the staff and to the Formby family. The Formby's neighbours were slightly more informative, but no one really spoke to the Formby's.

Nonetheless, the location of their new home was fairly common knowledge by the time they moved, and it was mentioned on the news in the neighbouring towns and counties—nothing too extensive, but enough to attract the Press Pests as Mark referred to them. It amazed Mark how even private affairs became fast-travelling news.

It wasn't like Mark could escape out the back door and down some alleyway; the house was detached, where land surrounded them at a three-hundred-and-sixty-degree angle. Any attempt at leaving the house would result in him being noticed by a loitering reporter or photographer. Amongst these journalists, people knew his face. People were nosey too and were sometimes as bad as the press although there were far fewer locals in the country.

He had no time for the media. Or people in general. But at least living in the country eradicated the disturbance of the general public.

Mark peered down gloomily at the handful of desperate journalists around the house. What could a sixteen-year-old do to stop them lingering? He didn't care about them talking to John, he was the one that caused it all. Now that he wasn't working all the time again—from home for the time being, Mark wasn't fooled into thinking that anything would change. John still remained passive and shut himself in his room. Locked. Not just to hide from the media outside, Mark thought, but also to shy away from anyone. It certainly wouldn't be a case of hiding away from guilt.

Mark was dubious whether he even had the capacity for guilt. He never seemed to show remorse. When Mark was younger, the excuses worked on him—he was even proud to tell people that his dad was a big, important guard. Proud! He felt disgusted that he hadn't seen through him earlier. He was a poor example of a man who only cared about himself.

What got to Mark most was that he was afraid he would turn into his father. Similar characteristics were there: inquisitive, shrewd, short-tempered. Heck, they even looked alike!

But he wasn't going to dwell on it. What could he really do about that? It seemed like he could do nothing these days. There was only one real thing he could do that was useful. Support Mammy.

She had been the victim just as much as Mark had been, if not more. It was her that had to lie next to him every night and had lived with him for nearly twenty years. Mark couldn't imagine how that must feel. To not be loved by someone you couldn't help but love, and who you wished gave you the time of day, or just appreciate everything you do, but doesn't. What kind of life is that for any woman? For any person?

He didn't understand why she couldn't just leave him so that it could just be Mammy and himself. He supposed it wasn't that simple.

But wasn't it? Could she not just pack a bag, get Mark to pack his, and leave? Or kick John out?

Surely there was an option?

However, there was nowhere for them to go. Mammy had no friends as such, and her parents were dead. Mark knew they would both be stuck with him, the man the media thought was some sort of hero for tackling so many druggies at once.

That just proved how misinformed the media could be Mark thought.

As he pondered over all this, Mark stared down at the press, sighing. He got up and walked out of his room in pursuit of Mammy. She was the only one he felt he could talk to. Not that he had many other options. It was either Mammy, him, or the pests outside.

There was really only one option.

He eventually found her lying on her bed, just gazing straight up at the ceiling. John had clearly wandered off somewhere. For a moment Mark stood to watch; she was barely blinking, breathing in a slow and steady manner in what looked like a form of trance. There was something in her expression that suggested that she wasn't lying there peacefully, but more strained. He walked slowly into the room, noticing as he got closer to her the dry, trickled tear trail on her face.

She snapped out of her trance when she realised Mark was stood over her. She looked at him with a tired smile, and a glint of life reappeared in her eyes.

"Hi sweety, are you okay?" she asked, feigning chirpiness.

"Yeah, I'm grand. Are you? You seem tired."

"I'm fine, thank you. Do ya want something to eat or something?"

As much as she tried to sound upbeat, it was more than obvious that she was not exactly in a great place. It was only for Mark that she had disguised her feelings, since he was small, but he was too old now to be reassured just by words. Mammy was suffering in silence.

And not just because of the situation with John. In her eyes, Mark could see there was something lying under the surface, something that made her feel trapped. He was good at reading human emotion, even if he didn't always understand or relate to it. Mark had assumed, initially, that she was just upset by John being hospitalised, and cried because Mark hadn't gone to see him. And at the time, that was probably why she did cry. Although now… now, Mark figured there was something else upsetting her. It had been several days since John had been allowed to leave hospital, and Mammy was crying on her own. Secretly hiding her emotions, or at least trying to: Mammy never did that. She never used to cry, particularly when he was growing up, but when she did, she never attempted to hide it from him—especially when he was old enough to understand why she was sad. She even talked to Mark about how she was feeling sometimes, as she was aware Mark was mature enough to talk to her about these things. She was just that kind of person. Maybe that was weird – for a mother to open themselves up to their child in this way – but Mark liked it and it felt normal for them. It made him feel valued, needed. And it wasn't as if she could rely on John to listen to her. Her tears never lasted for a long time either, no matter how bad things got. She hadn't even cried this much when her parents died—her recovery period was not that long, generally speaking, no matter how bad things got. But it had been several days now since John was realised from hospital, and Mark hadn't seen any signs that she still cried. Until now suggesting that whatever was making her cry now was ongoing.

A wave of pity washed over him. And guilt. He wished he had made more of a conscious effort to check how she was, but since arriving at the new house, Mark mainly kept himself to himself, occasionally talking to Mammy, but only when she came to check in on him. She was fine walking into his room for chats, and she had perked up increasingly each time they did talk, and so Mark assumed she was getting by okay, but now that he walked in on her for a chat, he realised just how wrong he was.

It was irritating him that she was pretending everything was fine.

"Mammy, is everything okay?"

"Of course, why wouldn't it be?"

"Because it's obvious you were crying"

Mammy looked at him in a way that said, 'I wish you weren't so shrewd', but it lasted a second before her face neutralised.

"Honestly, I'm fine honey. I guess I just need more time getting used to the countryside. I prefer the city," she replied, with a clearly forced smile. She glanced at the clock, looking for a distraction.

"Christ, you must be starving! I'll make you dinner." She got up, heading to the kitchen.

Mark's eyes followed after her with a troubled look. There was definitely something wrong.

Chapter Three

Eventually, the media left for good. It had been weeks before they slowly began to leave, but when they all left it was like a weight being lifted from Mark's chest. He had rarely left the house since he arrived, so he was fed up and eager to get some fresh air.

As Mark left the house, he saw John drive off in his black 2012 Mercedes-Benz C-Class, noticing Mark at the door as he swerved away. He stared through him, continuing forward. Mark stepped outside, already forgetting about John, and wandered away from the house. He couldn't care less where John was going.

The further away he drives, the better, he thought. He followed the road, plugging in his iPod as he did so. On shuffle mode, 'Everybody's Changing' by Keane came on, and checking that he was alone, Mark began to sing along softly.

It was one of his favourite songs from one of his favourite bands. The song explored frustration and pain, and the sense of time going by and feeling alone—all the things that Mark could relate to. Mammy too, he suspected. He appreciated songs that he could relate to and contemplate; he didn't really see the point in listening to songs he couldn't place himself in. Music generally allowed him to escape the real world, or at least act as a distraction. He wasn't really a gaming kind of person or anything like that apart from the odd game here and there, he was more of a reading kind of guy. And he often listened to music as he read.

Looking around at the scenery that was still new to him, Mark enjoyed the feeling of air running through his hair—it had been a while since he was able to be alone and away from his parents. It was a sunny day, and he was glad he had packed lunch and drinks; he climbed over an irregular stone wall and into a field that lay in between two country roads, the steady slope allowing him to look down upon the countryside. He didn't care if he was weird for having a picnic on his own. He was happy. Wasn't that the main thing?

He walked up to a tree, laying out his one-man picnic set, facing the scenery. It was the first time he felt good in a long time. Food, Pepsi, music, a book, and a good view. That was him sorted for the afternoon. He flicked through songs, thinking about how he was going to adapt out here, and how long until Mammy felt better... but he wasn't going to think about it right now. This was his time alone, where he could escape life and think about himself. He owed it to himself, he reasoned, particularly after being shut up in the house where the tension between the three of them was now building too much for him to cope any longer. At least if he had this spot to escape to, he would be able to cope.

Mark was just then aware of something or someone watching him. He paused his iPod, tossing away his earphones. He looked around and, initially, saw nothing. However, out of the corner of his eye, he saw a figure in the distance, on the other side of the field walking towards him. His sixth sense had not failed him.

After looking at the figure directly for a few seconds, he realised that the figure was a girl.

Suddenly his breath quickened. He didn't know how to talk to boys his own age, let alone girls. He couldn't really talk to anyone without feeling uncomfortable and awkward. He always ran out of things to say before he got past 'hello'.

Even saying 'hello' was considered too formal. He had no hope. What the hell was he going to do or say if the girl came to talk to him? He was on his own, and he wasn't an interesting guy. He knew that. He didn't need a girl to tell him that. This moment right here was Mark's idea of a nightmare. It sounded melodramatic but it was true. He could handle anything with calmness if it required him to be on his own. That was fine.

Mark was his own best friend. And that didn't bother him. It suited him.

He genuinely considered packing up his things in a panicked frenzy and running off. If it had been a boy approaching, he would have been able to stay fairly calm, despite feeling out of place. It would have been endurable. However, this was a girl. That was a whole new ball game. But what was the point of running off? What use would that be? He would feel even more foolish. Like a shy cat running away from a pedestrian on a street.

He turned away, prolonging the inevitable as much as he could get away with. Maybe that would deter this girl from approaching any more. He closed his eyes, breathing heavily, desperately hoping the girl would steer away.

He then opened his eyes. Was he really going to be this much of a coward? He could do this. She didn't know who he was—it was brave to approach a random person the way she was walking towards him. That took some guts. And he envied that. It was admirable. He had never been able to do that in school. Never. And she wasn't accompanied. If she had others with her, Mark may just have run off. Perhaps she was lonely. And if she was, maybe they had something in common, maybe they could talk about that, if nothing else. With this in mind, he realised he was not about to make her feel weird too, not if he could help it. It wasn't her fault.

He quickly glanced over his shoulder. The girl was definitely heading his way. Mark guessed he had about a minute before she reached him.

He was thankful he had some time to psych himself up, instead of her popping out of nowhere.

That would have been intolerable.

With his heart beating even harder now, he dared to glance over his shoulder again. She was less than a hundred metres away. He reckoned it would be odd if he looked away again at this point, so he stayed where he was.

Fifty metres away.

Twenty-five.

Fifteen.

Another surge of panicked thought hit him as he considered the possibility that she was wired up. Sent from a journalist… or maybe on someone's land he had innocently trespassed.

Paranoia was powering through him like an electric current. He was lost. As if he was a robot about to malfunction in a plume of smoke.

He looked up into her face, and almost forgot to breathe. He had not looked at her face before she was up close, but now he discovered that she was astoundingly beautiful. The most beautiful girl he had ever seen. He knew it sounded corny, but it was true.

Which was not saying much considering he hadn't looked at many girls.

Don't say that to her whatever you do! He thought furiously.

"Hi," said the girl, smiling friendlily. "Y'all right there?"

The slight look of confusion swept over her features then as Mark didn't answer back straight away. Mark supposed he must have looked unnaturally nervous. He was sweating despite it not being overly warm and his mind had hit a wall. No words came to him. His chest seemed to constrict from the stress.

Just say anything. Anything at all!

"Hi," he said, half croaking.

Oh God.

He cleared his throat.

"Hi," he said, more clearly.

The girl smiled again.

She had an amazing smile. Her lips were full without being too big and there was a slight hint of shine on them. It must have been that gloss stuff or something. Mark was clueless as to what all this girl stuff was. Whatever it was, it complemented her face. Her eyes were big, again made particularly lovely with a touch of make-up.

Was it eye-liner girls called it? He wasn't sure.

"What's the story?" she asked.

He registered the question much longer than was expected before stammering: "Erm yeah, I'm good. I mean, I'm new around here. I live down the road."

The girl smiled wider. She seemed to find his nervousness amusing.

"Calm down, I'm not gonna bite you," she said kindly, "I figured you were new around here. I didn't think I had seen you before. But I think I do recognise your face from somewhere though… I'm assuming you came from Dublin originally?"

So far, not too bad. He had sustained conversation for almost a minute now, so as far as he was concerned, he could hold his head up high. Besides, she seemed nice enough.

He was feeling slightly more relaxed. "I'm from Dublin, yeah," he replied. "You probably saw my face in the paper, my dad was the guard who was shot." Talking felt abnormally natural.

"Oh yeah, that must be where I recognised ye… how's he doing?" she asked with what seemed like genuine compassion.

Mark shrugged. "He's fine," he replied shortly, wishing to change the subject. "I used to live near Finglas," he continued, trying to drive the conversation elsewhere.

"Ooooh", she said, mockingly flinching away, cringing. Then she smiled again.

Mark laughed. "Yeah, not the nicest area in parts," he remarked. Then, thinking he was not really helping the conversation, asked: "Yourself?"

"I have always lived here. Born and bred near Dunshaughlin. Country girl, I am," she replied with ease.

She spoke with complete fluidity, and she boomed with confidence. More things to admire about her.

He allowed himself to look at her properly. She seemed to notice his interest and her eyes glistened in the sunlight, and it was then that he noticed her electric-blue eyes, sparkling like the sun shining on the bluest of seas. Her hair was blonde and voluminous, styled deliberately in a half-curly way that just added to her face, if that were possible.

She was already perfect as far as he was concerned.

Was he gawping at her he wondered, panicking that he would freak her out?

His sudden terror alarmed her a little bit.

"Are you okay?" she asked, slightly concerned.

Mark managed to compose himself. "Erm, yeah. Sorry. I er-I just… I…"

Nooooo! he moaned to himself. He was stuttering.

She smiled sweetly again. Every time, it seemed to take his breath away.

"You don't speak to many girls, do you?" she asked.

Mark looked up at her, searching for mockery and scorn in her expression. She wasn't mocking him.

He smiled back. "No," he said timidly.

"That's okay. In fact, I think it's kinda cute," she soothed.

Mark couldn't help but smile stupidly at her. It was probably a bit creepy, but no one had talked to him like this before. And then it hit him.

This girl was flirting with him!

At least he thought it was flirting. It wasn't as if he had ever experienced it before.

Was this really happening? An amazingly beautiful girl flirting with him! Surely not. Maybe she was just being friendly?

Something inside him niggled at him though. The cynical part of him recovered from the initial overwhelming sensation of talking to this girl. Why was she talking to him? What made her come over to him? She had never seen him before, so it didn't make sense for her to stride over the field just as he sat down. It seemed too timely, too predestined.

It was as if it was planned.

Like she was waiting for him...

The smile on his face waned, turning into a look of indifference. His normal passiveness with people was returning, and he felt the awkwardness coming back, assuming now that she wasn't there for a reason he would have liked. It was a defensive mechanism to act distant. He didn't know what to say next, but the girl seemed to sense the shift in his mood, so she broke the silence as if in answer to his queries.

"I never see anyone around my age around here. If that is what you're wondering. It seems like you are questioning why I am here," she said, looking at him carefully.

Wow. She reads people well, albeit bluntly he thought.

"Why are you here?" he enquired frankly. Immediately he felt rude.

She sighed, seemingly disappointed. "I just happened to see you walking by from the road opposite. You went into the field so I thought I would say hi. You seemed lonely." The tone in her voice had changed. "If you want, I'll leave," she said, beginning to step backwards.

He didn't mean to insult her. Perhaps she was just being nice. She probably was. It was the country, and her excuse was plausible. He figured not many people were around to talk to. Not that she needed an excuse to talk to someone roughly her age. It was just him being his weird self, thinking there was always a motive for everything. He tended to overthink things, after all.

"Wait, no. Sorry, I didn't mean to offend you!"

She stopped, looking at him seriously for a second. Then she smiled, still not fully assured.

Again, his heart skipped a beat.

"Erm, what's your name?" he asked, desperate to be the one to say something first for once.

"I thought you would never ask," she replied, winking. "Alice, it's nice to meet you…?"

She was so confident that it intimidated him a little bit. He knew he would never have it in him to wink so casually.

"What's your name then?" she asked, smirking. Clearly, she was waiting for an introduction.

"Mark."

"Nice name."

"Not as nice as yours," he replied.

It was out of his mouth before he realised. Had he really just said that? Thinking out loud was such a horrible thing to happen. Especially for guys. They always ended up putting their foot in it in some way or other, always screwing up with girls because they were corny. He had seen it in parks, in cinemas, and restaurants where he was able to overhear date conversations. He wanted to learn how to talk to girls, but instead he mainly learnt how not to talk to girls. He was glad he wasn't the guy on the date in those situations. Now, however, he was guilty of the same crime. A 'school-boy error' guys at his school would have said.

He was disappointed in himself. He was so careful to have safe answers and not embarrass himself, but it only took a second of a lack of concentration to turn the tide on how the conversation was going, and now he knew what that felt like. He didn't care for it. This was why he didn't like talking to people.

But maybe that put him in a good light? He chanced a glance at her, blushing furiously.

Alice beamed at him. "Thanks!" she exclaimed.

It was obvious she was not expecting that.

She and him both.

And… *and…* she seemed to like it!

"Do you mind if I sit with you? I have been standing for a while. I could do with a sit down." She smiled sweetly.

"Yeah, of course."

His heart started beating again. Faster by the second.

"It's okay honey, I'm a nice girl, I think," she teased.

Another wink.

"Are you this suave with everyone you meet for the first time?" he asked, smirking.

"Yes," she said with a serious face.

"Thought so," he said. What was the social protocol at this point? He was beginning to worry again; natural conversation was not his strong point. He was still going to have to get through this uncomfortably. Even if his corniness did actually work.

What was he going to say now? They had been staring out down the road, at the view he was looking at before she came over, for what seemed like hours. Without any words.

"Do you want something to eat or drink? I have snacks and a spare Pepsi?" he asked.

It was the only thing that came to his mind.

"Erm, yeah, go on then. A swig of Pepsi wouldn't go amiss."

"Here, have the whole thing. I won't drink two of these".

It was almost like he was being smooth.

"Thanks."

She looked at him then, a curious smile grew on her face. He felt a bit self-conscious as she sat watching him, particularly because she was stunning. He was not used to any attention from anyone, let alone from gorgeous girls. Was his lack of suaveness weirdly attractive to her or something? Or was he good looking to her?

He doubted that greatly.

There was naturally a reason why she was looking at him, but before he could think any more about it, she nodded and looked away. Smiling.

She seemed to smile a lot.

"You're a cutie," she announced.

Mark had no response to that. He could only faintly smile, wondering how on Earth he had managed to get a girl to say that. He had never been considered cute to anyone.

It seemed too good to be true that a girl like this would be willingly sat next to him saying he was cute.

She could know how to spot a target when she saw one, Mark thought. She definitely must know herself that she was a looker; the confidence in her walk, her talk, her general demeanour suggested that. He looked at her;

she tilted her face to the sun's gaze, facing ahead with her eyes closed, which made her face sparkle a bit with the touch of glittery make-up she was wearing. It was just right. Not too much nor too little. She was flawless.

She turned to look at him, catching him looking at her. Mark blushed and looked away.

"What are you thinking about?" she asked with that cheeky smile of hers.

Mark just blushed even more, unable to think of an answer that wasn't cringy. What had happened to him? He was always in control of his emotions, speaking when he felt necessary, but otherwise keeping himself to himself. He was a thinker more than a doer. He needed to come up with something to say though, and the effort of trying to think made his head blank.

"I ... I don't know. Not really anything," he said finally.

That was a safe answer, right?

She chuckled. "Yeah, okay. That's why you're looking at me," she joked.

"I guess I am just bewildered how I'm sat here talking to someone like you ... out of nowhere," he answered.

He had done it again. So feckin' corny.

"Things happen like that sometimes," she replied. "Besides, you seemed a bit lonely, and thought you could do with a friendly face." She smiled broadly then, revealing her perfectly aligned, brilliantly white teeth.

She has more than just a friendly face, Mark thought.

The whole scenario was surreal. It felt like he was in some sort of trance or had just taken some love potion and seemed unable to look away from her for more than a few seconds. He had never had a girlfriend and had rarely been able to talk to any girls, having been at an all-boys school for most of his life. Nursery was as close he came to proper female conversation, and even then, he was an outcast, playing with toy blocks on his own.

A phone rang. It snapped Mark's attention away from Alice's face, and towards her pocket where the call was coming from. Alice jerked a hand towards her pocket, and after looking at her caller ID, said: "I should probably get this, sorry."

Mark nodded, understanding.

After a bit of normal small talk, nothing which stuck out to Mark as being interesting or unusual, Alice hung up, and looked at Mark with an apologetic look.

"I'm really sorry, but I'm gonna have to leave ya. Da wants me back home, you know what parents can be like," she said frowning.

Oh yes, he knew what that was like.

"That's okay," he replied, unable to keep a note of disappointment out of his voice.

Alice picked up on this and leaned in to kiss his cheek. He could feel her soft lips on his face as she pecked him, which made him want to melt. They were softer than he could ever have believed possible. He found he could not move, save for a gulp in his throat, and sat thunderstruck. He could not quite conceive the reality of what had just happened. No girl had ever kissed him. His mother was the only one who had ever kissed him on the cheek.

Alice stood up, and began to walk off, calling behind her shoulder as she went: "See ya Mark, lovely to meet you. I hope we talk soon."

Mark cleared his throat, managing a, "See you. Great meeting you too."

And he meant it.

Chapter Four

Mark watched Alice walk off until he could no longer see her. He was tempted to follow long enough to see if he could see where she lived, but he didn't want to risk being caught. He figured that was a bit creepy, particularly after only meeting her once. There was no way he would let himself fall for her, at least not yet, until he could tell if she actually liked him. He may never see her again anyway, so what was the point on getting so fixated by her? For another thing, if she walked so coolly over to him –someone she had never met – how many other guys had she walked so casually over to? She said something about never seeing anyone in the countryside though, so perhaps she was making the most of talking to someone roughly her age? Thinking about it, Mark did think seeing anyone young around here would be rare enough. Talking to people on Facebook would help isolation for a while but not in the long term, not for someone as sociable as Alice seemed to be. This wasn't the city; any chance to socialise here would be grabbed for someone with such confidence. Maybe she was just being friendly; maybe what just happened wouldn't be considered weird to locals. Surely though, there were only two options to whether or not she liked him or not.

She either did, or she didn't.

At least it seemed to be that simple. Mark had heard that girls were often indecisive. If that was the case, Mark was baffled. He would never learn how girls' brains worked. Whenever a man does not know what a woman is thinking at any given time, the more power the woman has since, naturally, she will know what she herself was thinking without the man knowing anything. How on earth would he know what to say if he bumped into her again? Was it simpler than he thought? You could learn all the theory needed to know how to ride a bike, for example, but actually riding one in practice was completely different. You couldn't ride a bike just by knowing the method. Surely all girls weren't the same? Perhaps knowing how to talk to Alice differed completely from talking to another girl.

Mark frowned. Why was he bothered anyway? It's unlikely he would see Alice again, he even thought it would probably be better if he didn't for fear of embarrassing himself. He could see now why guys had so much trouble trying to figure women out. It seemed impossible. He no longer felt like he had the right to laugh at the guys on dates stumbling over chat-up lines in the attempt to impress girls. Especially since he was now guilty of doing just that. If he was honest with himself, he was hoping to impress her even though he was uncomfortable the whole time. Okay, his chat with Alice hardly qualified as being a date but still, he acted just as a nervous guy would on a first date so he now believed he could empathise. It confused him how much he was thinking about Alice already—he had never given girls an awful lot of thought before, probably as a defence mechanism. He had secretly found girls attractive, but he would never show it. He had always presumed he would be hopeless with women, so he had always asked himself: why bother thinking about them?

However, there was something in Alice's smile and the way that she looked at him that changed all that. He knew nothing about her, but he felt that he wasn't wrong, at least not completely wrong, in thinking that she liked him. Maybe not necessarily fancied him but *liked* him. That was not impossible. Not after how she was with him. It certainly could have been worse from his point of view.

I could do with having a friend, he thought sadly.

He looked towards the place Alice had walked, wondering if and when they would meet again. Perhaps if he came to this spot often, she would know how to find him. He wished he had thought to ask for her number or where she lived but he knew that he wasn't courageous enough to take the leap and go asking for numbers. Heck, he was contemplating running away for goodness' sake. Keeping a conversation comfortable was a demanding mission in itself. She may have forgotten about him already, Mark mused bitterly. Just because Mark hadn't thought his chat with her had gone terribly, she may have found the whole affair dull.

There was so much to think about from just a few minutes' talk.

It was only then that Mark discovered that he should be thinking about going back. His thoughts after Alice left had lasted over an hour, time which slipped by fast now he had a girl to think about. He wasn't disappointed he hadn't had much time to himself to think about nothing in particular. He gathered his things and headed home.

Chapter Five

Mark arrived to find his dad waiting for him by the front door. Mammy stood by the stairs straight opposite the door, looking tentative. It was clear that something was not right, something had happened since he had left. It was also obvious they were waiting for him to come home; John was probably looking out the living room window to wait for his approach; that explained why he was at the door when Mark walked in. Why he would do that though after ignoring him for the past few months struck him as being odd, and from the look in his face, it was apparent that whatever was on his mind was a pressing matter.

"Living room," he ordered, pointing towards the room.

Mark obeyed. He did not see the point in making a big deal out of it, although his father's tone did strike a nerve.

He slumped onto one of the sofas, making himself look as indifferent as possible to what he had to say. He was curious, however. But he wasn't going to let that slip on his face, so he remained neutral in his expression. A flicker of anger ran through him too and he was sure he wasn't stoic enough to hide that. What gave him the right to not talk to him or Mammy for months on end and then, when it suited him, suddenly order the two of them about to do his bidding?

John followed in pursuit, with an air of urgency in his manner. His eyes were more animated than they had been in a long time—they expressed more emotion. Genuine emotion. It was enough to make Mark forget about his anger. He was now curious to find out what had got him so agitated and was oddly enjoying his unease. Mark figured it was something to do with him in some way, but he had no idea why.

"What's wrong with you?" Mark blurted out. His tone sounded mocking, bemused.

John dismissed it, seemingly not noticing.

This must be important, Mark thought to himself.

"That girl you were with, what was her name?"

Mark was not expecting that.

"Why? How did you kn—"

"Answer the question!"

Mark's anger was back. He wasn't going to sit here like a dog and just spoon-feed him when he demanded it.

"No! Why should I? You don't speak to either of us for months," he cried gesturing his head towards Mammy's direction, "and you think you can just interrogate me like this?! No way."

Mark stood up, intending to storm off. He was already tired of this and wished he was back on the field talking to Alice. It was the only bit of pleasure Mark had had in months, and he was not going to let his dad take that from him too.

But as soon as he tried to leave the room, John grabbed his shirt, forcing him back onto the sofa with a rough shove.

"Quit it with the attitude, you little shit!" John spat. "Now you'll tell me what I want to know, and you can go off in your little hissy fit."

Mark had given up trying to leave, so he simply looked at his father.

"No."

John Formby slapped him in the face.

"If you don't answer me, you'll get more than that!"

Mammy came into the room then, whimpering.

She managed a "John… please…" in a frightened, upset whisper.

The mood changed then slightly. John bent his head down impatiently, thumb and index finger pinching the bridge of his nose:

"Sylvia, this won't take long," he said in a strained, forced patient voice. "Go along into the kitchen and make dinner or something."

He then looked at her blankly, expecting her to leave. There was also a glint in his eye that was challenging her, daring her to resist. Mark knew she would leave then, she was too weak to resist his orders, and he gave her a look that implied it was okay for her to go, along with a weak smile. She glanced at Mark, and then back to John. She was about to leave.

Except she didn't. A mother's instinct?

In her there was suddenly a fiery courage and Mark could not help but stare in amazement as she stood rooted to the spot, glaring at John with a determination that Mark had never seen before. Had she finally had enough? Had he pushed her too far? Mark was sure he would never forget this moment right now; never before had she stood up to him like this, and John himself looked stunned. They had had arguments but only when Mammy initiated it, never when John was the one who was angry. Certainly, she had never disobeyed him. Probably out of guilt, a means of repaying him for her infidelity. He seemed to disregard Mark completely as he glared back at her with a look of disbelief. In Mammy's face, there was fear. Mark was sure she had always been afraid of John—even though she was standing up to him now. It was only natural for her to still be frightened. Physically, John was an intimidating man. He didn't think John had ever actually hit Mammy but there was something inside Mark which told him that he was not incapable of doing so, not if she got in his way.

Like right now.

He had never faced any opposition from her until this moment, so that was probably why he hadn't needed to hit her to get his way.

The light from outside shone into the room then, casting a glow upon Mammy like some sort of Divine spotlight, and as Mark watched, he couldn't help but think just how proud he was of her, even if she did cave in and leave the room. Mark genuinely thought she didn't have it in her to stand up to him, but here she was doing just that! Mark was sure the surprise on his own face was clear to see. She glanced again at Mark as if to give her fuel to maintain her rebellion against him. She held her stance.

Mark suspected that him being in danger of being hit had something to do with her opposition.

Maybe Mark's surprise acted as a motivation to stay strong too?

Whatever it was, it was admirable. He knew that after all these years of taking the abuse and abandonment on the chin, challenging John Formby was a brave thing to do, albeit unwise.

As an answer to his assumptions, Mark's dad stood up, walking towards Mammy.

"I told you to leave," he snarled menacingly.

"I know what you'll do to Mark, you'll hit him properly. I-I won't allow that," she said in a terrified, but firm voice.

John laughed. Mark was touched.

"A beating never hurt anyone... well, not in the long run. My old man beat me, and I came out stronger for it. Great man, my Da. He taught me a lot."

It was his jovial tone which was creepier than if he had shouted it. There was something in his voice that made him sound a bit insane.

And even Mark had never thought his dad insane...

Maybe, the drug accident has tapped him in the head, Mark thought. *He hadn't spoken properly to anyone for months – was that the first sign that he was going barmy?*

And then Mark's mind switched to the immediate threat at hand. He stood up, ready to attack his dad if he so much as looked like he was going to hit her.

He would never stand by and watch her be hit without trying to do something.

"He didn't make you stronger. He turned you into a cold-hearted coward," Mammy replied forcefully. *What was she doing?* Mark wondered desperately.

A dark look washed over John's features then.

"You dare stand there and insult me? Do you think I can't handle my own son properly?" he hissed, his voice rising steadily. "He hasn't suffered much over the years. He just needs to tell me who that girl was that he was with, and he will be fine. Now, leave."

It sounded like a final warning.

"Well if it such a simple question, I don't see why I can't stay here," she countered. Mark could not believe what he was witnessing.

Was there something in John's demeanour that made her fear for Mark's safety? Why did he care who Alice was?

He had never thought his dad insane before now, not truly insane, but there was something in his eyes that was manic. He couldn't describe it in any other way than that, but he could definitely see it.

Was that why Mammy was standing up to him now?

"Don't get smart with me. Leave!" he roared.

And then he calmed down again, almost mellow. It was surreal, eerie, and scary.

"This is your final warning," he whispered coldly, his dark blue eyes set dead on Mammy like a lion waiting to pounce.

He was insane. He had to be.

Mammy still stood, closing her eyes, as if accepting her fate.

John straightened himself up to his full height, ready to strike.

He would not hesitate in hitting her, Mark could see that now.

Mark glanced around the room, panicked, looking desperately for something he could use as a weapon. On the mantelpiece was a vase with a few wilted flowers in, and instinctively, Mark darted over to it, grabbed it, and rushed towards his dad without hesitation, swinging as he did so in the direction of his dad's head.

Mammy opened her eyes slightly, just as John went in for the punch, and noticed Mark in the corner of her eye looking furious, swinging a vase at John.

Just as John's fist connected with her face, the vase hit John on the side of his head with the full force of Mark's raging adrenaline.

The impact of John's punch was significantly reduced by the blow to his own head, and he collapsed on the ground on his right-hand side, looking dazed and semi-conscious.

The vase broke into several large pieces, Mark holding the largest of them, standing over John with what was now a fashioned dagger.

Mark's adrenaline was still pumping, and he had an urge to bury the broken glass into his father. He resisted. To anyone else, the intention to kill or maim was probably very noticeable. A part of him *did* want to kill him for hitting Mammy though.

A voice in his head alarmed him as it whispered:

Don't stoop to his level. Don't be like him.

Both of Mark's parents had seen the intent. Mammy flung her arms around his body after seeing the look in Mark's eyes, kissing him on the temple, pleading with him to drop the glass. And John, despite being dazed and his vision blurred, had turned his dizzied head enough to notice Mark's intent from his body language.

Mark could see a glimmer of fear in his father's face or was it surprise? A lot of force went into his attack, more than John had probably anticipated.

After a few seconds, he threw aside the glass onto the other side of the room, and cried, turning to embrace Mammy, horrified at what he could have done.

It was a few minutes until John staggered onto his feet. All his aggression seemed to have been replaced with concussion, but he still stayed in the room staring at Mark with wonder as he gently swayed. It was a look that told Mark that he was never going to look at Mark in the same way—it was as if he had realised Mark was old enough to do damage if his adrenaline got the better of him. Obviously, in normal circumstances, there would be no competition. John was well-built, and strong. After all, he was a guard. It was a necessity for him to be physically tough. No sixteen-year-old would out-punch him, but Mark was stronger than most sixteen-year-olds. He wasn't the weedy boy that he had once been but was now broad and naturally muscular, just like John himself, and had a strong physical presence about him. That alone surprised John. He had previously only seen Mark as a wimpy kid, and he hadn't really paid much attention to him at all.

There was something else in his stare, Mark noticed, that wasn't immediately obvious, but it was similar to the look you would give something you recognised, or something you only just discovered—kind of like an epiphany.

And then Mark knew what it was. It was something Mark was already aware of. John had discovered that Mark was not unlike him; he had a short fuse, and he had the same kind of attitude about him, the same instinctiveness that John possessed in his character.

Between them, for the first time in years, an understanding materialised.

John had regained a little more consciousness now, but he was still dazed. He had seemed to have forgotten that Mammy was in the room as he focussed all his attention on Mark.

"I'm sorry," he choked, also side-glancing at Mammy to indicate the apology was aimed at her too. It was momentary though. His gaze, again, was rooted on Mark.

At least the manic look in his eyes has gone, Mark said to himself.

"You should be sorry," Mark replied blankly.

John went straight back to the point, though. No injury would stop him being like that. Except that the firmness in his voice wasn't as potent now, but raspier and more slurred instead. He was panting, struggling for air too. Mark wondered if he was about to pass out.

"Look, that girl... the girl you were with... you have to understand something about her... if it's the same girl I'm thinking of. She's trouble. What was her name?"

Mark's curiosity returned. He had no idea what was going on, but he was more willing to talk now that they had both calmed down. His father seemed in no fit state to get worked up again, he was finding it hard just to find his voice now. Mark looked at him. One hand rested on his head, whilst the other leaned rather shakily against the open door, turning his head towards Mark determinedly until he got his answer. Mark couldn't help but be impressed at his dad's willpower. He decided there was no harm telling him her name.

"Her name is Alice," he said evenly, carefully observing his reaction. "What has she got to do with any of this? Whatever this is," he said, gesturing around the room. "What is the big rush to know? How did you even know I was talking to a girl?" he asked all at once, desperate to get the questions off his chest.

John's pupils widened a little bit. It was clearly the answer he was expecting, even if the concussion was shadowing most of his body language.

"Don't talk to her again," he stammered.

Before Mark could question him any further, John's eyes rolled to the back of his head and he collapsed, out cold. *I must have really hit him hard,* Mark thought, secretly impressed with himself.

Mark glanced at Mammy, who gave him a blank look, but tears were streaming down her face. She was in shock.

Why had he wanted to know about Alice? Mark wondered.

But the question that was more on Mark's mind was how did he know her, and how did he recognise her from just the stating her name?

Mammy shook off some of her shock and went to check John's pulse and, after feeling the steady beat of his heart, looked up at Mark with a serious look.

And then she smirked slightly. It took a few seconds for her smirk to distract him from his thoughts. Mark was confused. He had not seen her smirk like that in a long time. She used to when teasing him he was growing up, and so to see it now was unexpected and rather facetious given the circumstances.

"What?" he asked, dumbfoundedly.
"So, who is Alice?" she asked with a wink.
Mark blushed.
Oh God.

Chapter Six

The first meeting between Mark and Alice was, in fact, staged. It was no coincidence that Alice had seen Mark walk up to the tree in the field and sit down; she was lurking nearby waiting, watching. She had not known for sure that Mark was going to be there at that precise time, of course, but from her house, which was on the opposite road to the field Mark was walking from, Alice had spotted him walking down the opposite road from an upstairs window. Ever since the Formby family moved to Dunshaughlin, Alice's father, Seamus Brenson, had made her watch for any signs of movement from the new Formby household and to approach Mark, in particular, if he left the house and went for a wander. Alice was sixteen, and her father knew that her good looks would be useful when it came to keeping an eye on the boy. As they were both the same age, Alice would avoid suspicion if she went about it the right way, meaning that Mark would hopefully be oblivious to the fact that he was actually being spied upon. And because Alice was attractive, her father had little concern that Mark would get too suspicious of anything—girls had a mesmerising effect on boys.

But he wasn't unconcerned enough not to rule out the possibility that Mark wouldn't be diverted by Alice's charms. He would have to think more about that if such a situation arose.

Seamus Brenson was untrusting of John Formby. For one, John was in the Gardaí; and two, he did not like the way he had moved so close to his home after his accident. Everyone knew he was a guard — because of his accident, John Formby was famous locally. The media in the neighbouring towns presented him as being a hardworking and reliable guard…but maybe it was a case of the press overstating his true detective abilities? Either way, John Formby was still a guard. He had an instinct for detecting criminal activity.

And now Seamus was right under his nose.

He did not want to draw attention to himself and give John Formby reason to 'observe' him. It was important to keep a low profile and not draw attention to himself or his work. For all he knew, Formby had already figured out that something wasn't quite right at the Brenson home or had a hunch. Perhaps he even had a tip off. Regardless, it was something that unsettled Seamus and presented a problem. If John Formby did know about his work, was staying low the best option? If he stayed quiet, and Formby did think there was something suspicious at Seamus' home, then staying in the shadows would, indirectly, alert him that he was hiding something. Likewise, if John Formby was oblivious and his moving nearby was a coincidence then keeping a low profile was just what was needed. Who knows what information he may have? Perhaps John would try to bug his home if he hadn't already. Thinking that, Seamus seriously doubted it, though, not with his security measures in place. But if Formby was as intelligent as he was depicted, it was something to consider. If Formby had somehow managed to bug the house, the only way he could have managed it was if he had persuaded one of Brenson's workers to act as a double agent, feeding him information of Seamus' business in secret, and even setting up the cameras whilst appearing to remain loyal to Seamus. He had checked the CCTV though and there was nothing, but all his staff would know, at least roughly, the blind spots…

Seamus was sure there was no breach in his workplace, and he realised he sounded paranoid. He was confident that Alice would get some information now that she had set a foundation to spying on Mark Formby. He was more relaxed now that Alice had made the first move. There was a risk attached to it of course. If John Formby was investigating Seamus, he would know he had a daughter called Alice. It only took Mark mentioning her name to John to make him aware that a drug lord's daughter was talking to his son.

He had never doubted any of his staff. He had known all of them for over twenty years save for his younger recruits; surely no one would betray his trust after that long? He was a good employer and treated his employees well. His recruitment process was almost unnecessarily extensive, he knew quickly who he could and couldn't trust. No one would go against him, he felt. They all knew the price of disloyalty.

Saying that, John was clearly a very clever, very practical man. His injuries came from a drug raid, he knew the game. It was not irrational to suppose that Formby had found out about one of his workers working for Seamus and blackmailed them. It was possible.

And Seamus hated possibilities. He only accepted facts.

The best detectives were just as good as the best journalists – it was not impossible that Formby had discovered something if he sniffed and dug in the right places – apparently his accident was as a result of his discoveries in order to bust a drug den, something no one else had thought of. Research told him that he was also responsible for a high success rate in solving drug-related crimes. But for all his seeming achievements, he had still nearly died. And just as everyone else seemed to ask: why was he alone at the time of the bust up? That fact was leaked, and it was that fact alone that Seamus thought kept the media interested. It seemed ironically stupid for a man of such high intellect.

However, despite his hatred for possibilities, it was Seamus' duty to assess every possibility. It was essential in his line of business. Therefore, if Formby had some form of evidence against him already without Seamus' knowing, to do nothing would be disastrous. It would be self-persecution. He would be a sitting duck. Seamus may as well just let Formby kill him if this was the case. On the other hand, if Seamus *did* intervene, he would be putting the spotlight on himself anyway, which would arouse suspicion when there may not have been any suspicion on John's part in the first place. That would make intervening pointless. Idiotic in fact. And he was not prepared to throw away his business, nor his freedom, so carelessly. It would almost certainly mean that his work would be discovered. That was the one thing Seamus wanted to avoid at all costs.

If he did nothing, and the guards knew something about Brenson's business, they may close in on him without warning, assuming of course that it was the case that the guards were actually suspicious. It was possible that they placed John Formby in the country on purpose, using his accident as a cover story to fall back on—that the country life was the perfect way to make a fresh start, which is exactly what he was quoted saying in the local papers.

He may be overthinking it, and it was just coincidence the Formby family had moved in nearby. Maybe they were just a family moving to a country home. If that was the case, then obviously staying low was best. But could he take the chance? Coincidences happen, and it was a small world for such a coincidence to happen.

But was the world that small? So small that an off-duty guard just happened to move to a house that was nearest the house which was involved in criminal activity?

It was a dilemma which deeply troubled Seamus. There was simply no way he could know for sure what the circumstances really were, meaning he did not know for sure how to act. It was a problem he had no answer to. And having answers was what Seamus was always used to. He was ruthless, decisive, practical. So not being able to be any of these things frustrated Seamus enormously.

It took weeks for him to come up with an idea that seemed to make sense. Fortunately, Seamus had a family of his own. His wife, daughter, and he lived in a Dunshaughlin country house, all of whom knew about his business. He knew it was still risky, but it was a 'damned if you do, damned if you don't' case. Every option presented a risk and if things didn't go well then at least Alice wouldn't have to try to meet Mark Formby again.

Another fortune: Alice was a confident, good-looking girl. It was at the dinner table one evening that he suddenly had the epiphany to use Alice so as to get a sense of John Formby's thinking. Mark Formby was Alice's age according to reports from newspapers and Seamus figured Mark must know something about his father in regard to the reasoning behind moving to the country. Maybe he didn't know why exactly they moved to Dunshaughlin specifically but sometimes the answer lay in the smallest of details. If Mark knew something, he might tell Alice of them if Alice played her hand well enough. A stab of guilt coursed through him about using his daughter, but he reasoned that it was a safe enough operation. He only wanted her to try and talk to the boy if she could.

However, the plan would only work if it was carefully laid out.

Firstly, Alice would have to make sure Mark didn't know where she lived. It was necessary that Mark assumed Alice didn't live where she actually lived. Just as a precaution. He could not risk Mark ending up in the house and exploring; heck, he may even have been asked to spy on them whenever possible. Mark Formby was not to know where Alice lived. If need be, a car would be on hand to collect Alice if she felt uncomfortable and needed to get away from Mark for whatever reason.

Seamus Brenson had thought of everything. He had even taken the measure of buying a house further down the country road, where two of his workers would act as Alice's parents if it was necessary. If Alice had different parents and John was investigating, it may put John off, making him think Alice was a different Alice.

Money was not something he was short of; he could easily set those provisions without a real struggle. However, he made sure the house was also within walking distance—if Mark kept meeting Alice in the fields near his own house, it was pivotal that Mark wouldn't start to wonder why Alice was walking so far to meet him. If Alice ever felt like she couldn't shake Mark off, she would ring them and ask to be picked up. If necessary, Mark would stay there for dinner at the second house. At least he would be diverted away from their actual house. In fact, Seamus wanted the latter situation to pan out. The closer Alice got to Mark, the better.

Everything had to fit. Everything had to seem plausible.

And everything was made plausible.

After Seamus had conducted his plan, he arranged a family and workforce meeting in the dining room. In all, there were ten people sitting around the dinner table.

Seamus then made his grand entrance. That was another thing he liked. He couldn't help but look authoritative.

"Hello ev'ryone. Thank you for your time. This shouldn't take very long," he announced with a genial look on his face, beaming around the room. He looked down at ten sets of plates and cutlery, with the mass of food in the middle of the table, made by Kirsten, Seamus' wife. Tonight's special was stuffed gammon joints with a variety of freshly steamed vegetables, served also with perfectly presented, identical looking fried eggs and pineapple chunks. She had a sublime talent for cooking. His hunger then took a hold of him.

"On second thought, what I have t'say can wait— food first! That should be a rule!" Seamus boomed. "Look at all this! Kirst, you've done it again, you beautiful woman!"

Kirsten Brenson returned the beaming smile, delighted for the recognition, and they all tucked in.

Afterwards, Seamus outlined every part of his story, starting from the potential threat of John Formby, to the finer details of the second house and acting parents. They all looked passive as they listened to his speech—no one expressed shock or acted like they were impressed, but their eyes were all shining with wonder, listening acutely to his every word. Seamus could tell that they all approved of the plan. He approached the end of his speech, and looked first at Alice, since it was she who would play the main part in everything.

"Alice is that all right with you?" Seamus asked.

"Of course, it'll be deadly," she said slyly.

"I hope not," Seamus frowned. Laughter rippled around the table.

"Now that bit is sorted. I now need one man and woman to step in as parents."

A slight pause. Then, one of his male workers spoke up.

"I don't mind. I used to do acting classes—acting as Alice's Da shouldn't be too hard"

Seamus looked in the direction of the voice and his eyes rested on Declan Doyle, one of his production assistants. He was a hard worker and a well-liked man. He was also trustworthy, reserved and fairly quiet. He had the qualities desired in a politician, Seamus often thought, meaning he was a valuable member of his team. He didn't think he would screw up a cover story—it was not in his nature to be careless and speak out of place. He was always careful with his words and he didn't speak often. That reduced the risk in Seamus' mind. Declan was forty-two and was tall, straight-backed, with light-brown hair and green eyes. He didn't look completely unlike Alice in appearance, and so he proved to be a viable option. He would do fine.

"Yeah, that would work," he confirmed. "That was easy. Now for the ladies—who wants to be Alice's mother?"

In his workforce of seven, there were only two women: Valerie O'Leary and Siobhan Connelly. They were more involved with the financial side of things, occasionally used as pawns to ensure a deal with potential clients. They were very attractive women, and, in their presence, Seamus could tell that Kirsten was not too happy. Most of these potential clients were men so the influence of Valerie and Siobhan definitely made everything much simpler. Valerie had blonde voluminous hair, with C-cup boobs and a size eight figure as he had found out when removing her clothes on one of the few times, he had slept with her. She had just celebrated her forty-fifth birthday, and there were wrinkles that had started to appear on her face. She was pretty enough, though, and her body was extraordinary for a middle-aged woman. In regard to Seamus' most recent requirement, she would probably be a better 'mother fit'. She had Alice's features and looked legitimately old enough to be her mother.

Siobhan, on the other hand, was a thin-haired brunette of twenty-seven he hired just over a year ago. She was naturally better looking than Valerie, at least in his opinion. She was olive-skinned, unlike Valerie's pale skin, which was more to Seamus' taste. Her double-Ds, hourglass curves and toned body complemented her seductive almond-shaped hazel eyes. She had everything going for her—plus, she had the edge of youth on her side. He had hit on her once but was rejected flat out. She was loyal to her partner in her personal life and loyal to Seamus in her working life and he respected her for it. He could tell she wasn't impressed about using her good looks to gain clients, but she was never asked to do anything inappropriate, so she accepted that side of her job.

He was a practical man: he knew it was easier to use Valerie, mainly for her age. As the two women sat at the table, Seamus observed both, picturing the two women as mothers separately. For the purpose of the plausibility of the story, Seamus preferred the idea of Valerie being the mother. For one, she was much closer to Alice's appearance with her blonde hair and dark-blue eyes. She was also of a more suitable age, which meant less hassle when setting up the pretend family.

"I guess I would be best for the role," Valerie answered indifferently, echoing his own thoughts, "I'm clearly older out of the two of us…" nodding her head Siobhan's way.

"Agreed," Siobhan replied. She did not want to act a part in a happy family situation. Her own relationship with her parents was complicated and she was also not keen on being someone she was not.

Seamus beamed in delight. "Exactly what I was thinking ladies!" he exclaimed.

He was always impressed at how they both knew what would work best in any operation. It was what made them so valuable in the business. It somehow added to the sexual attraction too, although only one of the two women allowed his advances. He was pretty sure Kirsten has no idea of what was going on, but he could tell Kirsten saw sparks flying every now and again.

Kirsten Brenson had always been uneasy ever since Seamus decided to hire the two women. She knew that it was necessary for the benefits of the business, but she suspected there was more to it than that. Kirsten disliked both women but was pleasant enough to Siobhan, perhaps because she could tell she was not interested in Seamus sexually.

As the discussion of the parent-acting took place, she could not help but scowl. It felt like they were taking over her role as Alice's mother—and a woman whom she did not trust too. She knew that it was relevant as a means of a cover story, but she could not shake off her discomfort. And what if Alice got into trouble? The idea seemed like a mockery to her as Alice's actual mother. She did not see why she could not play herself in the story; after all, she *was* Alice's mother. She had a maternal instinct that would be able to protect Alice should she need protecting. She was never asked her opinion about any of this kind of thing. It felt like a slap in the face.

"I am Alice's mother—why can't I be the mother?" she asked, with an edge of scorn to her voice.

Seamus frowned, picking up on her tone.

"Darling, y'know that can't happen. If Mark knows anything, he may know what ye look like since you are my wife. Y'never know, John Formby may have tutored Mark. It's a risk that is unnecessary."

She wasn't quite ready to admit defeat yet, though. "And if he has been tutored," continued Kirsten, "won't it look suspicious if, all of a sudden, there is a new mother and father?"

Seamus had to admit that she had a point. Even Valerie nodded as she thought about it.

But Seamus had previously already thought about it too.

"I see your point, Kirst. But Mark will only see Alice's parents once he is in the second house—if he ever goes there that is. If Mark has been tutored by John Formby, Mark will instantly know who Alice is. We would know then of John's intentions straight away. I know it is a big risk, but we need to know John Formby's mind here—at least we would know then if he recognised Alice. I've made arrangements for such an eventuality. This plan will work one way or another."

Kirsten looked away, clearly irritated. This did not go unnoticed by Seamus, Valerie and Siobhan; they glanced at each other, not knowing how to react. Neither Valerie nor Siobhan cared much for Kirsten—she was just the boss' wife in their eyes. They both thought she was a bit clingy. In Valerie's case, it meant that she felt no guilt when she got between the sheets with Seamus.

Seamus was quick to move on the conversation.

"So, Alice, if y'see Mark Formby walk away from his house, as he no doubt will at some point after bein' stuck in his new house for so long, you are to try and get close to him. Talk to him if you can. Text me immediately before you head out, and one of us back at the house will watch whilst you talk. One of us will follow from a distance if necessary. After a while, up to… let's say forty-five minutes, we will ring tellin' you to go home. Either Declan or Valerie will be stationed with a car if he begins to follow you home. He may want to be a gentleman or something and walk you home, you don't know. If that happens, your ma or da," he nodded his head towards Declan and Valerie, "will text you to let you know they will pick you up, so keep your phone in your hand. Otherwise, just go home as normal, looking over your shoulder every now and again to make sure he is not following or watching you. Sound, all right?"

"Yeah, fine. It might take time to see if I can sense anything from him about his Da, though," Alice responded.

Seamus smiled. He was happy she was thinking this through. She would make a great assistant in a few years if she chose to work with him. She was definitely a potential asset.

"I know sweetie, take as many meetings as you need. Enough to gain his trust. This might take weeks. You don't want to sound too eager about his father—we don't want him getting suspicious."

Alice nodded.

"And if he recognises you straight away, don't act alarmed. Introduce yourself with another name if necessary, to make him doubt," Seamus continued. "You could dye your hair or something if you like," he added as an afterthought.

"Everyone else understand?" Seamus asked, circling the room with his eyes. Everyone did, they all nodded.

The plan went ahead just over a week after the meeting took place.

She knew it was Mark. It couldn't not be. Alice felt like the talk was going well. She had been happy to leave the house finally, happy to actually be doing something after days of looking out a blank window, apart from one of the days when it rained. There was no point then.

It was clear that Mark liked her. She could tell that he was not used to talking to girls, and his hesitation showed. She thought Mark a sweet boy. Harmless. He was even not too bad looking, which was a bonus.

She didn't mind flirting with him. She was good at it. It was clear as day that Mark fancied her—Alice could tell from his nervous hand twitching, his dilated pupils, and his blushing. This was obviously the desired result, and so she left with a smile on her face when she walked away. She secretly enjoyed being marvelled at. What girl didn't? She wasn't sure whether she could know for sure whether Mark's dad knew anything, they had not really discussed parents. She didn't want to rush anything. She was right in thinking that she would need more time to ascertain John Formby's thinking. It was going to be a slow process.

As she reflected further on her talk with Mark, Alice remembered his occasional looks of questioning, as if he didn't know why Alice was there. She felt she addressed the potential problem well, convincing him that she was just bored with no one to talk to in a teenager-less countryside. But it was something to keep in mind when she looked to talk to him again. It showed he wasn't completely trusting of her, but it was definitely a good start she had made. She was pretty sure Mark would be thinking about her as he left, and for the rest of the day, next few days even. She was confident, particularly because it was clear he didn't talk to girls much. Talking to Alice was clearly a big thing for him.

Alice had Mark perfectly where she wanted him.

She had looked back a few times to check whether he was following. He didn't.

Alice then reported what had happened to her father in his private office. It was where all talks took place so that no one could overlook from the front window. Her dad listened closely to every detail.

Seamus was pleased.

"Good work," he said, kissing Alice on the forehead before walking out of the room.

Chapter Seven

John Formby woke up groggily the day after the incident with Mark and Sylvia. It took a few seconds for his vision to focus before he hauled himself out of the bed and stumbled into the shower.

He glanced at the digital clock on the way: 11:53 a.m.

Everything was blurred as he tried to think back, he figured he had slept fifteen or sixteen hours straight at least. He vaguely remembered picking himself off the ground shortly after Mark smashed a vase over his head – it was only around sunset then, he recalled – and then managed to climb into bed with a glass of water, badly concussed. It gave him a headache just thinking about it.

His numbed joints hurt as he stripped, and he reached out for the shower controls with a weak arm, the other rubbing his temple. The brightness of the day coming into the house still stung his eyes as he showered—concussion mixed with oversleep was a bitch, he thought.

Mark slamming a vase into the side of his head surfaced. John was surprised Mark had been able to strike him so hard. Usually, he would have been able to react to prevent being hit with his quick reflexes—he had had to many times in the past working with the guards, but given the situation, John's focus had been on Sylvia disobeying him. He did not anticipate Mark's assault until too late. Nor did he really think he could really do any damage.

He could not help but respect Mark for it though. He had always thought of him as a weak kid, but he'd shown backbone yesterday, a strength and determination he had not seen before. Perhaps there was some of him in his son, he mused

Knock, knock, knock.

John jumped slightly, the shock making him slightly more alert.

"What?" he called out in a tired voice.

"Come downstairs when you're done."

His wife's voice came through the door quite strongly despite the rush of water. He detected firmness in her voice, and it made him curious. *She'd better not think yesterday changes anything*, he thought moodily.

"Why?" he shouted so he could be heard.

No response. Sylvia Formby had already left.

Roughly twenty minutes later, John made his way slowly downstairs, and found Sylvia and Mark sitting at the kitchen table. They both looked serious and he could sense there was some tension there. An edge of nervousness hit them upon his entry into the room. He felt like the foreboding teacher about to grade them on their work.

"Drink this, it'll help," Sylvia said flatly, nodding her head at the fizzy orange drink on the table. The expression on her face was blank now. Mark too was looking at him reservedly. He could guess what was coming. He was too tired right then to really care about stamping down his authority, he'd let them both have this time in the sun. They all knew who was in charge.

"Thanks," he murmured, taking a gulp from the glass.

"We need to talk to you, John," Sylvia declared seriously.

John looked up at her, matching her blank expression, swirling his tongue over his teeth to wash away the drink from his teeth.

"Why?"

"Yesterday was the last straw," Sylvia began, "it made me realise just how aggressive you can be. Never before had you threatened Mark or me like that. There was no reason why Mark hittin' you over the head with a feckin' vase should've been necessary! We can't live like this." She looked at him keenly, hoping that he would see some sense.

John glanced at Mark. She was right that he had caused Mark to retaliate—there was no question in that. It was starting to come back to him how everything unfolded, and as he looked at Mark, he remembered Mark's murderous expression.

But more so, he recalled Mark being with that Alice girl. The concussion seemed to temporarily affect his memory. Out of nowhere, the same look of urgency reappeared on his face as he refocused on that. He had to know how Mark knew this girl, and where they had met. John had never seen Mark with a friend, let alone a girl. But his own curiosity wasn't the reason why he had to know.

His impatience cost him last time, though; John reckoned it would be better to hide his emotions and play the game. Bide his time and be more patient. Yes, that seemed best. He was more likely to get somewhere that way.

Neither Sylvia nor Mark seemed to sense what John was thinking.

As John watched them, he noticed they were looking at him expectantly.

Time to play the game.

"I'm sorry I have put you through such hard times." The lie made him cringe, but he remained outwardly stoic. "I guess I was just frustrated about not being able to do anything with myself for a while. It's been a tough time for me. Yesterday made me realise that I was out of order and of how I need to change."

That's it, agreement and sweet talk will do it.

Sylvia's face softened. Textbook. "I'm glad you said that, John. I'm really glad you acknowledge it. I don't want to be afraid of you hurtin' Mark like I thought you were going to yesterday. That's just not fair, and I won't stand for it. So, I'm happy to hear you say sorry."

Inside, John cringed again. He felt like a kid being told off for lying to their parents. He saw the disgust on Mark's face. He could tell Mark wanted him gone and that was fine by him. He was also far less naïve than Sylvia and he suspected his son would be a potential problem in the coming months. Despite enjoying making Sylvia's life uncomfortable, he had grown bored, like a kid outgrowing a toy. After all, it wasn't completely different, he was toying with his wife's emotions, which in its own way had become a plaything. Perhaps he would leave them once the business with Seamus Brenson was over.

"You're right, I can't keep going on the way I have been," John said, trying his best to sound convincing. "I was just wantin' to find out about the girl Mark was with. I have a bad feeling about her."

He had brought the girl up on purpose. By mentioning it casually, he hoped he could get the answers without too much pressing from his part. It was important not to act too eager.

He took another gulp from the glass and pretended to be absorbed by the drink by looking down at it with apparent interest. He noticed Mark purse his lips though and it was hard to hide a smirk.

"Why do you have a bad feeling about her?" Mark asked defensively.

John looked up, pretending to be surprised. From the look in Mark's eye, John could tell he wasn't fooled like his mother.

"Oh, I dunno. Just something about her," he said shrugging.

"You've never met her, what are you talking about?"

He shrugged. "What do you want me to say?"

Mark was silent for a few moments before asking why John even cared even if Alice was 'bad news'. He'd never cared about Mark's life before. Mark's eyes were angry, and the glaze that washed over his them reminded John of his own. The same green eyes with the same shape and sharpness. The same hatred...

Perhaps he would have to try a blunter approach after all.

"How did you meet?" he asked back, ignoring Mark's question.

"Just in the fields. We bumped into each other, why?"

Both of them were anxious to get answers from the other. It crossed both of their minds then that this was the longest conversation they'd had in a long, long time.

"No reason. I just thought it was odd, y'know, 'cause you don't really talk to girls. Or have friends," he added.

He didn't quite mean it to sound so harsh, just factual, but the look in Mark's eyes suggested John's words were cutting.

"Sorry," he added flippantly, not really meaning it.

"Yeah, well, Alice is nice. Not many people her age pass by so..."

"Hmm, makes sense. It is a bit of an empty land." He was getting bored. The conversation was getting stale.

He was trying his best to disguise his rush of thoughts with an apparently relaxed expression.

Mark studied him then. It didn't add up for him, but John had been frustrated and moody for as long as he could remember. Maybe the knock on the head did him good? His mood did seem lighter than usual. Or was that forced? Mark had a hunch that was the case. He was annoyed that Mammy was happy to accept an empty-worded apology, but he wasn't.

He was desperate to know about Alice yesterday. Why the sudden change of tack this morning? Mark wondered. Surely, he would be just as eager this morning if it was so important for him to know her name? This excuse of him just being curious was clearly a blatant lie.

It was all very confusing to him.

Mark observed John then, and he saw that he had that dazed look about him he got when he was deep in thought. Mammy was completely at ease again and had already forgotten about the conversation that had just taken place, and so was perfectly content to go about reading the paper, humming whilst she did so. She was such a free spirit at times that Mark had often got wound up by her; she was frequently oblivious to the present moment.

John, after a minute or two, realised Mark was looking at him, and smiled at him weakly, which to Mark always looked like a frown.

Probably is a frown, he thought.

"What are you thinking about?" Mark asked snidely.

"Me? Nothing," replied John.

Mark narrowed his eyes slightly, as if to reach deeper into his father's mind, but John gave nothing away. He had always wondered how his father could look so empty when he wanted to avoid people being able to read his emotions. He wasn't sure if it was a gift or a curse – he only knew he had the ability. It was only years of living with him – as well as Mark's shrewd, observant nature – that allowed Mark to know more about his father than most people did.

Perhaps it's something to do with me and him thinking alike.

It was only small, tiny things that occasionally gave John away. Mark daydreamed in a similar way when he was thinking about something or trying to concentrate. John knew Mark was observing him, which meant that he would be as resolute and unreadable as ever. And there was a dead look in his eye as he met Mark's gaze as if he was challenging him to play his hand in a poker game.

After making himself a big lunch, John swiftly left the kitchen. Mark could tell that there was something not right, something John was hiding— more specifically, something relating to Alice. This was what really grabbed his attention, and as much as John tried to hide it, the glint in his eye at the mention of her name was a big give away.

I won't get anything out of him though, he thought bitterly.

Mark looked back at Mammy, and she was still looking nonplussed about the whole situation, content just to read the local news. She had reverted back to her normal self it seemed.

Typical.

It never did take long for John to pacify her.

Mark left then, annoyed. What did he expect? For Mammy to now have the same suspicions as Mark? No, she didn't understand John's motives or his actions like Mark did. Mark inherited such traits. He knew when John was putting on a poker face but secretly keeping things to himself.

He often did the same himself.

Until yesterday, Mark did not think his father felt the same connection; it took a physical attack to make him see certain similarities between himself and John.

"Where you goin'?" Mammy asked.

"Havin' a shower."

Before doing so however, he searched for John to see what he was doing. Maybe he was up to something conspicuous, something which may point towards what he was withdrawing from Mark and Mammy. Just a small clue even.

If I'm not seen, who knows what I'll discover, he mused.

There was only one place he could be. Upstairs. Mark crept along the hallway and continued to creep up the stairs.

He made sure he was silent as possible as he sneaked up the stairs. Luckily, they were carpeted so that more sound would be absorbed as he walked upwards. When he got to the top, he had the decision of whether to go left or right on the hallway; the right end led to John's private study; the left John's and Mammy's bedroom. Mark's bedroom was on the left too—but he doubted he would be in there and he wasn't in the bathroom which was open. On his way along the hallway, Mark had glanced in the drive, and John's car was there. Mark would have heard the front door open and shut if John had left, though. As Mark teetered on the top of the stairs, he heard a noise—a footstep, and then the sound of a door opening. Quickly, he hid behind the wall of the top few stairs, peering over the edge to see down the hallway. What he saw stumped him.

John was leaving *Mark's* room.

Chapter Eight

Mark ducked down again just as John looked down the hallway towards him; he did not think John had seen him. What on earth was going on? When the coast was clear and John had gone into his own bedroom, Mark stood up on the top of the stairs and headed into his own room. As far as he could tell, nothing had been altered; had he not known John was just in his room, there would have been no evidence to suggest anyone other than himself had been inside his room. Nor was there any noticeable addition to his room, leaving Mark confused. Since John was in the next room, Mark did not want to arouse suspicion by checking all his drawers and cupboards right then and there—John may have seen Mark watching him, or knew by instinct that he was there so Mark did not want John to have another excuse to question him. Even though Mark was sure he had avoided being seen, it was best to act normal for now. He went for a shower. On the way, Mark looked over his shoulder towards his parents' room and wondered what his dad was up to.

After his shower, Mark saw that the door to his parents' bedroom was now open, and so he quickly peeked inside. Nothing different. No one inside. John had performed his disappearing act again. He then went downstairs into the kitchen and enquired of Mammy where John had gone.

"Out somewhere. He didn't say where," she said, shrugging.

Mark thought she was glad to have some time to herself after yesterday although it peeved him that a simple apology seemed to be enough to satisfy her. She left him to do his own thing.

After getting dressed he went outside to look for John. His car was still parked in the driveway. Mark desperately wanted to know the truth about his insistence upon knowing about Alice. He wasn't afraid to confront him. At the same time, he could not help but hope that he would bump into Alice again. Something about her intrigued him; she was gorgeous, she was cool, she was fun to be around. The feeling was strange. And confusing.

What guy wouldn't fancy her? he wondered, now daydreaming as he headed down the country road. Never before had Mark had such an infatuation towards a girl. After only one meeting too, it was extraordinary! He had never wanted to meet anyone, let alone a girl. Especially a girl. Girls intimidated him since he could never seem to read their feelings, not that he could really tell what Alice was thinking. In films, he could read characters better—but they were scripted and hyperbolic and relating to the viewer was much more essential than in real life. Mark could assess who was the culprit in crime novels and TV dramas nine times out of ten. That sort of thing was easy for Mark to work out. It was logical. Perhaps it was also that instinct for policing… like his own father had. He tried to push that thought away. Girls were unreadable in Mark's opinion when it came to romantic feelings. Most girls did not make it obvious they liked a guy; their passiveness threw him. Playing hard to get was a baffling and frustrating concept. Not that he had any experience himself at asking girls out.

Mark had been walking for five minutes or so when he came across a figure across the field standing next to a house, deep in thought. It seemed that whoever it was, they were examining the house, but he couldn't be completely sure. He knew the house was there, but Mark had not really given much attention to this neighbouring house since moving; it was one of several houses in the surrounding kilometre or so. Nothing about it really stuck out in any way, all the houses looked ordinary; this particular house, which was the cause of this person's keen interest was just that: ordinary.

A potential house buyer? Mark wondered.

As far as he could discern, there was no 'for sale' sign nor any pole in the garden which may have suggested there was, but then again, he could not see properly from this distance. For all Mark knew, there could be. Or the owner of the house was selling privately. Either way, there was nothing else that seemed to be going on to grab Mark's attention, just a vast mass of tranquil land with no signs of life except the shrills and tweets of the birds. He wasn't sure why he wanted to see what was going on, it was most likely nothing, but there was something urging him towards the person in the distance. All thoughts of his dad were temporarily gone. Who really cared about someone looking at a house? Despite being too far away still to establish, Mark was quite sure the figure was a man. Even from this distance he could tell the person was tall. If the man looked out at the field, he would no doubt see Mark walking across it towards him. Mark had intended to go back to his spot next to the tree if he was unable to find John, hence the backpack full of snacks, food, a book, binoculars to watch the scenery and any birds which no doubt would fly around. Perhaps he could use the same spot as a base to see what this man was doing. Also, in case he saw Alic—

Binoculars! The thought sprung out at him and he rooted a hand into his backpack. As he glanced at the figure, and noticing the man was still transfixed by the house as far as he could tell he found his binoculars and used them to see if he could get a closer look.

Magnified, the man was made clear enough to distinguish. His thoughts quickly spiralled into confusion. For a moment he lowered the binoculars, lost in thought, then raised them again to examine what was going on.

John Formby was the man examining the house.

Mark could not think clearly for a few seconds – he just stared at John through the zoom lens. What the hell was he doing? More relevantly, why was he looking at this apparently normal house? Was this linked to Alice in some way? Alice didn't seem to live there—she had walked away from the house as far as he could remember. He had to talk to him, get the truth if he could. He knew what John was like revealing information—like getting blood from a stone. Impossible. Almost.

He marched over towards John. His dad's attention was distracted after sensing that someone was there. His eyes widened in surprise on seeing Mark.

"What are you doing here?!" he asked incredulously.

"Erm, more like what are *you* doing here?" Mark returned.

"I don't appreciate you sneaking up on me. Were you spying on me?"

"Answer my question," Mark replied.

"Don't talk to me like that. Answer mine!"

John was always stubborn. Mark could be too.

"Don't flatter yourself. I don't go around following you like a needy dog," Mark retorted. *Two can play that game,* Mark thought to himself.

John's eyes narrowed. Mark kept his face expressionless, empty. He was good at that. Something else he inherited from John.

The battle of silence commenced.

It was like a Mexican standoff as they stood facing each other. The soft breeze rushed past their faces and all around them was silence, as if the world had stood still or time had frozen. Their hands hung loosely at both their sides, as if stroking imaginary guns in anticipation of a duel. Instead, their eyes were shooting the bullets at one another in the attempt to break the other down. It was a surreal moment for Mark who had never gotten so much undivided attention from his father, save for the interrogation yesterday about Alice. The curl of John's lips formed into a kind of scowl as if Mark was an unpleasant barrier getting in the way of him. Had it not been for Mark's curiosity about why Alice was linked to John's mysterious theory… whatever that theory was, Mark may not even have been outside to spot John here in the first place.

"Why are you looking about this house then? Is it something to do with Alice?"

Upon mention of Alice, John's eyes flickered, just for half a second but it was there. *Did she live here? If so, why was that relevant? How did he even know who Alice was? Why did he care?*

"Have you been inside this house?" John finally asked.

"No, why? Does she live here?"

John acted like he hadn't heard the question.

"What are you on about?" Mark asked, getting irritated.

"Nothing. Why are you here?"

"I went out for a walk, noticed someone looking at this house, so I got curious. Then I realised it was you, so I came over."

Mark was now forcing his patience—he didn't think he could control it for long. He was clearly getting in the way of whatever John was up to.

John surveyed Mark as he spoke. He came to conclude that Mark was at least being partly truthful—he knew Mark was a curious character by nature even though he did not know much about him. He often saw him wandering around the streets around their house in Dublin, coming and going like no one's business.

"I fancied a stroll too," John finally responded, "I saw this house and I wondered if it was derelict. I've never seen anyone come in or out of the place, so that's why I was looking around the place." John offered a faint but icy smile.

John felt that he had to lie. He needed some cover story to explain why he was at the house to avoid constant, annoying questions from Mark, and Sylvia too as Mark would no doubt mention it to her. He hated how close they were. Most teenagers shut everything up inside. The derelict story seemed most plausible on the top of his mind—he may get rid of Mark pestering him that way. The less confrontation he got, the better. He hoped his smile was convincing enough and masked his unease at the way his son was badgering him. Mark was more likely to lay off his case if he was 'pleasant' about the whole thing. Or would that make him more curious? John didn't know. Or care. He just wanted him to go away.

It also annoyed John how inquisitive Mark had become as he got older. He did not like how sharp he was. He wondered if Mark would see through his cover story. Mark, admittedly, was surprised by the explanation. John never explained anything. Mark wished he could read minds; he still didn't sense truth. He was utterly baffled.

Then another question occurred to him, which hopefully would stump John and he felt an anger swelling inside him again. The way John was looking at Mark made him feel small, like a pesky little kid. And it ground his gears.

"So... Da," he began sardonically, "why were you in my room earlier?"

John looked taken aback and Mark couldn't help but be pleased with himself.

"Excuse me?" he asked, acting dumbfounded.

"You heard. Why were you in my bedroom?"

"I wasn't in your room..." he replied.

His dead face was impossible to read.

"Don't lie. I saw you leave."

Mark was at the tipping point. He did not appreciate being treated a fool. He was too old for that. And the smirk certainly was wiped off John's face.

"So, you *are* spying on me," he said firmly.

"I don't have to explain myself to you," John hissed and turned away, powering down the road.

"Yeah, that's it! Walk away like you always do!" Mark shouted after him.

John said nothing and did not turn around, but Mark saw one of John's fists clench. Mark could only shake his head as he watched him stride away.

Seamus Brenson watched the whole confrontation between John and Mark Formby take place outside his house. From the CCTV camera, which was hidden on the outside of the house, Seamus paid his utmost attention to the debate that took place as he sat in his security room. His security employee, Dennis Waitfield, had alerted Seamus of a man examining the house. Seamus immediately recognised the man as being John Formby. It implied John was, after all, suspicious of the house or was aware who lived inside. Either way, surely, he was investigating him. But why be so obvious about it? The appearance of Mark Formby was definitely an intriguing development. Unfortunately, the sound system attached to the camera was too far away to pick up their voices in enough clarity to distinguish what they were saying (he made a mental note to get better audio systems in place), so Seamus relied on body language. It was clear that there was tension from the way they conversed, and even something hostile between the two of them. An argument was very much on the cards, and the cold storming off from John seemed to epitomise the relationship between father and son.

Seamus stroked his beard as he thought deeply about what this could mean. He watched Mark give the house a quick look before walking off down the road himself. He then sat back into the chair, spinning slowly as he did so. After a minute, he left the room with a quick thanks to Dennis for his good work, and then went off to find Alice.

It was time for Alice to spy again.

He found her in her bed listening to her music whilst her laptop remained open on her lap.

"Oh, hey," said Alice, taking out her earphones.

Seamus got straight to the point. "I want you to invite Mark over for dinner whenever you can. He was just seen outside this house having a debate with his Dad. John was examining the house, meaning... well, it suggests that he does suspect something. It seems that he is not living in Dunshaughlin coincidentally. I want you to press Mark about his relationship with his father in a subtle way – use your charms. Men often fall for women's charms. Believe me," he said with a smirk.

He hated the thought of using her daughter as bait in such a way, but he couldn't think of a better course of action. Besides, Alice had made a good start with Mark Although, if he sniffed danger at any point, he would tell Alice to stop.

"I didn't need to know that. But okay I'll do it. Just ask about his Da?"

"Well yeah, mainly. But make sure you don't just talk about that as he will more likely be wanting to know why you are so interested in him if you only ask questions about his Dad. Talk about other stuff, general things, everyday stuff. Flirt with him, invite him to dinner. I want Mark's attention drawn away from this house if at all possible. Which will surely happen if you bring him to another house: the cover house. John will have, to some level, now given the boy cause to give this house a second look even the boy doesn't know why. And that's far from ideal. If your name has ever been discussed at their home, it is possible that John knows you are my daughter. He is a professional detective after all. Even if he and Mark don't see eye to eye, it still remains plausible that the two of them have talked about you. If John does know who you are, that can't be helped. However, by bringing Mark to the other house we talked about, he might drop his attention from this house or at least make him think he's got something wrong."

For some reason, the concept of being talked about made Alice blush slightly. She wondered if Mark talked to his Mammy about her if not with John. Seamus didn't seem to notice. His plan was too fixed in his head for him to be distracted. He thanked Alice and left the room before an afterthought hit him and popped his head round the door. "Maybe we should dye your hair actually. That way, if John ever does see ya, he won't know who y'are straight away—it may confuse him if he is looking for an Alice Brenson. His memory has apparently been affected by his accident according to the press, so it may just be enough, even if his memory has fully come back. Tell Mark your surname if appropriate, obviously not Brenson. Make it up. I hope ya haven't already told Mark your surname?" he asked seriously. "That would complicate things"

"No, I haven't," Alice replied.

"Good. If y'do meet Mark's parents at any time, whether today or another day, make up your surname. Preferably a stereotypical Irish name so that it can't be inspected too closely by John if he went snooping."

"Okay," she sighed, now getting a bit fed up with his waffling.

"Good girl. Now get yourself ready," he said with a wink. "Not that my princess ever looks anything other than beautiful," he added.

Alice smirked, shaking her head. "Okay Dad."

"Oh, and dye your hair before you go out. That will take time, but your Ma is great with that sort of thing. Get her to help ya."

"Where should I ask to meet him?"

"Where you met him last time? He had a backpack anyway. He won't be too surprised seein' ya at the same place as the first time anyway. You should get his number this time. Make sure you woo him! The worst-case scenario would be him to say no to your invitation!"

"I'll make sure of it," Alice promised with a smile.

Seamus may have been a big businessman, but Alice was still his little girl. And she knew it. He still found lots of time for her.

He smiled back at her. "I'm dead proud of ya, Alice. You've turned into such a fine young woman!"

And then he left the room in a happy mood.

Alice began getting ready. For some reason, she was nervous seeing Mark again. Was it because a lot relied on her next meeting with Mark? Yeah of course that is it, she reasoned.

Except she wasn't so sure.

After Kirsten helped her with her hair, Alice looked in the mirror, looking at the chestnut-brown colour which had replaced her usual blonde look. She had to admit, her Mammy did a great job—she thought she looked good with her new hair, if Alice did say so herself.

She smiled, satisfied.

Kirsten gave her a quick smile. "You are so beautiful," she remarked.

"Thanks Mammy," said Alice, beaming.

This feels like a first date or something from the way she is looking at me, Alice thought bemusedly. Something inside her felt weird at that thought. A feeling that was alien to her. Alice was perplexed. She had done this before, many times, she had been on many dates. Why was she feeling odd? Why was there a twinge in her side? Was she ill?

"Knock 'im dead," her Ma said.

Alice looked at her mother as her words intercepted her thoughts. There was an air of confidence on her face.

"Bitch please, have ya seen how good I look?" she replied.

Kirsten laughed.

Alice walked out of her room, and down the corridor, passing her father on the way, who stopped to look at her, flinging her hair gently with a finger and then twirling her as if she was still a little girl.

"Looking good, princess!" he called.

Despite his business reputation, she always did think him a brilliant father. He always said the right thing at the right time.

"It's true, I hook all the guys who look," she said, acting matter-of-factly.

Seamus guffawed in delight, beaming at his daughter. "Oh, your humour is just as charmin' as your smile!"

After looking back at the CCTV, and Alice peeking outside to check if the coast was clear, Seamus boomed from upstairs:

"Time to go m'darling. Good luck! Oh, and ring when you want collectin' and Declan will drive down with Valerie to get you!"

"Will do, Da!" she shouted back. "I have to find him first!"

"I'm sure you'll find him," Seamus replied confidently.

And off she went.

Alice was pretty sure Mark would be found where he was last time like Seamus said, the distance from the house was about five minutes but she was quick about slipping outside, looking out for any signs of John Formby as well as Mark. No one could be seen. She relaxed slightly.

Chapter Nine

A part of her felt bad for Mark. She was being sent out to lead him on – he was a nice enough lad, and it was clear he was not used to talking with friends – or anyone in fact. Mark trusting her would be a lie, a false trail. She suspected that even after only one meeting with Mark, he was not one who easily trusted someone, meaning it would be a big thing for him to trust her. And that made her feel guilty.

Maybe that was why she felt odd beforehand: she was misleading Mark on purpose, which was totally unfair to him. What had he done? He was innocent. The guy did not deserve to be used—What would it do to him psychologically if he found out? Alice tried not to think about it. She had a job to do whether or not the Mark did deserve it or not. It was a shame, because Alice would have liked to have been genuine friends with him; he may be a bit socially awkward and perhaps slightly nerdy, but he was nice, you know? Someone who you could trust even if he didn't necessarily trust you. That troubled her a little bit, made her uneasy; it was supposed to be the other way around. However, she knew that it was necessary for her to get to know Mark. John Formby clearly suspected that something was happening at the house – the CCTV footage her dad talked about had confirmed that. Any information that she could find out from Mark would protect her father, and she was eager to help him in any way she could. Seamus was a great father, a great man. She often wished he did not do what he did, but things were as they were, and there was nothing Alice could do.

Regardless, she still felt a pang of… what was it? Sorrow? Pity? She wasn't sure. There was more to it than guilt.

She was now approaching the field Mark was in last time she came across him. It was time to get into character—even if that character was still herself. Her spy mode, if you like. Could she become friends with him as well as spying on him? Alice wondered if that was possible, forcing a friendship based on lies.

Alice looked over the hedge into the field. In the distance, as predicted, a figure was sat by the tree, looking out at the scenery. It was as if everything was perfect. She thought she would have to look for him, she even expected not being able to talk to him today.

What an unusual guy, Alice thought.

Alice continued to walk down the road until she was on a one-hundred-and-eighty-degree angle to the right of Mark, intercepted by a high hedge. Mark sensed someone walking by, and as he did so Alice pretended to do a double take as she spotted Mark.

"Oh hi!" Alice shouted over to him happily.

Mark waved coyly, not saying anything.

Alice climbed over the hedge and made her way towards Mark. Somehow, she managed to climb over it gracefully. Mark was sure he would not be able to do that. As she got closer, Alice helped to aid conversation, expecting an otherwise awkward encounter.

"Ya really like this spot, don'tcha?" she asked, smiling down at him.

The pupils in his eyes widened slightly as he smiled up at her. Alice knew what that meant. Every girl knew what that meant.

"Yeah, it's a decent place to escape. Ya... ya changed your hair?"

It was early days, but it seemed Mark was more relaxed this time. Slightly. He was still quite shy, but that was normal when a guy liked a girl. Mind you, Alice for some reason could sense that Mark didn't talk to many people his own age from as far as she could tell, so maybe that was the cause of the nerves, and his pupils dilated because talking to people was out of his comfort zone. Was she being cocky assuming it was all because he fancied her?

"Yeah, fancied a change, you know? I was fed up with the blonde hair anyway if I'm honest. What ya escaping from by the way?" she asked, as if ignorant of the fact he had just argued with his father.

"Just home really. Parents are a nightmare. I... I do like your hair by the way, it–it suits you", he added finally.

He was supposed to compliment her, right? That's the way things worked, wasn't it? He blushed all the same, looking away in embarrassment.

"Aw thanks sweetie! And I feel that. Mine are a pain too. Anything in particular on your mind?" with a concerned look.

Mark glanced at her, then smiled weakly.

"Nothing much. Just a bit tense." He didn't want to talk about the weird stuff his dad was getting up to, nor did he want to mention the fight which involved Alice herself. He didn't want to freak her out

"Erm... do ya mind if I ask what? You've got me curious now," she asked kindly. "You can talk to me about anything—obviously don't if yer uncomfortable or anything like that. Besides, even if I wanted to – which I don't – it's not like I have anyone to gossip to."

Alice played that card with expert sensitivity. Her eyes were big and kind, and her smile was reassuring. Her tone was almost a purr as she spoke and was having an obvious soothing effect on Mark. She was gently easing him out of his shell.

Mark knew he liked Alice; she was a sweet girl. And he would be lying if he said he didn't fancy her—she was gorgeous, funny, and not a dumb blonde... well, not a dumb brunette now. This was the second time now that she had seen him as he happened to be sat there. Did she like him too and decide to wait for him? If she was, then logically, this was where she would wait to find him. But she did seem surprised to see him, albeit happy too. Maybe she was a good actress – most girls seemed to be. They had only met once, and she was an intelligent girl. She would know where to find him. Where else could a teenage boy hang out in these parts? He decided that Alice was someone he could trust—it's not as if she had many people around the area to talk to, as she said. Even if she didn't like him in that way, she was definitely a potential friend. Someone he could talk to other than Mammy. Okay, it was not a definite, but it was definitely maybe.

Also, for whatever reason, Mark found he could speak fluently with her, like he instinctively knew how to talk to her aside from a few stammers; at least better than he would be with most people. Alice was right. He *could* talk to her—it somehow felt natural, which was an alien experience for him. After deliberation, Mark decided to tell his tale.

"My Da's the main cause of it. I don't know where t'start exactly, but I guess it all started when he had his accident." He glanced at her to see if she was still interested. "Actually, I lie, the problems were always there. He has never been what ya would call a good father—his job was everything and being a dad took a back seat. And then one day he got involved in some bust up and got shot a few times by some gang of some sort, I dunno. Have ya heard? You probably saw the media a few weeks back camped down there?" Mark pointed in the general direction of his house.

Alice could tell his father was bothering him from how passionately he suddenly talked. People speak passionately about what is meaningful to them, or what is on their mind. Her own father had taught her the basics of psychology. It was important to understand for his own business.

"Ah yeah, o'course. I heard about it in the news. What's the name? F something… something Fromby? Fromby?" she said vaguely with a thoughtful look. She didn't want to risk coming across as knowing more than Mark expected, especially now he seemed to be opening up to her. She made it sound convincing.

Mark smiled. "It's Formby, and yeah, he's quite the hero in these parts now, is my Da. Well, for those who remember, I suppose. Probably old news now, right?"

His smile faded a little then. *Another small shift in his facial expression spoke all there was to say about how he was feeling towards his dad,* Alice thought. She was surprised when he spoke further.

"No one really knows what happened exactly the day he was attacked—o'course the guards know, but it has been quite hushed up. Only the names of the gunman and a few other names were released."

Mark noticed Alice was listening diligently, which he appreciated. He didn't expect to go into so much detail but now he had started, he couldn't really go back. He had not been able to pour out what he was feeling to anyone, not even Mammy, so the more he spoke, the better he seemed to feel.

"Anyway," continued Mark, "since moving here, he has just been the same. He has been short with me and Mammy all the time and there have been quite a few fights. As ya can imagine, it has been quite tense. So… here I am getting away from it all."

There was a minute or so of silence then. He did not want to talk about the incident with the vase. Mark saw Alice look thoughtful, seeming to digest what she had heard.

I guess there is not much ya can say to that, he thought.

73

"I had no idea things were so tough, Mark," she said delicately. Which was true. She *had* had no idea until then. "I'm glad y'feel like ya can talk to me—it makes a nice change actually. No one is ever around for me to talk to. Y'know, erm, if ya like, ya could erm… come to mine for dinner tonight? I—I mean if you like?" She thought she slipped the question in very well. "It might allow ya to get away from it all? I don't wanna be all forward and make ya feel like I'm pressurising ya or anythin'."

Alice put on an embarrassed tone of voice so as not to seem over-confident. Another little trick. Too much confidence with a shy character like Mark, and he would feel overwhelmed. A touch of shyness herself meant he was more likely to react better. Plus, as well as she was acting, she could not help feeling a genuine burning to her cheeks.

Mark had not been anticipating the question at all. No one had ever spoken to him long enough for them to become friends, let alone be invited for dinner. Kids realised quickly he was no fun, not a normal kid, so he grew to learn to not be disappointed when it happened. There were no second conversations with Mark, in his experience. He had never passed the first stage of friendship—the initial interview. However, within two meetings with Alice, she was already the person Mark had spoken to most apart from Mammy. And she was not just anyone. Not an equally pathetic, vulnerable boy like himself. She was beautiful, way out of his league.

It was almost impossible to him that this was happening.

More surprising still, Mark noticed that Alice's cheek had noticeably become a touch pinker than it had been. With amazement, Mark realised she was blushing.

Blushing!

It took a few seconds for this to sink in. Alice wasn't mortified to be around him and had asked him to eat dinner with her. For her then to blush was surely a good thing? He had never had someone who genuinely wanted to get to know him better, someone who actually liked talking to him and being around him. For so long, he had been used to abuse, bullying, and if someone in school did have to speak to him, they blushed in a way that meant they were embarrassed to be there. He knew what that blushing meant, but Alice's blushing now was definitely different. She chose to be there at that moment talking to him, 'chose' being the operative word. It was not a big deal to most people, certainly not to most teenagers. They would not even think about it. But it was a very big deal for Mark. Acceptance was huge since he had never been accepted.

He could feel a sharp pain in the back of his eyes.

Don't cry! That won't make her want to be your friend! For the love of God, keep it in!

After managing to control his emotions, at least his tears anyway, he accepted the invitation. A surge of something rushed through him.

Excitement. Nervousness. Anticipation?

This must be what it feels like to be invited somewhere, Mark thought inwardly.

"Hang on, better let Mammy know I won't be home for dinner."

Alice could not help but smile. He was so cute. She watched with what was actual affection as he phoned his Ma, and it wasn't until after the call he noticed her looking at him in that way.

"What?" he asked smiling back at her.

She shook her head.

"Nothing."

Alice and Mark waited for Alice's parents on the pavement next to the field. She had rung to ask if they could be picked up. Mark stood staring ahead of him, not quite believing what was about to happen. The whole scenario was crazy to him, the feeling of being wanted surreal.

"Whatcha thinking about?" Alice enquired.

"Nothing... I guess, I just didn't anticipate this," he replied, looking down with embarrassment.

"Yeah, me neither," she said smirking. "I don't really ask many boys around for dinner." She gave him a wink.

Mark laughed.

"Bet ya don't," he said playfully.

As he said this, he attempted to wink back, he didn't plan it. It just happened. He hoped he would look suave—suave for him at least. Instead, he did more of what can only be described as an exaggerated awkward blink and instantaneously, he felt foolish. Alice laughed heartily but it was clear she found it more funny than idiotic.

But still a bit idiotic, probably.

"Ya can't wink?" she teased.

"Apparently not. I swear I can wink normally!" he burst out, feeling ridiculous.

She raised her eyebrows. "Ya sure? Ye're getting kinda defensive there."

"It's because I can actually wink!"

"Okay, okay, ya can wink!" she exclaimed, raising her arms in the air as if surrendering to something.

She could not prevent a fit of laughter at how ridiculous the conversation had become—she hoped he wouldn't take her laughing the wrong way. To make sure, she pinched his cheek with her thumb and index finger before giving a soft slap on his face and telling him he was cute.

Alice was glad to see him return the laughter then. It was clear that there was a sense of humour in him that no one seemed to notice, and that he wasn't just a serious guy who had no personality. The loveable blabbering in his speech as he spoke to her, and the way he laughed was pleasant to the ear, and Alice noticed that indirectly, Mark had a certain undefinable charm to him, as if his lack of awareness to his own cuteness was drawing her in towards him. Alice could not help but think that Mark was a bit of an unintentional and endearing fool. As she looked up at his brown hair and energetic, bright, green-blue eyes, it was at that moment that he reminded her a bit of Hugh Grant, just a bit more ginger. In his younger days that is, like in *Love Actually*. Foolish, but quirky, and charming in a babbling, rambling kind of way. Even though he was awkward, there was an easiness to his laugh that was full of happiness and his smile was crooked in a nice way as it emphasised Mark's strong, still developing jaw. Now that Hugh Grant came to her head, she could not shake off the comparison. Not that she minded that much.

Hugh Grant was on her list.

She kept this to herself, however. She did not want to display too much affection at once—even for Mark, she would have to subtly play the hard-to-get card. Not that she really wanted him; she had a job to do. Nonetheless, she did like him. She would flirt with him just enough to keep him interested, nothing more. That way he would stay in contact with her, and Mark would begin to trust her increasingly more. She had merely set the foundations.

This was no romantic fairy tale. She was a spy.

As this was going through her mind, a black Vauxhall Astra approached. The whole thing was arranged so that even the vehicle used for the operation would not suggest any extravagance of wealth to make Mark ask questions. It was an obvious thing, but an essential one all the same. It would not add up for Alice's 'parents' to live in a standard, semi-detached house whilst flashing about a Jag or Bentley. You couldn't really go wrong with a mass-produced Astra. The car pulled up right beside them.

Both parents were in the car, so Alice and Mark sat in the back seats. The first impressions of the parents were positive. They seemed very enthusiastic and very friendly.

"Hey guys, how are ya? I see Alice has found a friend," Alice's ma commented, turning around to face them, beaming.

Mark could not help but think that Alice's mother was also very attractive for a woman in her forties. Many of Alice's genes clearly came from her.

"Hey Ma. We're good—"

"Nice to meet ya, young man. What's your name?" Alice's father interrupted, extending a handout from the driver's seat for Mark to shake, smiling kindly.

He too, looked like he was in his forties. A touch of grey was spreading round his otherwise almost black hair, and he had big dull green eyes, but they were kind and warm. His tweed suit suggested he had just finished work. Alice's mother was wearing office trousers and a dark pink blouse, also indicating she had finished work for the day. Normal family, Mark thought.

"Mark. Nice to meet you too," Mark replied, politely smiling back.

"F'Christ's sake, at least wait for Alice to introduce Mark to us herself! For the love of God, man!" Alice's ma chimed in, smacking him over the head. Everyone laughed and Mark was relieved that there were no awkward silences.

"I get this abuse all the time," said Alice's dad. He put his hands up in the air, light-heartedly pretending to act the confused victim.

"Ah just drive, you eejit!"

More laughter.

"Anyway, it's lovely ter meet ye, Mark," the mother said. "By all means, call us by our first names. I'm Vanessa, and this fecker driving is Eoin," she said pointing towards Alice's father.

"Nice to meet you," was all he could manage, trying to contain more laughter.

Ye've already said that! Goddammit say something else—you're not a feckin' parrot!

Alice smiled sweetly at him. Mark suspected she could read what was going through his head. Vanessa and Eoin seemed not to notice, however; both were talking to each other again, probably so as not to overwhelm him too much. This suited Mark. He would probably have to think of something to talk about in his head before he got to their house to have dinner. They had been in the car for about five minutes, Mark guessed, and he had no idea how much longer they would be travelling. Probably not long if Alice was able to walk through the fields to meet him, and they had already passed the only house that seemed nearby – the house John was inspecting earlier. There were so many questions he wanted the answers to: why he was there? Why that house? What was he suspecting? Clearly something was on his mind for him to be focussing on this one, ordinary house. Was something going on in the house that he had noticed as he was travelling on his own? Had he seen something odd during the times he disappeared from the house hours upon end? It was frustrating Mark. It was in his nature to know everything that was brought to his attention, regardless of whether or not it was someone else that had made him aware of something or he himself. He still had a feeling it linked back to Alice in some way.

But he wasn't going to think about that now. It was much more productive now for him to think of things to say.

"Ya' all right?" Mark looked up; his thoughts disturbed.

Alice was looking at him. Mark realised he probably looked deep in thought, which made sense. He had been.

"I'm fine. It's just a bit surreal all this."

"I know, you sorta said already," she replied, placing a hand lightly on his arm.

There was a kindness to Alice's voice as she said this but she no doubt thought it a bit weird to say so little and yet repeat himself. That was a schoolboy error straight away.

"Sorry. I – I erm…"

"Calm down, it's okay. We won't bite ya."

Then she made a biting motion.

"Or maybe I will," she said with another wink.

That wink again. Mark felt himself melting.

He looked outside—he didn't want Alice to see him blushing, as he no doubt was, considering he felt a sudden touch of heat surface in his cheeks. He noticed they had just gone past the old mill, about a ten-/fifteen-minute drive away from his own house. It was not long before they turned into a drive where a semi-detached house awaited them.

"Don't be too impressed, it's quite small," Alice piped up.

"Oi! We're not made of money," Alice's mum said hotly.

"She's showing off," Eoin said with a nudge on Mark's arm.

They went inside.

Mark was not surprised to discover after Alice's comment that the house was indeed small. The staircase was about three paces away from the front door, and in the middle of the two was a small shoe-cabinet. Alice kicked off her converse sneakers and shoved them onto one of the shelves.

"I've loads more shoes in m'room, I'm not exactly a girl for one pair o'shoes," Alice asserted eagerly. Then, noticing Mark was unsure what to do with his own shoes said: "Oh, you can just put your shoes anywhere."

Mark smiled, placing his own shoes on the bottom shelf. He then observed the hallway. He found that it was narrow, with a room halfway down, which Mark could only presume was the living room.

"Dinner will be about twenty minutes," Vanessa informed them. "I always make extra food so ye'll get a decent plate, Mark. I've been usin' the slow cooker today."

"I'm sure it'll be lovely, Mrs – erm…" Mark stumbled, looking at Alice for help.

Alice sniggered. "O'Riley," acquiring a glance from her parents.

"I'm sure it'll be lovely Mrs O'Riley," Mark finished.

Valerie O'Leary, who of course was playing Alice's mother, joined in laughing.

"Mark!" she said pretending to be annoyed, "call me Vanessa."

Mark smiled, relaxing slightly from her kindness.

"Now you kids go in ter the livin' room and watch something if ya like," Valerie instructed before heading towards the kitchen.

Mark and Alice wandered into the living room and decided upon *'The Simpsons'* as a suitable enough show for background noise.

"I love *'The Simpsons'*," Mark declared, trying to ease conversation.

"Ah me too, can't go wrong can ya?" Alice replied. "Personally, Ralph Wiggum is my favourite. He's so dumb!"

Mark laughed lightly. "True."

He noticed that on the double-seat sofa that they both sat on, Alice was sat not too far away—enough for him to smell her perfume and create a warmth inside his body.

He hoped Alice felt it too.

"Yer Mam n' Da seem nice," Mark commented.

Alice twitched the corner of her mouth.

"Meh, alrigh', I guess. Always embarrassin' me."

"At least they're both nice."

Alice looked into his eyes. She could see a sadness in them that suggested he meant more to it than one would initially have thought. Alice guessed he was indirectly referring to his father. She could deduce from Mark's outpouring earlier that he and John Formby did not see eye to eye, and as she looked at him, there was more than a professionally faked concern in her face.

She genuinely felt sorry for him. His whole demeanour screamed neglect and loneliness.

She put a hand on his arm, gently squeezing.

Mark looked down; he could feel himself blushing again.

He was secretly over the moon that she had touched his arm. It was nothing, of course, but for Mark, getting some form of affection from someone other than his mother was a miracle. Particularly with someone like Alice—a normal girl in a normal family. He was sure Alice thought nothing of it, but for Mark, that one touch meant a lot to him. It was not often he was given such treatment.

The twenty minutes flew by, and Mark soon found himself eating with Alice and her family. It was nothing fancy, but it tasted wonderful.

Cauliflower and parsnip soup, followed by a beefsteak, vegetables, and mash, covered in homemade beef gravy. Mark was not usually a beef fan, but the tenderness of the beef blew him away. Mrs O'Riley was clearly a talented cook.

Her family obviously agreed, silence ensued during the meal. Mark was happy that Alice's mother was a very fine cook, otherwise it would have been an effort finishing the meal; he would have hated to force eating something just out of politeness. But he had no concerns there. He devoured everything.

"This is delicious, Mrs- I mean, Vanessa. Thankin' you!" Mark exclaimed. Addressing her as 'Vanessa' seemed better since she told him to do so.

Vanessa looked at Mark, wondering if there was any falseness to his manner, but she could see from his expression and empty plate that he was genuine. Her face lit up.

"Ah Mark, ya charmer, thank you!" she exclaimed. "I jus' chucked this into a pot and waited."

"Mammy always says a modest cook is always the most professional," Mark politely returned.

Alice smiled at Mark, giving him a subtle nod of the head. She realised he was a smooth talker when he wanted to be. It was attractive.

Evidently, this was a good thing to say. Vanessa turned her chair to face Mark fully, with a smile that stretched to both sides of her face and remarked: "Well ya can tell your Mammy she seems a very wise woman, from me! And ye're too kind! Quite the gent, ain't he, Eoin?"

"Indeed," Eoin, who was really Declan, agreed. "She always seems to cook better when guests are around. Otherwise, it's mediocre," he added with a grin.

Mark could tell there was nothing malicious in what he said. Alice laughed, Vanessa scowled, giving him a swift punch on the arm.

"Gobshite! I did not know we would have company!" she shouted bemusedly. "Better than that shit you serve… ya know Mark, I got food poisoning the last time this fecker cooked! And I was already ill that time, before eating. Can ya imagine! Poisoning an already ill woman!"

Mark laughed. "He has stiff competition to be a good cook to be fair," he stated, inclining his head towards Vanessa. He was amazed how at ease he felt.

"Oh Mark! Can I keep ya? Ye're much nicer than him," said Vanessa, tilting his head towards Eoin.

"Hey! Hands off my mother!" Alice shouted, pretending to be shocked. She was surprised to see Mark flirting with his 'mother' – at least, who Mark thought was her mother –Alice could see in him a cheeky, charming side to his character, which even out of politeness, was quite an attractive feature in him.

Mark seemed alarmed now though. He didn't want to appear like he was flirting with Vanessa—he was merely being nice. At least in his eyes, he was. Alarm grew in his eyes.

"I--I didn't mean… I…"

Alice intervened.

"Mark, it's okay, I know ye're were only being nice," Alice butted in. "Flirt with me next time though mister, got it?"

Alice made out she was being serious whilst somehow showing that she was only teasing. Mark was fascinated how she did this—she must have done some sort of acting classes or something, because she would be a natural at it, Mark thought. Now however, he could not hide his red cheeks from Alice and her parents at once.

"Oi! None of that flirty business in the kitchen, madam!" Vanessa declared.

Bet you're not saying that when you're screwing me Da on the kitchen counter, Alice thought to herself bitterly but painting a smile on her face for appearances' sake.

She had known her real father slept around, but she kept it quiet because it seemed easier. She didn't know if her real mother knew about it for one, and so she didn't want to hurt her. Plus, by saying something, she risked problematising the business. But that didn't mean she didn't secretly hate Valerie, who of course was pretending to be Vanessa in this scenario. At first, Alice resented Seamus because of it, but she realised that a guy like him would probably sleep with employees he was often in contact with, particularly since he was a man who did whatever suited him. And even though she did not want to think about it, she could not help but notice that women would be attracted to his boisterous, confident nature, not to mention his rugged good looks. A lot of women gave him the once over as they passed by.

Alice's acting in character for the case of the cover story slipped for a second. She could judge that by the way Mark was looking at her with a curious expression. She snapped out of her thoughts, annoyed with herself.

"Are we excused?" Alice asked.

She felt it would be easier to be in character once she was away from her 'parents'.

"Sure," Vanessa replied, also looking at her strangely.

She and Mark left, and she took him upstairs instead of the living room. She knew that by wooing him a bit more, she may earn brownie points with her dad. Alice knew, too, she would have to tone it down here; she did not want to overwhelm Mark. She noticed the troubled expression on his face as they walked up the stairs, which was a sure indication that he was not used to being with a girl. Not that this surprised her much. She was pretty sure Mark was a virgin, so she was in no way expecting to have sex with him—not that she would even if it was clear that he wasn't. It was adorable how innocent he was. His vulnerability made her smile. And it was not like they had to have sex; it was just what was normally what happened amongst teenagers. And, she thought, it was often part of the job description for spies. Not that she had had sex much anyway: twice for a sixteen-year-old wasn't horrendous. The first boy, Tommy Duncan, had been her first when she was fifteen, but she was going out with him at the time so that was okay, and he was the same age. The second time she was sixteen and it was at a house party. She felt embarrassed by the whole affair and definitely wouldn't have gone along with it sober. The guy was no looker.

But Mark was not a typical teenage boy who was fixated by sex. Alice guessed he did not even consider it; for him, the possibility of even making a friend was huge.

Maybe it was his humble nature she liked about him most. All she could tell was that she liked his personality. He was harmless, yet she could tell he was not stupid. No. He was sharp. Observant. He had noticed Alice's change of expression at the table, for one. Despite his humility and vulnerability, he was not childish or immature in many respects. Anything but immature.

For the purpose of the cover story, the Brensons had played safe. The walls were painted magnolia and a few band and TV posters hung on the walls of Alice's fake room. They were the only thing in the room which she had owned previously: a Paramore poster, a large poster of a posing Ryan Gosling, a field of flowers, and the Eiffel Tower canvas were all fairly old. She had grown bored of them years ago, and they were lying around in the storage in her real home, but since she found them to be in good enough condition, it was agreed that the posters were usable. Everything else in the room was cheaply bought, such as a second-hand bed, a cheap duvet set, and second-hand furniture. It was clean, tidy, respectable—just girly enough to be a typical girl's room. Alice noted something of surprise in Mark's eyes as he surveyed the room—perhaps he thought it would be more girly. Or maybe it was just the surprise of being in a girl's bedroom. Either way, the basic nature of the room did not contradict the 'not overly wealthy' aura Alice's fake family was attempting to portray.

A basic precaution, yet nonetheless essential.

Mark looked uncomfortable. He stood just inside the doorway with his hands in his pockets. He was fidgety, and his eyes were moving to and fro, looking around the room as if looking for any distraction. He could not focus them. She had felt that when it was just the two of them, he was more open—more willing to talk. Less concerned with being himself… or not himself, if being himself was what he felt was the issue. She had, after all, found out something about John Formby and Mark's relationship already. As soon as they were accompanied by Valerie and Declan in the car, Mark's shyness returned for a time and even when he spoke more, it was small talk. In public, alone, he was more talkative.

Was that the best way to effectively extract relevant information from him? To keep them away from each other's houses.

Only time would tell. If so, she would have to consider getting Mark to go on days out to Dublin city centre or something because they couldn't keep meeting in a field.

Alice star-fished backwards onto her bed.

"I've always wanted t'be Lenny Henry. You know that guy on telly? Gettin' paid for lying on a bed!"

"You're odd, you are," Mark remarked. He was smiling. She smiled in return but could not help but think 'you can talk'.

"Join me, it feels great."

Mark hesitated, but after a few seconds he walked over to the bed and fell back but in a rigid, prosaic way.

And then Mark gasped in horror.

Mark had gone down heavier than he had hoped, and the upward force of the mattress threw Alice off the edge of the bed.

"Oh my God! Are ya alrigh'? I-I-I didn't…"

"That's it! You're gettin' it!"

Alice jumped up from the ground and leapt over the bed at Mark. He found himself pinned down and unable to move in an instant.

No wonder I'm terrified of women, he moaned in his head but another part of him was enthralled.

The look on Alice's face showed anger, but it was an act, Mark realised, when she released him and laughed merrily. If anything, this made the whole situation all the more terrifying.

I really don't get girls, he thought.

Had he seen another side to her? Or at least a side to her that he hadn't thought existed before? They had only met twice but it still caught him off guard. Was she annoyed? Wasn't she? Did she even know?

Because he sure didn't!

Mark just stared up at her.

"Ah chill ya big baby," Alice said casually.

"Sorry," he replied flatly.

Alice decided to tone it down. Clearly, the whole jumping on him hadn't gone down well. Scared him off like a pigeon flying away from a chasing child.

She smiled at him gently, putting a comforting hand on his leg.

This sent him into overdrive.

He jerked away.

Alice slowly retracted her hand away looking confused.

"Sorry," Mark announced, "I'm just not accustomed to this kind of thing."

'Accustomed'? Who even says that? Alice thought.

"I didn't mean anything by it," Alice said seriously.

Mark looked at her, feeling more out-of-place than ever. He hadn't been sure whether she was coming on to him or not, but now she could see she was just being playful. Surely, she could see this was alien to him. Mark supposed she did… and that this was how she was attempting to coax him out of his shell.

"Why are ya so interested in me?" he suddenly burst out. "I'm not confident like you. I must be really dull to you, surely."

He could not help sounding self-pitying and paranoid. He realised it sounded pathetic, but Mark was genuinely confused now why Alice was paying him so much attention. *Was this how someone normally acted with someone they had barely met?* He didn't think so.

And then another thought hit him.

Was all this attention linked to John in some way?

It sounded crazy, but if John could know Alice, was it not possible she knew him too? Just a little bit? Aside from what Mark himself said about him.

There was no way he could find that out without asking her directly about his own father. If there was no link, which he doubted considerably, he would look even weirder. And he did not want that.

Alice shrugged. "You seemed nice," she said neutrally and straight away he could tell he had hurt her feelings.

He was so close to making a friend. Instead, it may be more prudent to say something else to get an answer. And then, without realising he had stated out loud:

"My Da saw us together."

He hadn't meant to say that, but he wanted to change the subject.

Her reaction completely shocked him. She shot upright with a glint in her eye.

Was that panic? Mark wondered. It seemed an odd reaction to have.

Alice recovered, but that flash in her eye was enough to suggest something was up. *What the hell was going on? How did Alice know John? Did she know him? Was she just unprepared for what he just said?? Was he reading far too much into this? Probably. Even so, it seemed an exaggerated response for expressing embarrassment.*

Or was it? Did girls react over the top with weird stuff?
Before he could think more on it, Alice replied.
"Oh yeah? What did he say?" She was trying to sound calm.

Mark could not help but denote some edge to her voice. Again, he wasn't sure, but she had definitely appeared to appear casual rather than actually *being* calm. There was a definite alertness in her manner now.

"Nothing other than he saw that I was talking to you."

He didn't want to bring up the whole interrogation incident where he almost struck his father unconscious with a vase. Mark's eyes narrowed a millimetre or so to study her, and there was that flash again…

It was like she was unsure of what to do or say.

Alice looked like she was anxious.

Chapter Ten

Mark was unable to sleep that night. The comfort of his bed was usually enough to achieve sleep from him, but his mind was too active with wonder to even consider it tonight. He had not once considered before this evening the possibility that Alice had known John in the same way John seemed to know Alice; that flash in her eyes must have meant something. If Mark had more experience being around girls, perhaps he would be able to read her better. But he could read expressions, and it didn't take a socialite to notice that he provoked a response, a reaction from her. Girl stuff aside, that flash could not have been coincidental. It was instinctive, something that can't be acted away or disguised. Mark talking to Alice had provoked a reaction from both John and Alice alike. Something was niggling inside of him. Something didn't add up.

Saying that, as corny as it sounded, Mark had never felt about anyone the way he did about Alice. He was emotionally numb to everyone usually. Yes, he felt empathy, anger, and whatever else towards certain people but nothing that scratched more than the surface or made him think in ways which were alien to him. He had never cared about how someone felt towards him other than Mammy – he had no friends to think about. Perhaps that was a coping mechanism, a way in which to live without the help of kids his age. Was how he was feeling towards Alice misconstruing his thoughts, and making him irrational? Was the whole 'flash of the eye' merely a result of Mark saying something unexpected? Wasn't that most likely? He had not considered that Alice might be reacting because she was embarrassed that the father of a boy, she liked had caught them talking. That was mainly because he did not think she did or would like him. She was still young enough to blush at that kind of thing, Mark thought. Assuming she did like him, of course. Was not thinking about how Alice may or may not be feeling towards Mark a barrier? A defence mechanism to protect himself from getting hurt. That was also possible.

None of it made sense. It felt like his brain was a court case, as if the prosecutor and the defence were fighting each other for the truth and he was the jury who could not decide which argument held the most water.

Mark's head started to hurt; a dull, aching pulse in his temple which stemmed to the back of his head. Realising he would not figure anything out tonight, he admitted temporary defeat and tried to fall asleep.

Annoyingly though, the headache prevented sleep.

"Just what I need," Mark mumbled.

Mark stared at the ceiling when a soft tap on his door stirred him.

Mammy stood in the doorway.

"I noticed you weren't asleep. Y'all right, hun?"

"Yeah, just can't sleep," Mark whispered back.

Mammy sighed, then came over to sit on the edge of his bed.

"How was dinner with Alice?"

She smiled that mischievous smile Mark knew all too well—particularly when she teased him. Mark anticipated where the conversation was going. She was going to go on about Alice, making a big thing about this dinner being the first date he had ever been on. Mammy's curiosity irritated him at times—she always wanted to know his inner thoughts like any mother, a maternal instinct. Sometimes she would even try to tell him how he was feeling which was annoying. Mother knows best and all that. Besides, Mark wasn't even sure whether or not tonight's dinner had been a date or not. As far as he was aware, dates did not include parents having dinner with you. But that wasn't why he was unable to sleep. Not strictly speaking anyway. He didn't want Mammy to know what he was mainly thinking about though, so he was going to play up on the whole date thing.

It was the easiest approach.

"Dinner was fine," Mark replied.

"She must like you to invite you for dinner," Mammy whispered encouragingly.

"I suppose," Mark answered, "I don't know if it was a date though. She was probably being nice and felt like she had to invite me or something."

Mammy rolled her eyes. "Stop thinking no one likes you, Mark," she said with a serious edge.

"I'm just being realistic. I'm not exactly filled with friends. Hardly attractive."

Mark said this to play up on the 'date' keeping him up, but he realised there was an element of truth to his words. Actually, it was more than an element, it was the whole truth. Sooner or later Alice would notice just how undesirable he would be as a boyfriend or even just a friend, and she would soon lose interest. They had exchanged mobile numbers just before he got out of Alice's parents' car, but he was dubious as to how much contact the two of them would make.

He wasn't exactly suave, cool, easy-going, or talkative—the necessary characteristics to have a chance at making a friend… or more. He was annoyed how he had acted in her bedroom too.

"What isn't attractive is thinking it's not," Mammy responded.

Mark supposed that was true.

However, he wasn't even sure whether he actually liked Alice or if he was just infatuated with a pretty girl giving him some attention.

Let alone worrying about how she might feel about me, he mused.

"Don't lose sleep over it, sweetheart," Mammy piped in, "girls can be hard to read and often don't know how they feel themselves, let alone wondering if the guy likes them."

Mark looked up at her. He had never really thought about talking to Mammy about girls. It had just never occurred to him. He had never liked a girl in the same way as Alice, so it was probably for that reason that they had never really spoken about it. But Mammy was a girl, or at least was a teenage girl once upon time, surely, she must have some insight into girl behaviour.

"Really?" he asked, genuinely interested.

"Well yeah," Mammy said, as if it were obvious, "don't be fooled into thinking girls always know what they want. But no girl invites a fella over if they don't like them. Polite or otherwise. They just don't," she added with a smirk.

Mark laughed.

"Nice way to make me feel more comfortable around girls, thanks," he said jokingly.

"Anytime."

Mark reckoned his mother had been quite a mischievous teenager judging from her quirky, humorous comments every now and again. There was an impish, childish side to her, and it was that which often allowed her to be all the friends he could possibly need in one person.

Of course, it wasn't quite the same as having friends his own age.

But then again, he had never really sought any. As much as it would have been nice to make a close friend, he survived fine without one.

She could relate to his thoughts. She listened. She answered. She cared. It was enough for Mark.

As they shared a smile, Mark wondered if he could tell her about Alice reacting strangely and he nearly said something after a minute's deliberation when he noticed John standing at the doorway. Mammy noticed too and an alertness swept across her face like a child being caught doing something they shouldn't by their parents.

Even though she was one of the parents.

"What's going on here?" John asked. His eyes were steady and emotionless.

Mammy hesitated. She was uncomfortable. The teasing expression in her face had been rudely replaced with one of alertness... and fear? It troubled Mark.

"I went to get a glass of water. On the way, I went to check on Mark, but he was awake," Mammy replied.

John stared her down. His face remained still and in the faint moonlight that shone upon Mammy's face, he was peering at her, looking for something in her expression. The shadows emitted from the night sky outside wrapped around John's figure as he loomed over the doorway and for a moment, Mark felt something unsettling and almost monstrous in the way he stood there—so still with his eyes unflinching.

Mammy had shrunk away from his ominous gaze; she must have felt it too. There was something threatening in his demeanour, but Mark was determined not to appear shaken by this.

"Yes?" Mark quizzed him, attempting to sound confident.

John's eyes flickered towards Mark's face and the directness of his stare remarkably resembled that of a giant arachnid fixating on its prey, trapped in its web. Not that Mark knew for certain how spiders stared at their food-to-be, but if the monstrous spiders in *Lord of The Rings* or

Harry Potter were anything to go by, John's stare was strikingly similar as he stood in the darkness. The tone of Mark's question stirred a look of anger, and maybe even hatred as he glowered down at Mark.

Silence. Only the steady tick of the clock could be heard.

It was almost deafening.

The same searching glare was given to Mark, clearly hoping that this was enough to expose what might be hidden from him, Mark assumed. Ever since seeing Alice, and after the incident with the vase, John had been even more reserved despite the argument in the field. Mark wondered if he had continued to spy on him and Alice talking. It was evident that Mammy had not talked about where Mark was for dinner. Even John would have noticed his absence if it was commonplace for him to eat with them. He would come in, take his dinner, and bring it out to his study without fail. He would have noticed Mark not being there. He had no idea what Mammy would have told him. He felt selfish for not considering Mammy's position before now.

Could he have followed us to her house too? Mark suddenly thought.

Quite possibly.

John must have guessed that Alice would arise in a conversation between Mark and Mammy. Particularly if he had followed Mark and Alice. Was that why he was there listening in? Mark had not seen him approach the door. It suggested that he had walked slowly to the door, tiptoeing even, so as not to be heard. Which meant he was trying to eavesdrop. Which certainly meant he had heard something about Alice. Surely. Who knew how long he had been lurking in the shadows behind the door?

It was possible that he had heard everything.

If so, Mark was relieved he had not mentioned Alice's weird behaviour. He did not want Mammy mixed in with something that tested *his* patience. Furthermore, Mark had a hunch that if he described Alice's behaviour, John may interrogate him even more aggressively than the first time he demanded to know about Alice.

Clearly, Alice was bad news in John's eyes.

But why?

Nonetheless, John remained statuesque, but he had finally looked away from Mark's steady gaze. He took another scanning glance towards

Mammy and then looked back at Mark for a second before slowly turning to walk away.

Relief surged through him as he watched John's shadow in his room disappear. Mark then glanced towards Mammy, noticing that her face had turned ashen white.

And not just because of the moonlight shining on her face.

Across the field, Seamus and Alice sat discussing the meal in his office. Declan and Vanessa had described Mark's behaviour from their perspective and then Seamus had dismissed them. Seamus was distressed to hear that John had seen Alice talking to Mark.

He was wondering whether John recognising Alice led to him inspecting the house. If he did, it was evident that John knew about them. Recognising Alice could only mean one thing, surely, for a guard? And if he knew about them, it was safe to assume that he also knew about his lab.

Which would suggest that he had been sent to investigate, using his recuperation as a means of a cover story just like Seamus had considered.

But did he know all about his lab? Did John know he was running two businesses in his lab, one perfectly legal, and one unquestionably illegal?

Again, the fact John was a guard inspecting his house suggested that he knew enough.

It was two a.m. and Seamus sat in his leather swirly chair, gently swaying right and left. He ran a hand through his carefully groomed beard. It was a typical habit of Seamus'. Whenever he was deliberating something, he gently stroked his beard with a glazed look over his eyes. Alice often thought he looked like a philosopher or a highly academic professor when he did this as if he had a look of a man who was contemplating life's meaning.

He was logical. He was wise. He was decisive. And most importantly, he was ruthless and clinical. Difficult decisions were never made on the spur of the moment with Seamus; he profoundly considered every area and every viewpoint, so it was rare that he ever needed advice when making a decision.

Perhaps that was why Alice thought of him in such philosophical ways.

However, this particular predicament was unique even for Seamus. He was a man that took great care in going about his business. Every leak that could possibly expose his secret was sealed before anyone could get suspicious and start investigating him. It was his financial situation which was a massive factor in his ability to hide the truth in his home.

He had registered his 'lab' as a chemical factory which provided specialised medicines for local hospitals, veterinary practices, and small pharmacies in Dunshaughlin. He even provided chemical solutions to local high schools for their science laboratories. Health and safety officials, and specialist medical and chemist inspectors visited Seamus' lab each month to check everything was running as it should, given how important it was to the wider community.

And of course, it always did.

A government investment had allowed Seamus to convert his once large cellar into a huge underground laboratory. His business played an important part in the storage and accessibility of necessary drugs and medical equipment. The house attached was his own home but the entrance to the laboratory was so secure that virtually no sound could be heard from inside the house itself. It was completely soundproofed. On the far side of the laboratory was a large electronic shutter which led to a long parking bay. This allowed for the transportation of such medical equipment. As for Seamus' own car, he generally kept his main family car in the small garage on the outside of the house. He had cars such as an Astra and a Peugeot 306 in the parking bay, but no one asked questions. Anyone would assume they belonged to employees.

In the laboratory itself, there were many wide steel pipes which connected to the ceiling for any gasses and a large number of stowaway metal containers full of all kinds of medicines and substances.

He was always given notice of visits, meaning he had time to ensure that any evidence of cocaine and heroin production was hidden, and as such, he had a concealed, secure room in which he stored the evidence. Converted secretly himself so that the entrance looked like part of the wall. It was a remarkable set-up. So much so that he had workers who knew nothing of the illegal side of his business. There were a team of qualified chemists who dealt only with the registered chemicals. After they left, his other team began their work. They too all held PhDs and doctorates in chemistry, including Declan. And they were bloody good at what they did. They had to be. Each batch was over ninety percent pure.

But were they all loyal?

Seamus now doubted it.

Because of his registered laboratory, the guards were never involved. There was no cause for suspicion. How could there be anything dodgy about a registered laboratory that provided for hospitals, employing only specialists in chemistry? Any doubts would lead to a full investigation. And with all the necessary checks to the facility, who could point the finger at Seamus for the increase of cocaine and heroin consumers around Dublin and Meath?

Never before did Seamus Brenson have a guard come inspecting secretly. He had planned everything down to the last detail. Everything. His employees were only employed if he was certain they could be trusted. He paid them for the silence as much as anything. No leaks should have gotten out. No guards should be circling the house, sniffing around like a dog for clues. Now that Seamus knew John recognised Alice, he was certain that John Formby was suspicious of him.

No guard would know what Alice looked like unless they were investigating them. Seamus himself was not that well known either. If he was, that may have answered how Alice may be recognised as his daughter. But even that was unlikely.

Meaning someone in his trusted workforce told him.

Seamus stopped stroking his beard, troubled, and looked at Alice.

"You're sure Mark said John recognised you? I need you to be absolutely positive, Alice. This could be the most essential piece of information you can give me. If you're sure, it means someone in my team has exposed us." He paused, timing his last sentence with dramatic emphasis. "It would mean somebody being severely dealt with."

Alice gulped.

Seamus was usually a placid man, but for something as serious as this, all calmness was stretched to his limit, all his light-heartedness extinguished. Alice could see the dark and dangerous power in her father's eyes. There was something in them that chilled her to her core.

And she was truly scared of his potential at that moment.

In fact, more than potential. She knew what he was capable of.

"Mark said 'My Da saw us together'. If he is inspecting the house too... well it does suggest he knew who I was, don't you think?" Alice reasoned.

Yes, he did think. Seamus' pupils dilated. He was angry. Furious!

He was going to kill one of his employees.

And Alice knew it.

She looked at him with fright. She desperately did not want someone to be killed. No one deserved that. Especially when her being certain about something was the last nail in the coffin—she didn't think she would be able to cope with the guilt.

What if her dad was wrong? What if there was another explanation for John Formby knowing what she looked like? *I mean, he is a guard,* Alice thought, *isn't that his job to learn stuff?* She had to admit, there was something weird with John knowing exactly who she was, guard or not. There had to be a reason why John had learnt it after all. Her father's conclusion of a traitor was consequently understandable.

But was it incontrovertible? Alice was sure there was an alternative. She thought hard. She remembered John's name being all over the news, papers, and radio for a while. She herself had heard bits of the heroic story. She had even seen a Mark in one or two of the reports.

And that is when she thought of how she herself had been in the paper. Once.

When Seamus opened up the facility now known as BrenSolutions Ltd.

"Da, what if John knew me from his own searching rather than someone giving him a tip?" Alice asked desperately.

The anger that had set his eyes fixed on the floor whilst he thought furiously of who could have betrayed him now slowly turned up at Alice, his anger shifting more to a sternness. He could not remember the last time he had laid angry eyes on his sweet Alice.

However, he could not hide the impatient scepticism from his voice as he asked her: "What the feck are you on about?"

A thread of guilt trickled through him snapping at her, but he was too busy in thought right now to let that concern him.

"Do you remember when you set up BrenSolutions and one or two of the newspaper fellas came round to take your photo?"

"Yeah, what about them?" he asked impatiently.

"Well, I'm pretty sure I was in one of those pictures. I remember seeing myself in one of the newspapers. Well, what if that is how John knew my face? I mean, he's a guard, isn't he? He could easily have found that article if he dug hard enough. I don't know why he would, but he could have done."

Seamus' look of fury subsided slightly. Instead, he was looking away from Alice, eyes flickering as he thought this through. It was true—Alice's picture had been published in one of the local newspapers, he remembered. *The Meath Chronicle*, wasn't it? Yes, it must have been that one. Indeed, John could have searched local newspapers to search for the Brensons. But the same worry still lingered. Why did John Formby first get suspicious of something in the first place?

There had to be a reason, a motive. Surely a breach in his own workforce was the most likely and possibly only reason? You only search articles as a guard if you are unsure of something or investigating something. And to investigate something, there was always a reason why you did so in the first place. Nonetheless, it gave Seamus something to think about… and now, even Seamus himself was less convinced he had been betrayed. As ruthless as he was, he was fair. He would not punish until there was a one-hundred-percent certainty. If not, how could he be successful if he was seen to distrust so easily?

He looked up at Alice, nodding gently.

"Maybe, clever girl," he mused with a half-smile. There was no one who could call Seamus Brenson's daughter thick.

Despite her perhaps being right, Seamus knew there was still a strong possibility that she was wrong and that he had, in actuality, been betrayed.

In his head, it still seemed the most likely.

Chapter Eleven

John walked away from Mark's room feeling as though there was something that did not fit. He knew that the Brensons lived in the house he had managed to look round. That was a fact. Seamus Brenson was the name attached to the laboratory which administered medical supplies to hospitals; he was the owner of the facility. He had managed to fish out several newspaper articles with Seamus' picture in to prove that he lived there. In one article, there was an image of Seamus standing outside his house – clearly the same house he had inspected – shaking hands with a hospital representative. Another showed a small image of part of the laboratory, described as being 'underground from the main family home with a drive-through at the back of the laboratory for vehicle access to and from the facility'.

It wasn't exactly first-page headline news, but the fact that a medical laboratory existed there was known to the general public of Dunshaughlin. But John knew too there was a secret being hidden. A secret disguised by the legal storing and administration of medical drugs.

He knew that Seamus Brenson also produced Class A drugs. Not in particularly large quantities maybe, but definitely enough for his product to raise a fuss. Producing too much would be risky.

It was no coincidence that John Formby had moved to a home so close to the Brensons. As soon as he recovered in hospital, he searched for a house that was 'for sale' in or at least nearby Dunshaughlin and he happened to come across a house directly opposite. The chances of that happening were slim, especially in the Irish countryside where houses were scarce. It was almost unbelievable.

And yet, it came to be.

But for John, it also felt like fate.

However, he had never anticipated Mark talking to Seamus' daughter Alice. Mark never spoke to anyone his own age, at least that was what he had always assumed. If it had not been Alice, John may even have allowed enough room for respect towards Mark; he'd had similar social problems himself all his life, although he always seemed to mask it. He could appreciate the hardships of talking to a girl.

But it *was* Alice.

And that changed everything.

John had not believed what his eyes told him when he had seen Mark and Alice together. He had been driving along the country road, just out of Mark's sight, when he saw two teenagers talking in the field. He could tell it was Mark even from that distance. Call it paternal instinct. In the passenger seat compartment was an old pair of binoculars he used sometimes whilst on duty. That was how he could spy on the two of them. He could discern enough of Alice's face to know it was Alice. He held a few pictures of Alice in his possession at home, in a place where neither his wife nor son would find them. This was his task, and his alone to know about. Again, they were cited from newspapers and e-journals, particularly the medical-related publications.

Considering the two of them were so close to the Brenson household, as well as the fact that you just did not see kids hanging out in country fields - particularly those who looked like Alice - it was clear that it was Alice Brenson. There was no mistaking it.

The whole situation became complicated from that moment onwards.

After the failed interrogation attempt of Mark in regard to Alice, he decided to take a much quieter approach.

Silence was often the best option.

It was a trustworthy friend.

So, he had decided to follow Mark, unobserved, when he could. Being a guard, this was not a demanding task. It was pretty much common sense too as well as second nature to him. Hire a car if need be, wear a hat of some sort and remain at a distance. They were the basics to spying on suspects from afar, although hiring a car was not always necessary. John knew this, but obviously Mark knew his Mercedes well. If he saw a Mercedes stopped in the distance at any point, he would be alerted. Fortunately for John, he knew the manager at Hertz Car Rental at Dublin Airport. Niall Kiely helped him in an immigration-based case five or six years ago, whereby a dangerous ex-convict who had been suspected of being involved in an armed robbery, rented a car from Niall under a fake name to make an escape. CCTV confirmed the man they were looking for was the man assuming another identity. Kiely had made an instant impression upon John. He was a small, middle-aged man, with receding grey hair, and at first glance there was nothing particularly striking about him. However, there was something in his eyes that was shrewd and intelligent. John's habit of reading a man's eyes before listening to his words led to John's appraisal of Kiely, straight from the off. John could not to this day put his finger on what it was but something about Niall's demeanour pleased him. He remembered their first discussion relatively well.

"This man here," said John, pointing at the man in question taking up a whole piece of A4 paper, "you're sure this is the same man who signed his name 'Nigel Murphy'?"

Niall had given him a curt nod with curious interest lighting up in his eyes. "Yes," he said, "there is no denying it. Look, four and a half minutes later he is signing his name." He pressed fast-forward on the CCTV recording before pausing when the man signed his name. "I record every name and time of every customer I get, Inspector Formby, and I am quite sure my own recordings match that to this man. I can go and prove it for you if you like?"

"Please," John had replied, quietly impressed. He had eyed Niall with interest—he liked anybody with pristine organisational skills when not entirely necessary. It added a quality. A kind of classiness that demanded respect, he thought. It said a lot about a person, and he liked being able to study people with ease.

Niall was his kind of guy.

It was his liking of Niall that was John's main reason for approaching him. If he explained to him that it was work-related, he was sure there would be no problems.

So, after seeing Mark and Alice talking to each other, and when he decided what he was going to do, he had gone to pay Niall another visit.

"Hello Mr Kiely," he said after opening the door of Hertz, "long time, no see." A rare, genuine smile stretched slightly across his face. John did not hand them out easily. He could see Niall busily organising documents into assigned folders.

Niall looked up. Watching Kiely sort documents seemed to clarify for John his contentedness in approaching Niall.

"Inspector Formby!" he exclaimed, smiling pleasantly, "this is a surprise. Come in, come in! How are you? Would you like a coffee?" John could see there was wariness in his smile, and his eyes did register wonder but that was probably due to John being a guard. It was understandable for him to be on edge.

John chuckled at the effect he had.

"I'm grand Mr Kiely, yourself? No on the coffee front, thanks. Never been my cup of tea if you will excuse the pun," he remarked.

Niall sniggered. It seemed to relax him. He had a pleasant, monotone-sounding laugh but it suited him, John thought. "Not too bad, Mr Formby—sorry, I mean Inspect—"

"Call me John. This isn't a formal visit."

Niall's face visibly relaxed this time and John thought it knocked several years off his face. He figured he was having a stressful time of things recently. But as much as he respected Kiely, a rare thing for John, he was not here for a social call. Not really. It was still business as usual.

As if reading his mind, Niall asked: "What can I do for you, John? Wanting to hire a car?" He laughed as if he had just told an inside joke. A patient expression met John, and Niall's grey eyes twinkled kindly in the daylight.

John smiled. "Yes, in fact I am. That's exactly what I'm here for."

A surprised, curious look swept across Niall's face. "Oh, right!" He was silent for a few seconds as if trying to work out an answer to a hard question. "Well we have a large selection for you to choose from. Was there any car in particular you had in mind?"

"Any. It's for a case I'm working on," John lied. "A rented car would be best for this particular case. Preferably the cheaper the better," John added vaguely.

"I see," Niall replied, looking at John with keen interest now. "How come you chose here if you don't mind me asking? Surely there are car rentals closer to where you live than the airport. I hear from the radio you're living in Dunshaughlin, if I'm not mistaken? I'm sorry to hear about your accident. All better now, I trust?"

Again, Niall demonstrated shrewdness in his character. Even if it was a known fact where John lived, it showed Niall did not miss much. Not many, for example, would still remember he lived in Dunshaughlin.

"You're right, Niall—may I call you Niall?"

Kiely nodded. "Of course."

"Well Niall, I guess you could say I am not really a fan of strangers. At least I know of you—I'm more comfortable about that. And it was only a twenty-five, thirty-minute drive down. No biggie. Plus, I admire how organised you are. I guess that is comforting for me. I can relate to that," he added.

Niall's cheeks turned light pink.

"Why thanks. That's very flattering."

Inside, John began to squirm. He had said too much. It was uncharacteristic of him to be so kiss arse. This was getting a bit too clichéd and complimentary for his liking, and he hated smushy-ness.

On the outside, John smiled politely.

"Well," continued Niall, "you say it is for a case you require a car?"

John nodded.

"Then I would suggest choosing a well-known, branded car. Something mass-produced. No one is likely to get suspicious of your everyday car. I have a range of models from Ford, Vauxhall, Peugeot, Toyota, Renault and others for you to choose from."

John thought about this a moment. He had not really thought in that kind of detail; he was good enough at spying at a distance that an 'unsuspicious car' did not really matter. But he figured that for this case, such a car would be ideal; particularly since he was hiding from people who might be looking for him, and not necessarily just him looking for them.

"I like your thinking, Niall," he said.

Twenty minutes later, John drove away in a three-door Ford Fiesta.

A day after getting a car from Niall, John had followed Mark all the way to the house he had dinner at with Alice from a distance. And that was the most confusing part of the whole thing.

He knew that it was Alice Brenson. And he knew that she lived in the house opposite his own. So why was her 'house' different? In other words, why was she being driven to another house which he knew for a fact was not the Brensons' house? The two adults who got out of the car at this other house did not seem familiar to John either. He could only get a second's glance at them from a distance, but they did not seem to match the description of Seamus and Kirsten Brenson, especially Seamus. John's records showed, as did the pictures in the newspapers, that Seamus was roughly six-foot-three, maybe four. This man who got out was nowhere near that height—even from the couple of hundred metres away where John was, this man was at the very most five-eleven/six-foot. A wide, medium tree got in the way of John's view after that point. But it was enough for John to know that this was not Seamus. And that meant something was being set-up. And this worried him.

This worried John a lot.

Clearly, whatever was happening was planned. And if it was planned, Mark was a part of said plan somehow. Was Seamus Brenson trying to divert attention away from himself? Did Seamus know John was on to him?

That night, John crept behind Mark's bedroom door to listen, if he could, to Mark and Sylvia. Without too much thought, he went off to the kitchen to get a drink, and after bringing a Jack Daniels bottle with a few cans of Coke with him, John wandered off to his study. He always worked best there with a stiff drink.

Mark talked about Alice to Sylvia as if he and Alice were dating. The fact he mentioned the name Alice reaffirmed what he already knew: that this girl was definitely Alice Brenson. His confusion slowly shifted as he began to think about it... and suddenly, things began to become clearer.

John figured he was known for his 'heroics' from the media—and if someone like Seamus Brenson knew that he therefore knew John was in the guards. That much was not hard to work out.

But if he knew that, which he inevitably did, he would subsequently know that Mark was his son.

So, when Mark went for a walk, it was likely that Seamus was on the watch. It was therefore not too much of a leap to suggest that Alice, his daughter, could be exploited. A man like that would have no trouble demanding that from his daughter. Because she and Mark were of a similar age, was it not possible that he had gotten Alice to spy on Mark in order to extract information about John himself?

It was certainly a possibility. A strong possibility.

But why bother? Just because he was a guard who had moved into the area. Was it a kind of precautionary thing?

Then the answer to that came to him too. It was the most obvious answer so far. So obvious that John swore at himself for not thinking about it as soon as he did.

He had been seen inspecting the Brenson household. Obviously, there was CCTV or something—why had he not thought of something so blatant?

Because you're reckless, John, a voice told him. *You act on instinct too often without thinking it through. Just look at the reason why you landed in hospital with five bullets in your body. You're so obsessed with the idea of your plans that you don't actually plan it out properly. Not all the time. Obsession makes you careless, and some day it will cost you.*

He was furious with himself... and angry at the voice in his head. Mainly because that voice made sense.

And the more he thought about it, the more *everything* seemed to make sense.

He downed a shot of JD before adding Coke to his next serving.

He sat down, gathering himself in a more rational, less panicked mode of thought.

He realised investigating the Brenson house was a stupid move. But he had to try to get something on them somehow. He wasn't going to do much sat around. Nonetheless, he grabbed a notepad and pen from a drawer and began to write down what came to mind. It was a force of occupational habit that helped to channel his thoughts more clearly.

He continued writing: ... *Let's assume Brenson is onto me, or is at least wary of me, how much does he actually know? Mark will know nothing. That is good. Seamus won't know what my plans are necessarily, even though I made the stupid move of wandering around his property. There is no information Mark could accidentally pass on to Alice... surely not. He is probably still being encouraged to voice his thoughts about me, though. He is probably being wooed by Alice on Seamus' command. And who knows? Maybe he has told her that I'm a 'useless father' and all that shit. I don't care about that—he can bitch all he wants. But Mark did spy on me whilst I investigated the house. That means he knows I'm up to something, right? Shit... what if he was told to spy on me? ... Working against me. Watching my every move. It would make sense—is he really that intuitive that he would follow me just out of curiosity? I doubt it. He must have been influenced... he must have been persuaded to do it...*

The thought that Mark was being used to spy on him chilled John. But that was just paranoia talking. Right?

Things were never supposed to get complicated. Everything was supposed to be running smoothly. Just him and his old binoculars, waiting.

Waiting to make his next move.

But what about before that? he wrote. *I only inspected Brenson's house after I saw Mark and Alice, and after questioning Mark about it. Alice would have been spying on Mark to get to me before they knew the two of them were even talking. So... if Seamus was using Alice to spy on Mark to get to me, surely me inspecting the house has nothing to do with Seamus' plans? Yes, I was an idiot—it probably made him all the more determined to know what I am thinking, but his reasons for wanting to know what I am thinking are not as a result of me lurking about his place... not if Mark and Alice's first set-up meeting was before I went inspecting...*

Could Alice have convinced Mark on that first meeting between the two of them to watch me? Seamus would know I was a guard, that's definite. That news was not exactly secret with the media about. Is it simply the fact I am a guard then that he wants to check on me? Check the road is clear.

This is also possible.

So, what does he want? What is he up to? More importantly, what is his goal? And what about this second house?

John laid his pen down then. He read and re-read what he had just written, satisfied that he had managed to express his thoughts as clearly as he would have liked to. He decided writing was a good idea for the future—for a guard it was a good idea anyway, but in order for him to avoid being reckless again, writing down his thoughts seemed like a good way to control his actions.

And, John realised, that that could ultimately make all the difference.

Chapter Twelve

Mark awoke the next morning later than usual. As often happens when falling asleep much later than expected, Mark overslept, meaning it was almost noon by the time he awoke. He groggily got out of bed, groaning with displeasure when a cool rush of air hit his body after removing the duvet. It was a horrible feeling, particularly when Mammy came in on those days when he was in danger of being late for school and so removed it for him unnecessarily quickly, as if she were performing the 'remove table cloth from a set-up table' trick. That was Mammy at her cruellest. Mark figured he could live with that, though.

Today, however, Sylvia had not disturbed Mark. Perhaps because she knew he had trouble going to sleep last night, and so wanted him to catch up on sleep. Still, she must have heard Mark getting up because by the time he had staggered sleepily to the kitchen to make himself a morning (almost afternoon) coffee, she had made fresh toast with a cute little jar of raspberry jam beside it. He smiled because he knew it was homemade—making jam was one of Sylvia's weirdly obsessive hobbies. And also, he appreciated the effort she had gone to; the toast was still warm, meaning she was waiting for him to get up.

"Someone's up early," she murmured.

Mammy raised a playful eyebrow at Mark from her seat at the kitchen table.

"Cheers for the toast," he replied, not awake enough yet to respond with something witty.

"That's alrigh'. Sleep okay in the end?"

She was more serious now, a tone of concern in her voice.

He nodded, looking down at his breakfast.

Sylvia looked down again to read her newspaper after giving Mark the quick motherly once over. He noticed a small frown on her face though as she resumed reading, and he knew she had spotted something in him that didn't satisfy her. It wasn't from something she had read.

"Someone's grumpy."

There it was.

"I'm fine," Mark replied, getting irritated. She was annoying when she was like this, especially when he was sleepy. He was not a morning person.

"Have you any plans for your half day?"

She was smirking now. Mark knew she was doing it on purpose to wind him up, but he didn't want to be rude either. He rolled his eyes, sighing at his need to be nice.

"Not really, I might go on the Xbox for a while but other than that, nothing. What are you doing?"

"I was goin' to ask if you wanted to go to town, Dublin like, and go for a bit of a shop. 'Suppose you're fed up here all the time, right?"

It was true. He was finding it increasingly more tedious in the country with very limited things to do. Even fewer things when it rained. Maybe a day out with Mammy would be good for both of them, especially now she was also getting bored being trapped indoors all day. She had a car, but she didn't really have any reason to go out other than for the weekly food shop. Mammy had been working in an office – something to do with finance – before John's accident, but since then she had not returned to work. She explained to her boss that John was in a critical way and that she would be of no use to the company in the state she was in. In hindsight, Mark figured she was right in that assumption, but he secretly wished she would return now that things had settled down. Mark worried that she was being held up by John's treatment of her and thought that going back to work would at least give her a daily routine, something to think about other than life at home. As far as he was aware, her work was pretty reasonable about the whole situation, saying she could return whenever she felt ready. But in the meantime, a trip with her seemed fun—that would be what he would focus on. Mammy would know when she was ready to go back.

"Sure. I'll get ready. When were you thinking of leaving?"

"No rush, sweetheart. Within the hour okay for you?"

"Yeah."

"We haven't done anything for a while, it will be nice to get out for a change."

She seemed genuinely happy now she was going somewhere. It was nice to see.

Half an hour later, Mark was ready. It was a blustery August day, so he stood in the hallway with his windproof North Face coat, shoving his hands in his pockets to check he had his wallet and phone. His phone, as of last night, had become his most prized possession as soon as he and Alice had exchanged numbers. So far, he was too cowardly to text her, but he still felt he had to have his phone there with him anyway, as if he might take the plunge out of the blue. Or if she texted him. But he was pretty certain she wouldn't text first; he was dubious if she would ever text him full stop. Mark didn't know much about girls, but he knew enough to know that it was like a rule that girls didn't text a boy first.

He was disturbed away from these thoughts when Sylvia appeared at the top of the stairs. Her mousy brown hair was tied up in a neat bun, and she had applied a light coverage of make-up on her face—even a light shade of red lipstick on her lips. It was a strange thing for Mark to see, particularly since she had not worn any make-up for many months; this had allowed Mark to grow accustomed to her lack of make-up as being her normal look. Mammy winked as if it was her sole intention to show off the neatly drawn on eyeliner to make her eyes bolder. There was even a touch of greenish glitter. It oddly suited her, without looking tacky or trampy. It was just the right amount to keep the look classy.

It only occurred to Mark then that this trip meant more to her than he realised. She was making a real effort. She was ready to go out shopping, something that was an everyday thing for most women, but it seemed an occasion these days for Mammy. She was acting like a typical woman and Mark could not help but feel a mixture of relief and happiness as she used the staircase like a catwalk.

"And here, ladies and gentlemen, is Sylvia Formby strutting her stuff in the latest summer look from the Next collection," Mark announced into a pretend microphone.

Mammy pouted, turned around dramatically, walking up one or two stairs before stopping again. She whipped round to face Mark in an over-the-top way. The hyperbolic pouting reminded him strongly of Ben Stiller's performance in '*Zoolander*'. He chuckled to himself.

"Hang on. Next? You mean the young Next models right, not the sixty-odd-year-old ones?"

"I'm saying nothin'," Mark grinned.

"Oh, you lil…"

She ran down the stairs, towards Mark, but he was too quick, hurrying out the door for her to catch him, laughing on the way out.

Today is gonna be a good day, thought Mark.

"Oww!"

Mark looked back – Mammy had run out after him forgetting she didn't have shoes on and the gravel under her feet pricked the soles of her feet, which sent Mark into another round of laughter.

Eventually, both of them got into the Kia Picanto that had not been driven in for almost a week. They were shortly on the country road and as they drove past, Mark absentmindedly glanced out of the window, coincidentally looking at the house John had inspected. Mark did a double-take at the top window. He was sure he had spotted something in the window. As he looked properly, squinting to try and see from the distance, he saw something move. Was it a person in the window? A curtain? A cat? There was definitely something. Sylvia had noticed him looking closely at the house.

"What you looking at?"

Mark looked away, but still not that focussed on what she had asked.

"Nothing. I just thought I saw something in the window of that house Dad was looking at yester—"

Shit!

"John was looking at that house yesterday?" Sylvia asked, confused.

He had not wanted to tell her anything about the events yesterday, other than about Alice. But he would not be able to get out of this one now. Reluctantly, he told her about John investigating the house the day before and the argument they had afterwards. Instead of his explanation making everything clearer, Sylvia only seemed to be more perplexed.

"Why would he do that?" she asked, momentarily distracted from driving and having to swerve sharply back onto the stony road.

"I don't know," he replied honestly.

He was still annoyed at himself for being careless, so he stared at the road immediately ahead as he answered her questions.

Silence followed. Mark waiting – and hoping – for a change in topic. He looked at Sylvia as she drove and saw she was staring ahead in concentration, not from driving since they were still on a straight, empty road that required minimum focus.

He shouldn't have said anything. He didn't want her to get involved; the last time she had; John had tried to hit her. Mark had no idea what he was going to do next in his attempt to work out what John was doing and why in particular he was so invested by the house across the field, but he did know that the less Mammy knew, the safer she would be and the quicker Mark could go about his own investigation. It would be more difficult to monitor John when he started school at the end of August – even he couldn't avoid not going to school – but he would still be able to do some digging, despite having less time to do it. That was why he was diligent in getting to his set task now, so he could use his time well while he had a lot of it. He wondered if Alice would be at the same school. It seemed unlikely but he did wonder all the same. Not that he would think about that too much right now; he could always ask her himself whenever or if he texted Alice.

"What do you think your da is up to?"

Mark snapped out of the blank stare he was giving the road ahead, looking at Sylvia. He had not expected the question, or rather he wouldn't have had he not told her about the house incident. He presumed that she had not been paying too much attention to what John was doing, but thinking about it, Mark thought it didn't take much to notice that John was acting peculiarly. Mark didn't want her to pry or asking any questions. It was better that way. Safer, too.

But what is better doesn't always work out does it, Mark?

No, he supposed not. He didn't want to lie to her either—that would just make her more determined to find out what was going on. But that was the trouble. He didn't know the truth.

"I couldn't tell you. He's up to something though, I know that much," he said carefully.

Better to appear like he was answering her questions as best as he could otherwise, she may start taking matters into her own hands. Sylvia was like that. Once she suspected something, she wouldn't stop until she got the answer. Ironically though, she didn't bother with John. She just left him to it without resistance. She must have thought it was easier that way after years of probably getting nothing out of him anyway. Probably wise, thought Mark.

They were stationary at a traffic light, the first car behind the line, waiting for the green light. Sylvia was drumming her fingers slowly on the steering wheel, appearing to not be thinking of driving.

"I don't like it, Mark," she suddenly announced.

"What do you mean?"

"Everything. He is acting weird, more reserved than ever, and now you're telling me he is going off searching empty houses for no reason. And I don't like how he is with you. I saw the way he looked at you the other night when you hit him—it scared me. I don't want to think you're in danger around your own father but—"

"Mammy, it's fine. You don't have to worry about me. He barely even looks at me, anyway."

She looked sad, concern clawing at her from inside.

"I know. I'm sorry about that," she said, looking at him sympathetically.

"I'm fine with it. Got over it long ago," he replied. He gave her a reassuring smile.

He was hoping the conversation would stay as it was now and not about what John was up to. He much preferred the idea of a conversation about a neglectful father rather than what that neglectful father might be doing.

"It's not fair for him to punish you just because of me cheating on him," she moaned, "that's when he started acting indifferent towards you. It's not like you're not his son. You're definitely his son. And it was a long time ago."

Mark looked down quietly. Even with Mammy, this was an awkward conversation to have and she rarely spoke of the affair.

"Everyone makes mistakes," he said, hoping to be helpful.

"Yeah..." She stared off into space then, troubled.

"I just thought he would be better by now. I... I thought he would be back to the way he was before we were married."

That stirred Mark's curiosity. He had never really thought about what John Formby might have been like before he married Mammy.

"Was he different?" he asked.

She smiled a little. She began to nod slowly, lost in a world of nostalgia.

BEEEEEEEEEEEEP

Sylvia started. She whipped her attention back to the road and noticed the green light was glowing. The car behind gestured at her impatiently. She drove onwards, raising a hand in apology.

After a minute she spoke again.

"Yes, your da was a different man back then, before we were married. Even before he joined the guards. I met him in a bar in Galway. I was there because I went out to see my friend from school. Maura Lane she was called. She moved to Galway soon after she got her exam results to start a new job in some office. Anyway, your da came over to me at the bar and started talking to me. He had a brilliant smile and his green eyes glistened…"

Mark coughed loudly.

She laughed.

"He was a friendly person back then, always making jokes and not taking life too seriously. I was only nineteen at the time, your da twenty-one. He was working as a pizza delivery boy in Galway on his gay little moped that he had."

Mark burst out in hysterics. The idea of his father being some spotty pizza boy, just out of adolescence, delivering pizza was outright bizarre.

Sylvia laughed along for a few seconds.

"He was living with his nanna—he moved in with her after being chucked out by his parents when he was nineteen. He lived in Kildare growing up, as you know. I've only ever seen John's parents once. At our wedding, and that is only because I made John invite them. John wasn't happy but back then he was a different man—he didn't dispute what I had to say."

Mark raised his eyebrows in surprise. Sylvia noticed.

"I know," she said with a bemused huff, "who would ever have thought it? But yeah, we were surprised that they actually came to the wedding, and I kind of regret inviting them. They were an unsmiling, unfriendly couple and were very strict Catholics. I think they came because of some religious duty, perhaps because we married in a Catholic church. It was John's choice—he wasn't religious himself, but he was a man who believed in church weddings. Again, you wouldn't have thought that of him, would you? He convinced the priest, Father Brennon he was called, that we were both practising Catholics, otherwise I doubt he would have married us. It's weird how one experience at a bar can shape the rest of your life."

She looked at the traffic ahead thoughtfully.

Mark did too. Mammy had never gone into how she had met John, other than meeting him at a bar—he vaguely remembered being told that once. But the rest of it, Sylvia never discussed. Mark never thought to ask, which he found slightly odd. He supposed he had just taken it for granted that John was always as he was now, even if it did seem surprising how Mammy could ever have wanted to go out with him. At least some things made a little bit more sense now.

"What were his parents called?" Mark enquired.

"His father was John too, his mother's name was Anne," she replied.

Mark was curious about his grandparents. He had not met any of them and it was something which occasionally made him feel sad, but less so as he got older. It was something he was used to; he couldn't miss someone he had never met, after all. Sylvia's parents had passed on before Mark was born, and he had never asked about John's parents—Mark did not think he wanted to, judging by how John was with him and they never made the effort to form a relationship with him either.

The rest of the journey drifted into a natural silence with the hum of the radio being enough to substitute conversation. He couldn't think of anything else to ask about his grandparents. After Sylvia had parked and the two of them had wandered in and out of several shops – none of which were of any interest to Mark – they wandered into Starbucks. With a hot chocolate and a latte in front of them, Sylvia brought up the topic of Alice. The café was not particularly busy, but it still made Mark embarrassed; he did not like the thought of someone overhearing about a girl he fancied. He couldn't deny that he fancied her either; he realised that he really did like her. It was a strange feeling since he had been so indifferent to girls before now—they were nothing but an added complication to his already well-established belief that he could not socialise. Never had he cared about his mobile either; the only person he called or texted was Mammy yet he still found himself in possession of it most of the time. The old-fashioned reason of 'take your phone in case of an emergency' rang through his mind if he ever had to go anywhere, which he knew that was not normal for a sixteen-year-old, and it did add to his feelings of being an outcast. He didn't mind most of the time, but it did make him feel lonely occasionally.

Now that he was a little less lonely meeting Alice on more than one occasion, he didn't know what course of action to take. He knew he wanted to text her, but he was afraid of rejection and humiliation. Was that fair to say? Mark thought so; again, he could not deny the desire to talk to her… and he never had that urge.

All of this was in his own mind, of course, and that was how he liked it. There was no possibility of being laughed at that way, no way anyone could see what he was thinking and judge him. It made his own thoughts a powerful weapon, something that he could cling on to. His private nature did not appreciate being prodded to speak its mind, but that was what Sylvia was doing right now. Behind closed doors, Mark sometimes chose to reveal his feelings to her – he knew it was still important to tell someone some things – and he knew that Sylvia would not gossip. In a way, that's what made them so close; neither of them had anyone else to talk to, no friends to confide in. Despite this, he had never talked about girls. He had never seen the need, as if it was a subject he could keep putting off until he was forced to deal with it. After the last conversation about Alice, where he first talked to Sylvia about a girl, he knew that that time may be soon.

But he did not want that time to be in public.

He looked around the nearby tables. No one sitting down until two tables away either side or behind him, although on his left the couple who sat there were about to leave. Mark could not help thinking that it was unusual for a place like Starbucks to be so quiet and it was 1.30 p.m., which meant a lot of people would still be having lunch.

"So, Alice," Sylvia announced confidently, "you like her, don't you?"

Mark did not like how she asked this so casually—anyone could be listening in.

"Yeah, she's all right," he replied, trying to act nonchalant to disguise his profound discomfort.

"I didn't get to speak to you properly about the dinner you had at her house. Did everything go okay? I know you haven't really done that before."

"Ma!" Mark hissed.

"What? No one can hear."

"Of course, they can!"

Sylvia looked around quickly without really paying attention. "Nope, no one's listening," she said.

Mark sighed, annoyed, and sat back in his chair, making it clear he was not going to talk about it in the middle of a coffee shop.

But Sylvia persisted.

"Why would anyone want to listen to a teenager anyway?"

"Why wouldn't they?" Mark countered.

"Well, they don't know you for a start. You're only talking about a girl who they almost certainly don't know either."

"I still don't want to talk about it here," he murmured so that only she could hear. He was getting increasingly hacked off by her now. He wanted this to be a nice day out, not one where he would be shown up and made to feel uncomfortable.

"Sorry, I just want to help," Sylvia said quietly.

Mark was unsure whether she was pretending to sound hurt or whether she actually was you never could tell with Mammy as her sensitivity was unpredictable. She sipped at her latte, cupping it in her hands for warmth and looked gloomily to one side. Mark studied her for a minute. If she was pretending, she was making it last. He rolled his eyes, glanced around him again, and caved.

"Fine, okay. Dinner was fine. Nothing really happened. We ate. We watched '*The Simpsons*'. I went home. Nothing more to it."

He was moody but it was a truthful answer, not much happened other than that. He didn't want to mention he was in her bedroom—he could not be bothered explaining that one. There was a natural silence after that.

Malcolm May sat at a table from behind Mark, pretending to read a copy of *The Irish Times* he had found at the table, with a pot of tea for one and a blueberry muffin. Every now and again as he sat there, he wrote something down in a little notebook. Everything he jotted down was in shorthand, a skill he acquired during his studies for his journalism degree that ultimately came to nothing. It was useful to write in shorthand as most people did not know or care what you were writing even if they noticed you were doing it. He sat not too close to the Formby's that they shied away from the conversation they were having, yet not too far away that he could hear nothing of what they were saying. He was writing down, in short bursts, what they were saying. That was why he was using the newspaper as a distraction. If, by the off chance, one of them noticed he was writing stuff down with nothing other than a cup of tea, it may draw attention to himself. At least with a newspaper, he could always say he was taking notes from the national newspaper to come up with fresh stories in the local newspaper he worked for if one of them confronted him. He would come up with a name of a local paper on the spot if needed; he was good at improvising. He would say something like 'why else would I know how to write in shorthand?' and 'why would a stranger care about a mother and son conversation whom he's never met anyway?' But he was quite sure neither of them would. And they didn't. Twenty minutes had gone by when Sylvia and Mark Formby stood up to leave and left without so much as a glance behind them.

Their conversation was an interesting one, Malcolm thought. At least it was decent for a discussion compared to the drivel you sometimes heard at a bus stop. And now he could hear more conversations had between the mother and son, thanks to him having applied a GPS tracker and recorder on and inside Sylvia's car. He had done this as the car was parked in the public car park in town. Malcolm had parked his own car next to it after trailing the Formby's from a distance on the way into the town centre. The Formby's were well away from the car when he parked into the space beside them. It was an old black Renault Clio he drove, nothing to cause neither Mark nor Sylvia to pay it much attention. He clipped the magnetic GPS under the wheel of the Picanto when he got out, bending over and pretending to do his shoelace. That was all he was going to do initially but as he stood up, he noticed the car park was empty. Nothing stirred, no one walked to or left their cars, no one to see him. He thought that at least someone would be around and so would only just have enough time to apply the GPS without anyone noticing. Nonetheless, he had a micro-recording chip system wrapped inside a bit of paper so that it looked like an insignificant bit of paper on the floor in his pocket if he was able to put that on the car inconspicuously as well. And he was able to. The front passenger seat window was open ever so slightly – about five millimetres – and it was just enough space to cram the small bit of paper through the window and onto the floor, where it rolled gently under the seat where it could not be seen. Provided it was not stood on and crushed accidentally, the recording system would pick up any conversations had in the car. He had checked again, casually, that no one was around and then took out the device from his pocket. In the action of putting on his coat, he tucked his outstretched fingers through the gap for a split second. It was that easy. He was chuffed he was able to do that as well as apply the GPS—the recorder was an added bonus. It was the only safe time he could get to the car; when it was in the Formby driveway in Dunshaughlin, it was too risky. The house was detached on a country road, and there was too high a chance of being spotted being in the driveway, even at night.

Following the Formby's around town as they went from shop to shop was not too much of a concern. All the shops they went into had the entrance as the exit, so it was not too difficult to spot them come out again so long as he didn't get distracted. But he never would; he was sharp and observant, and that was the reason Seamus had texted him that morning to trail the Formby's; Seamus had watched the mother and son leave from an upstairs window. He was already in the factory underneath the Brensons' home when he got the text as he was at work checking CCTV. He was one of the security guards who worked for Seamus. From the factory exit, it did not take long to catch up to the Picanto. Malcolm smirked when Sylvia had been beeped at from the car in front of him when she failed to notice the green light, but it also got him thinking that something was distracting her… or could be. He had hoped that he would find out something about that, as the Formby's spoke in Starbucks but what he did get was just as good. Alice's job at inviting Mark in was working; it was clear that the kid fancied Seamus' daughter. As soon as the Formby's were out of sight of the coffee shop, Malcolm wrote more vigilantly in his notebook about his thoughts toward what he had seen and heard, including the traffic light incident. He would talk to Seamus about it when he got back, but after several minutes of writing some more notes into the notebook, he figured his job was done. He had inserted the sound device in the car. He sat back and took more notice of the tea he was drinking and decided that he had

had better. He munched through the muffin with more pleasure and by the time he had left, he was in good spirits.

Back at home, Mark played FIFA on his Xbox, not really paying attention to the game he was playing. He was deep in the first half of the game, playing as Chelsea, and was 1-0 down to Spurs. The noise of the game was therapeutic to him. Listening to the commentary of the match was a pleasant background noise to have on as it drowned out everything else, allowing him to think about Alice whilst appearing to be preoccupied. He hated it when he lay on his bed in silence thinking, as he so often did about something, and Mammy came into the room. She always seemed alarmed that he lay there with his thoughts. "It's not what normal boys your age do, Mark," she had once said to him.

Well, I'm not a normal teenager, am I? he thought.

That was the time he decided to get an Xbox 360. He was relieved EA still produced FIFA games for the 360. When Sylvia came into the room, he would be playing a game or taking advantage of the apps installed in the Xbox and watching Netflix. It was a good disguise because it was more normal, whatever that was. Now when she came in, Sylvia smiled at him, pleased that he wasn't just staring into space, and then left again.

Half-time came and he made a substitution, getting more into the game now that he knew he had a task of coming back into the game and at least getting a draw out of it. It was the start of the new season in career mode, so he wasn't too concerned if he did lose, although he hated it when he lost after deserving at least a draw. He was daydreaming during the first half, thoughts turning back to Alice as they so often did now, again torturing himself about whether he would text her or not. Maybe he would finish this game, send a text that sounded as casual as he could make it, and then forget about his phone and play another match. It was as good a plan as any. It was times like this that he was thankful for the distraction his Xbox gave him. The second half was in full flight and he was gaining more and more possession of the ball, weaving the ball to and from players around the Spurs penalty box, trying to find a way through. However, he was struggling to find a decent goal scoring position, and the time was ticking.

65th minute.

70th minute.

80th minute.

Mark was getting frustrated. It would probably be still 0-0 had he been paying attention, and he was playing at home too, which made losing worse. But it was no use. The Spurs defensive line was too strong, committing even the attackers to defence. All thoughts of Alice were momentarily gone as he hissed a curse at the screen. He deserved an equaliser. He had 70% of the possession last time he checked.

Then, on the 87th minute, he found a small gap for a shot. It wasn't a brilliant position but still worth a shot. The goalkeeper, Lloris, got a strong hand to the ball, lifting it away from goal for a corner, clipping the crossbar on the way.

Mark sighed; head raised as if to the heavens.

He took his time taking the corner, longer than he normally would. He spotted one of his players unmarked now in the box, and sent a drilled cross, low and hard into the box. Pedro was first to it, but the ball bounced unfavourably away from him; however, Mark managed to get Pedro to get a strong foot on the ball, defeating the defenders closest to him, and aiming a relatively fast close range shot towards goal. Mark fist-pumped the air when the ball ricocheted off Lloris' hand and into the bottom left corner of the net. He was feeling much better and was almost upbeat now he allowed himself a half-thought about sending a text to Alice. If he managed to find a winner now, he would almost be able to project enthusiasm in a text which an hour ago freaked the hell out of him.

But sadly, for Mark, his contentedness was short-lived.

On the 93rd minute, Mark managed to get Willian to send a beautiful cross to Giroud who was running into the box, and Mark thought he had sneaked a winner for a split second when Hazard took the cross first time, whacking a volley towards goal but instead of finding the net, the ball thumped cruelly against the crossbar at such a pace that it flew past all the Chelsea strikers, and found a Spurs defender. Many of Mark's blue men were committed in the Spurs defensive half, and now Spurs were on the counterattack. Mark glanced the clock flicking into the 94th minute, praying for the final whistle to blow. It didn't come. And before Mark knew it, the Spurs striker Harry Kane was one-on-one with his keeper. It was going to be the last kick of the game and Mark's feeling of hopeless dread rushed through him horribly, and he groaned with disgust when

Kane slid the ball smoothly past the goalkeeper and into the bottom right-hand corner of the goal.

Full-time arrived. Chelsea 1-2 Spurs.

Mark scowled at the screen and threw his controller down onto a beanbag in disgust after flicking back to the home screen. He then shut off the Xbox, temporarily annoyed at the game he knew he would probably come back to later.

For the time being though, he was finally going to text Alice, channelling his trivial anger from FIFA into a text that was not glaring with nerves and hesitation.

But he found it harder to do that when he came to pressing the send button. Currently the text read:

Hey, it's Mark. Was wondering if you were free tomorrow to do something?

Mark stared at the text, wondering if it was okay or not debating whether or not to add a risky kiss. But he decided that if he was going to make a fool of himself, it was better to go down knowing he had not looked desperate by sending one. Ten minutes had gone by, he noticed, and he was still struggling to send a one-liner to a girl he barely met. It all felt surreal to him, almost overwhelmingly suffocating, to even think he was doing this. The last text he sent that wasn't Mammy was about a year ago, when he texted a boy from his class, Simon, about an assignment they had to do together. Mr North, a new geography teacher from London, had set the groups, completely unaware that Mark didn't speak in school other than maybe the school librarian, Mrs Hevshan, an elderly yet agile woman from Lebanon who always took the time to speak to Mark. He had laughed amusedly when she made a comment to him about her surname being a fitting one for a librarian due to the fact that it was remarkably close to Miss Havisham's in 'Great Expectations', only she was much more outgoing and more pleasant to boys than the cruel Havisham. So, he spent most lunchtimes in the library, where he could blend in surreptitiously and not be picked on. Mrs Hevshan always kicked out the troublemakers by lashing her walking stick against a nearby table when school bullies came into the library just to hassle the kids like Mark. Not that this happened often, because of the terrifying threat Ms Hevshan posed to them. It was really quite remarkable how lovely she was to the quiet kids who were not a nuisance.

Mark's hands were beginning to shake. His anxiety was getting the better of him. He almost wanted to trip up and accidentally hit send so that he could do nothing about it. Purposefully hitting send was a big deal, possibly too much of a big deal. Was he being pathetic? He certainly thought so.

Come on Mark, it is one click of the button. The worst she can say is no, and if she does you don't have to speak to her again. You're not going to see her at school, very rarely out here, so who cares what she says really? Either way you'll survive.

"Yes, well it is easy for a voice to say that, isn't it?" he muttered aloud. "You don't have to press send!"

But he knew there was sense in the voice inside his head. Did it matter?

Of course, it matters, the conscious part of Mark argued. *This is an issue of pride we're talking about here. And didn't you think she liked you? She invited you to her house,* his inside voice enquired.

That was true. But it could also just be him desperately hoping she liked him because he liked her. It was like Freud's wish-fulfilment argument that religious people believed because they wanted so much for a God to exist that they then *made* Him exist.

Then, out of the blue, his phone bleeped and vibrated in his fingers. At first, he thought one of the apps on the phone was beeping at him to remind him of an update, which Mark found really irritating. But it was not the same noise. He realised after a second that it was a text message alert. He was just not used to hearing texts go off on his phone. It was rare.

Confused, but curious, Mark pressed on the text message icon. The phone took its time loading the text menu, as if also confused why Mark was trying to access his next to non-existent messages.

It's probably just an ad or a reminder to put credit on my phone or something, he thought glumly.

It was neither.

It was a new number. He clicked on it and read:

Hiya! It's Alice. I hope you're okay, I was just wondering if you wanted to go to the pictures tomorrow or something? I haven't been in a long time and I figured I'd see if you were free? Totes okay if you're not (or don't want to lol). Soz for short notice, I don't wanna interrupt any plans. Anyway, let me know. A. xx

Chapter Thirteen

Malcolm had enjoyed telling Seamus about his successful venture. Indeed, Seamus definitely seemed pleased, even though he had not managed to hear anything about John specifically as he would have liked. Malcolm thought the fact he had managed to insert a recording device into the car made up for it. Seamus asked if he had been seen by the mother and son, to which Malcolm said perhaps only in Starbucks but only as someone drinking his tea in peace without an agenda. There was nothing that made him stand out. Seamus had looked at him long and hard then, as if trying to detect a lie in what he was saying.

But Malcolm knew that was exactly what he was doing.

Malcolm met this scrutiny with a cool, neutral expression. After all, what he told Seamus was true. There was nothing to hide or be shifty about.

Seamus nodded slowly, smiled after a second, and then nodded again.

"Good," he said, "good to hear."

Malcolm smiled, bowing slightly.

"You may leave now. Tell Alice to come in, would you?"

"Of course, sir."

"And have a beer or somethin'. You deserve one."

"I will. After I drive home. Good evening, sir."

Seamus huffed, bemused at Malcolm's formality as he left the room.

Seamus had instructed Alice to meet up with Mark again, but not in the fields this time. If they met there often, there was a bigger possibility of John noticing. Keeping a low profile was the safety net that Seamus was always anxious to put in place.

The text was sent half an hour later and according to Alice there was no reply for a good hour. Was the boy trying to look casual towards a girl texting him? If he was, fair play, he thought. Seamus did not like people who rushed into things, putting their feelings on a silver platter without as much as a second thought. But what surprised him more was Alice's reaction. The normally cool exterior that Alice showed was now set with a troubled expression on her phone as her impatient hands constantly flicked towards her phone to see if he had replied. He did not know why but he decided it did not matter. All that mattered was Alice getting her claws into Mark Formby, working away at him until the hook was immovable from his adolescent, acne-ridden mouth. He didn't really know if he was a spotty kid or not; he had only ever seen pictures of the boy or seen him from a long distance. Either way, it looked like Alice had done a good job so far. The more Alice met him, the more he would tell her to ask Mark about his father, slowly getting information so as to get inside John's head without him even realising it.

As of yet, there was nothing to report from the recording device in the car. There had been little conversation between Mark and Sylvia Formby as they made their way back home, and when they did speak it was of no interest to Seamus. And judging from the number of times the Kia Picanto left the driveway, hardly at all, Seamus doubted that he would get anywhere with this. However, it was better than not having anything at all even if nothing resulted from it. It would be so much more beneficial to have bugs around the Formby house, but he knew that would be very difficult to accomplish without big risk. Could Alice do it? Probably not. She would have to be there only when John was not in the house. Alice told him that John had seen the two of them together, it was unwise for Alice to be in their house. It may be that John knew who Alice was then, and that that was the reason behind him lingering around the house the other day. No, he couldn't risk Alice being in the house. There was no way of knowing when John would enter and leave the house; there was no pattern up to now; his actions were unpredictable. And even if Alice was in the house, how would she bug Mark's bedroom without him noticing? The classic 'can I have a glass of water, please?' that gave you the time to snoop around quickly? That was possibly the only way. Plus, if Alice was given more of those micro-device thingies' that Malcolm had this would be easy for Alice to apply them. That could work, even with Mark in the room at the same time, it could work. It took less than a second to slip something small under, in, or on top of something when the other person wasn't looking.

John was the issue though.

Seamus considered too how odd it would be if Alice never went to Mark's house after meeting him fairly frequently, particularly when he was just across the field. Questions would be asked eventually.

It annoyed Seamus that Alice's picture was in some newspapers regarding the medical factory he had under his house. If John went snooping, finding such articles would be easy work for a guard. The question really was: had John Formby been snooping? If he hadn't, Alice could go to the house. If he had, Alice simply could not be in the house in case John did come. Perhaps down the line he would get Alice to tell Mark to get into his father's study (if he had one) when John was out and look for anything strange. But even that was risky. Very risky. Even a boy besotted with a girl would find being asked to do that fishy. It was a lot to think about.

But it was nothing he couldn't handle.

Seamus fiddled with the papers, pens and documents on his desk, arranging them so that everything was where it should be so that by when he left his study, every inch of the room was spotless. He was going to relax in front of the TV after a relatively successful day with a Magners. As he entered the living room, he saw Kirsten sat on the furthest away armchair, reading. It was another of those sickeningly pointless novels about 'real life' that had some romantic tie to it. He wasn't going to stop her reading that shite, but he certainly didn't see why she did. He frowned.

"More of that drivel?" he asked with a hint of disapproval.

She looked up, saying that that it was a decent story if he actually gave it a chance. He supposed to a degree he could not really judge her, particularly when he himself did not read; he had liked some James Patterson and Dean Koontz when he was younger, but only if the spine of the book was short enough for him. He had little patience for long novels; he did not see why they could not just get to the point within three hundred pages and he was not a patient man anyway. Doing something, by choice, which required a great deal of patience did not appeal to him at all. *Do not get me started on a series of books*, he thought, *they're the worst of all.* He grudgingly conceded that it may actually be good for him to read more so as to help him be more patient, but there was no motivation in that, and nobody was ever going to do something they didn't really want to do. That was why New Year resolutions failed nine times out of ten. People didn't set targets they wanted to achieve more often than not, but rather what they thought they needed to set for themselves.

Kirsten bookmarked her place and closed the book on her lap. She looked at Seamus with a raised eyebrow as if waiting for him to continue with his 'why do you read these kinds of books?' lecture. Instead, he just shook his head, planting himself down onto the leather sofa and reaching for the remote control. Glancing at the digital clock on the DVD player he saw it was nearly eight p.m.; it was midweek so there was unlikely to be anything on that was worth watching, although he did enjoy the odd TV drama that had only several episodes to the story. They were nice and short, and they got to the point without too much messing around, Seamus thought, and that was why he liked it. There was always Netflix if there was nothing else on. Alice had pleaded with him to get a subscription and after the free trial he too had been won over by the service.

"So, Alice is meeting this Mark fella again, I hear?"

"Alice told ya?" Seamus said, looking at her briefly.

"Yeah."

There was a tone in how she said this that made him look over at her again. She had some sort of issue. She always had an issue.

"What is it?" he asked, scowling.

Kirsten looked at him then for a few seconds before answering. Like she was trying to use her eyes to answer the question.

"It's not fair that you are using her like this, Seamus," she eventually replied. "What if this John Formby knows her? What if he's dangerous? Surely that would be putting her at risk?"

"There is a risk in everything," he answered calmly.

"But this is your daughter! How can you just put her in danger like that?"

He rolled his eyes at her. She was getting paranoid, just like she usually did when something had to be done to keep the drug production thing secret. It was really fecking annoying when she was like this. The only thing that was preventing him from raising his voice now was Kirsten in her pyjamas, her breasts exposed under the loose material of the T-shirt that was cut to accentuate them. The short bottoms also revealed most of her legs even though there was a dressing gown which did cover them somewhat. But as inviting as she was as she sat there, he was not going to take criticism.

"She is not in danger, Kirst," he said reassuringly, walking towards her. "They're going into the town centre tomorrow. Nothing will happen in a public place." He planted a kiss on the top of her head.

"Yes, fine. Tomorrow will be fine. But what about the time after that? Or the time after that? One of these times she could find herself in a bad situation that you put her in."

"Nothing will happen to her. She's a smart girl. I would never ask her to do anything that I thought was too risky or too dangerous. I know you're her mother and you are just looking out for her, but I am her father and I, too, happen to care about her."

Seamus added the last bit with an edge to his voice that Kirsten did not appreciate. It was condescending and it made her feel small. She bit the inside of her lip, as she usually did when something didn't sit right. An argument was brewing.

It seemed Seamus could read what she was thinking because he quickly added: "You don't have to worry so much about her. I know I am asking a lot from her, but she wants to do this for me—she is not being forced. Besides, I think she likes the kid. It isn't a chore for her. And if John has a problem, I'll know about it and will act, as necessary. He won't harm her in front of his son and wife even if he wanted to; heck, we don't even know for sure whether he knows who Alice is other than a girl who Mark has gotten to know over the past week or so. The only reason so far, we think that he may even know who I am, let alone Alice, is because of him searching the outside of the house. And even then, he may just be curious generally. It is possible he saw a house with all its windows covered by curtain and went to look around. He is a guard; all of them sniff around even when they don't smell something suspicious. It's their nature. It's in their blood. Alice is just trying to get a sense of what John might be thinking more than anything. If she can find out specific things, then great, but otherwise just a feeling will do. After all, I am only doing all this in the first place to keep us safe. People like me doing what we do have to look over our shoulders every now and then. It's a precaution more than anything else. Alice will be safe. The minute I think she will be in trouble, of any kind, I will make sure that she is removed from that situation—and you know better than anyone that I can make that happen. I plan for all events."

Kirsten's face softened increasingly the more he spoke. The more he said, the more of a spell was put onto her. It was like a weird, hypnotic trance that instantly made you feel like everything was okay even if your conscious self would normally tell you different. Seamus could see her mood changing as her face was less contorted and tight with anxiety, and he flashed his winning smile that he knew would make her feel jittery. He was still able to make her feel like a teenage girl when he wanted to; he was aware that he had a sex appeal about him, and this was what he was doing now. He had done it with Valerie, and he had done so with his wife for years. It was easy for him.

And boy, did it help.

As Kirsten relaxed, she breathed in and out slowly. When she breathed in, filling her lungs with as much air as she could take, her breasts swelled to their fullest and as Seamus focussed on this—also taking in her legs and what he knew lay underneath the folds of her dressing gown, he felt himself get excited. She then walked past him towards the door, and for a horrible moment, Seamus thought she was walking out of the room and leave him alone with his urges.

The cruellest of sexual nightmares for a man.

To his delight, Kirsten stopped at the door and closed it. She then locked it with a little key that lay in a little bowl on the windowsill closest to the door. She knew at this time Alice would be on her laptop and completely preoccupied with it. Any noises Seamus or herself made would not be noticed by Alice either; she was inseparable with her Beats at the moment, listening to all the garbage teenagers listened to.

Twenty minutes later, Kirsten sprayed the room with Febreze that she had in a cupboard to eradicate the smell of sex, and only then did she unlock the door and head up to the bedroom to take a shower.

Kirsten smiled all the way up the stairs.

Chapter Fourteen

Mark stared at his phone as if the object in his hand was something much more valuable than a budget smartphone, something that made his jaw drop in disbelief. He read and reread the text from Alice many times without any emotion other than an overriding feeling of disbelief. The world had stood still, or at least that was what it felt like for Mark as he sat on the edge of his bed. All his movements seemed to be in slow motion as the shock had yet to pass through him and allow him to function normally. All thoughts of his Xbox were now forgotten; instead, he tried to think about what had just happened. A swift reply to the text was impossible and Mark would later be pleased about that as it made him look less affected about the text than he actually felt. He wanted to play it cool, whatever that meant. He was aware that texting was political despite his lack of experience. He knew you had to know the game and know what exactly to write.

Particularly with girls.

And he didn't. So, the large amount of time in which it took him to reply was really a blessing in disguise, although he doubted that anything, he put in his reply would adhere to its politics. No amount of time would help him, he feared. There was only one thing for it, and that was keep the reply as short as possible.

He wanted to show Alice that he really did want to go to the pictures with her because that was true, and he didn't want to hurt her feelings either. There had to be a balance which was short, but not so short a message as to look indifferent. Girls who sensed that in a boy would soon move on. Even though he was terrified at the prospect of a date – which this surely would be – he did not want to deter her.

Despite his fear, he wanted this.

He checked the time Alice's text had come through. 7.57 p.m. Now it was 8.13 p.m. How long could he afford to take sending his reply? An hour? Two hours? He didn't know. Timing of replies had never really been at the forefront of his mind. He didn't need to before having no real friends to text.

Mark dropped the phone on his bed before heading off to use the bathroom. His room had suddenly made him feel claustrophobic and his phone had begun to feel heavy, like it was a medium for everything weighing down upon him. The walls were closing in on him and his phone was the only thing he could see in the room, taunting him, luring him in, and daring him to reply. He knew he had to but not just yet.

I know you can see me so stop trying to look away. Alice is just a girl who likes you and you like her. It's simple. A few words. Maybe a smiley face. A kiss or two. You see, simple – so, so simple. Go on, it is easy Mark. Use me. Press my buttons and type the words. Don't think, just type. I see the beads of sweat on your face as you watch me – just get it over and done with.

Mark took a deep breath as he stumbled into the bathroom, trying to compose himself. How was this happening? He was just a boy who had no friends and no ability to talk to girls and yet he was about to agree to a date with a beautiful girl. Not just beautiful to him, but also to anybody walking past her. Without bias, he knew she was truly breath-taking. She was one of those girls that everyone of a similar age did a double-take over, a 'wow, she's stunning' kind of stare as she walked by. Mark gripped both sides of the basin, looking at himself in the mirror. He looked worn and almost afraid. Was he ready for this? To take a girl on a date. He wasn't sure. He didn't think he would sustain any conversation with a girl like Alice, yet he had managed it, even with her parents. Perhaps it wasn't impossible. Maybe he should trust himself, but he knew that was risky. A date was a whole different ballgame to meeting a friend (not that he had much experience of that either) and he could only hope that he could act stoic in front of Alice in a date situation. Even if it was just a friend-date, he would still be nervous. It was at this moment that he realised just how immature he still was: he had no life experience, not even the basics like friendship. He was usually in control of himself, so it was unnerving to see such helplessness in the face that met him in the mirror.

Mark had calmed himself down after a few minutes, so he went back into his bedroom with the full intention of replying to Alice. He knew that the text would take a long time to send because he would no doubt type and retype various forms of saying 'Yes, that sounds great' so that he went mad deciding which would be the best reply. And that is just what happened. By the time he was happy (well, not happy exactly but happy enough) about what he was going to send, over an hour had passed since receiving Alice's text. What eventually was sent was:

Hi Alice, I don't have any plans tomorrow so that sounds great. Looking forward to it. What time do you want to meet and where? Mark. X

He debated about whether to include the 'looking forward to it' comment but he wanted her to feel like he wanted to see her as well as being polite; and he decided that it was more of a nice thing to say rather than cringy. Wasn't it? Either way, it would do. It had to do; it was now too late to change it. The more time that went by the more he just wanted to send anything that was not embarrassing. Anything at all. He didn't want to keep her waiting, after all. However, part of him couldn't help but be pleased that he was the one in control by making her wait. The kiss at the end made him feel the most uncomfortable though, particularly since he never sent any before. He only added them now to show he knew something about texting, and he also wanted to make her feel less awkward about sending kisses by returning them, even if girls did send kisses as meaninglessly as if they were handing out sweets at Halloween to kids in over-the-top costumes.

Mark made sure his phone was not on silent and turned his Xbox back on. FIFA had certainly become a blessing for Mark when he needed a distraction. He played four more matches in career mode, bringing the number to played league games of the season to ten matches. After his disappointing loss to Spurs, he had won three and drawn one. He was currently in second position in the league, trailing by two points to Man United. Annoyingly, the match he drew out of the four he had just played was against Man United, sacrificing a victory and a place at the top of the league by a cheap penalty conceded in the 82nd minute. He was about to play a fifth match – a Champions League match against Lyon – but then his phone bleeped. He wasn't anywhere near as nervous reading this text than he was in sending his reply, but his heart still raced as he reached for his phone.

Great! I'll see you tomorrow. Shall we say meet at your house at one p.m.? Mammy will drop us to town, she wants to go shopping anyway ☺ *xx*

A smile crept upon Mark's face. It was the first one he allowed himself. Now that a plan had been made and had been made concrete, happiness surged through his veins for the first time in a long time. After the dinner he had with Alice, he had been pleased but he was still too concerned about whether or not he made a decent enough impression upon Alice to feel raw happiness. It seemed that he had, and relief mixed in with the happiness to make him feel excited. He was looking forward to spending the day with a girl, which was something he never thought he would feel… or want to feel. At least not so soon and out of the blue.

But it was happening. The seed of anticipation had been planted in his brain now, allowing his continued thoughts upon the matter to blossom and grow.

One p.m. is fine with me. See you then!

He had hit the send button before realising he didn't add the kisses back this time. Quickly he sent a second text with two kisses, making it obvious now that he was sending kisses just because she was. He hoped she wouldn't mind.

She probably won't notice. She's probably feeling a pang of regret now, he thought.

Mark forced the usual negative thoughts out of his head this time. He was going to enjoy this moment, not taint it with doubt. It was nearly ten p.m. and he wanted to look as fresh as possible for tomorrow so got ready to go to bed earlier than usual. As he was brushing his teeth, Sylvia appeared at the doorway of the bathroom looking at him curiously.

"You okay? You don't usually go to bed this early."

Mark could not help but smile as he looked at her, despite having a toothbrush and a load of toothpaste in his mouth. He spat it out into the sink and answered: "I'm great. I'm meeting Alice tomorrow. We're going to Dublin to see a film."

Sylvia beamed at him, feeding off the excitement which must have been radiating from him like a toxic chemical.

'That's great!' she exclaimed, 'I'm so happy!'

She put her hands over her mouth and nose in delight, and for a horrible moment Mark thought she was going to well up with tears. It was bad enough seeing her cry when she was feeling down, but this would be cringy if she cried with pride. Fortunately, she did not, but instead made a weird squeaking noise to express her feelings.

"So yeah, going to bed early so I look fresh," he added, hoping that this would be a cue for him to go to bed and avoid being fussed over further.

"Yes, yes of course! Don't want my handsome man looking tired and being cranky because of a lack of sleep!"

She ruffled Mark's hair and then grabbed the back of his neck, pulling him in to kiss his forehead with a relative amount of force.

"Har-har," he said ironically. "I'll see you tomorrow. Night."

"Night, Mark."

As he walked into his bedroom, Mark could hear another happy squeal - quieter this time - but he could hear it, nonetheless. He rolled his eyes and got into bed, picking up his copy of *Nineteen Eighty-Four* to read before going to sleep. Fictional worlds fascinated him, and no matter what he thought about his own life, he was grateful that he was not being watched twenty-four-seven and that there was no such thing as the thought police in reality.

He had no idea that he was, in fact, being watched very closely.

And often.

Chapter Fifteen

It was a long night and Mark got less sleep than he would have liked. He was pretty sure that at the very most he had gotten no more than three to four hours' sleep. It was impossible to know for sure as people often drifted off to sleep despite a seemingly sleepless night. Nonetheless, when he looked at the clock on his bedside reading six a.m., he knew there was no chance of sleep; he was wide awake, and his brain was in overdrive thinking about the day ahead. The more he lay awake, the more agitated he became, tossing and turning in the attempt to kill some time until it was a reasonable time to leave his bedroom. After a while, he conceded defeat and reached for the small lamp by his bed, intending to read more of *Nineteen Eighty-Four*. However, after a page or two he could not switch his attention enough to the book in his hands; his thoughts inevitably wandered, meandered and raced to Alice; her smile, her text, the unknown of the day in store for him. It was a strange feeling for him. The only time he had been this nervous and upright awake so early with his thoughts racing at a hundred miles per hour was last year when he had an interview for a part-time job at Dublin Central Library. He had been fourteen then – almost fifteen – and it was an interview that his mother had organised for him. He was a frequent visitor of that library growing up and it was the one thing that Sylvia had been determined to make Mark go as often as possible, particularly since he was not interested in any sport clubs or anything like that. Not that Mark minded; he liked reading and the library was a home of escapism in more ways than one. As much as he loved his mother, their home in Dublin was a hostile, unwelcoming place. It wasn't a home as much as it was a house. A place to live rather than a place to make family memories, although there had been times with Mammy that Mark had enjoyed and treasured. Small, usually forgettable memories but they were as strong as any because of the bond between the two of them and Mark learnt to savour the good moments. The interview was agreed upon mainly through Mark's interest in reading. The staff knew both mother and son on a first name basis and it was as close to social interaction Mark had ever gotten other than his mother.

However, the interview did not go well. Despite them knowing Mark well and liking him well enough for being a young reader in a world that seemed to not have time for reading, it was still a job, and at the interview Mark stuttered, hesitated, perspired too much for them to realistically consider him for the position. He had no people skills, and even though they had pitied him and encouraged him to take his time, he still crumbled under the pressure of the situation… or rather, the pressure he had put upon himself. It didn't help that whilst he had the interview, his mother was perusing blurbs in the romance section. Her hopeful smile dropped when she saw Mark approach her looking ghostly white and staring at the floor. She could see beads of sweat around his hairline, and she did the only thing a mother could do. A comforting hug with a 'never mind, Mark, there's always next time. It was your first ever interview after all'. He was glad of her consoling him, but it had done little to raise his spirits.

Mark was afraid he would crumble today too but he had at least improved a bit since that interview. In front of Alice and her parents, he did not sweat – at least not visibly – and he was too focused on being polite and eating everything that was on his plate to think of anything else. He hadn't even been this nervous with exams; he always revised thoroughly, so he pretty much always felt relatively relaxed, which for his usually nervous disposition, was unusual.

At 6.45 a.m. Mark was too restless to stay in bed. He robed up warmly in his housecoat and slipped on his slippers before heading off to the kitchen to make himself breakfast. No one was awake apart from him. He made himself some toast and coffee, then went into the living room to watch TV at a low volume. Nothing was on that was of any interest so he settled for the news channel, half-listening to the news reporter talking about the disappearance of a woman in the Killarney area; a report about the Royal Family, more specifically about Prince George's first day at nursery; a report about an elderly man being mugged by a gang of youths in North London, before going on to the news and weather, from which Mark zoned out completely. The background noise of the TV was comforting though as his thoughts raced.

The next few hours dragged by at an excruciating pace. It was not until 11.30 a.m. that time seemed to speed up again; suddenly there was no time to compose his thoughts, to remain calm. More importantly, there was less time to get ready. It felt like he was rushing by the time he had changed into clothes he thought were good enough for a date, shaved, and applied his cheap supermarket aftershave. Looking presentable was a pain at the best of times, but Mark was utterly unprepared to present himself for a date. Without realising it, he had brushed his teeth so thoroughly that he may as well have gone through the process of brushing his teeth three or four times. Then, ten minutes before he was due to meet Alice outside his house, Mark was staring at himself in the mirror, trying to place his hair in the position he wanted with an increasingly shaking hand. This had happened at the interview too but there was something more pressurising about this somehow. At least with an interview you could leave whenever you wanted to if things were going badly but this was a whole day in Dublin with a girl. It was harder to run away from that, and if he did, this would only upset Alice and that was definitely not what he wanted to do either. So really, there was no backing out. The fact that it was a whole day made it an even bigger deal for Mark. He thought that it would be daunting to anyone. If they lived in the city, a coffee would have been enough. It would have been normal. It was not the case for those living in the countryside; they could visit in the nearest town but since they were driving and Alice's mother was going to Dublin anyway, there was not much option. But then another thought came to him. What if everything went well? He had survived previous meetings with Alice without frightening her away so he could do today too. Right? It was something he had not thought of before now.

But he wasn't going to think on that too much either. It was unnecessary pressure to put on himself and he wanted to learn from his mistake in the interview of overthinking. Not that that was easy to do. Whatever would happen was going to happen and that is all there was to it.

Sylvia appeared at his bedroom door just as he was about to leave. He could see that motherly smile on her face as she took him in with her eyes, giving him a satisfied look up and down. In another situation, Mark probably would have cringed and moaned at her to stop being a soppy mother but her look of approval settled his nerves a bit. It meant he didn't look too bad, even if it was just a casual checked top, black jeans and Converses.

"I'm so proud of you, Mark!" she burst out. "My boy, on his first date. Oh, you look so handsome!"

The near torrent of emotion was on the tip of her tongue but all that came from her now was an excitable squeal with her hands wringed around her mouth as she looked at him. Mark smiled, secretly pleased, but quickly turned his smile into a grimace, telling her that she was embarrassing him.

But she ignored him.

She strode over to him, squeezed him into an over-the-top embrace and kissed his forehead. Mark wiped it.

"Ma, stop it!"

"Okay, okay… but you'll tell me all about it, won't you? What are you doing with her? Are you back late? Oooh and what about dinner—are you eating out?"

"Ma… I don't know. It's probably not even a date. I'll text ya, okay?"

"How can you not kno—"

Mark interrupted. "Mammy, please."

He hated it when she bombarded him like this. How was anyone supposed to answer ten questions at once? He told her that he was in a rush and that he didn't want to keep Alice waiting. He figured it better to be the one waiting rather than her to be the one waiting for him, especially since Alice's mother was driving them in. He didn't want to give a bad impression. He had only met her once, after all.

Thankfully, Mark was the one who was waiting. The weather was pleasant enough, dull but not threatening to rain but, of course, that could change. Nearly ten minutes later, Mark spotted a black car in the distance, and suddenly his heart raced. The familiar black Astra was hurtling toward him and now there really was no way back. This was it. His first ever maybe-date was less than a minute from beginning. He wasn't one hundred percent sure that this was a date, but at least if he treated it like one, he would be careful about what to say and do; he would be regardless, but more so if it was a date.

The Astra stopped in front of him. From the back seat, Alice beamed. He presumed she had not taken the front passenger seat so as to join him in the back and be polite and he smiled at her consideration. It took about half an hour or so to get into the city centre as the traffic was not too busy. The journey there was comfortable and not awkward to Mark's relief. Alice's mother dropped them off near O'Connell Bridge and then drove off in search of a parking space. Apparently, it was her sister's birthday the next day and she wanted to get something decent for her.

"Thank God we got rid of her," Alice huffed, giving Mark a smile.

"Ah, she's nice though. I was glad to get away from my Ma this morning." He frowned, attempting the 'you know what mothers are like' look, but Alice must not have understood this intention as she gave a slightly concerned look.

"Oh yeah? Why's that then?"

"Fussing over me like I was a dressed-up puppy"

Alice's face relaxed, and she laughed. Mark liked her laugh. It wasn't irritating or too loud like you sometimes hear but pleasantly cheery. He really hoped he wasn't staring at her too much.

But naturally, he was. Or at least enough for Alice to notice. She didn't seem to mind though; there was no creeped-out look in her eyes, only friendly and inviting. She was so easy to get on with. Even for Mark, who didn't like talking to people normally, he didn't feel intimidated at the prospect of talking to her. Instead, it felt like he could talk to her without trouble for hours. It was a surreal feeling.

"Is there anywhere you need to go?" Alice asked.

"No, I don't really shop," he replied.

"Well, I'll have to teach you then, won't I?"

Alice clearly knew how good she looked, and it was only then that he really saw what she was wearing. A flowery, tight-fitting dress to show off her curves. She looked elegant and was so full of life that he wouldn't be surprised if her aura gave life to the flowers on her dress.

"Good luck! I really hate shopping," he murmured, trying to distract himself from her.

The corners of Alice's lips lifted up as if she were trying to solve a puzzle. Then, out of nowhere, she grabbed his hand and ran off down the high street, past the old post office and on towards Pennies Stores.

"Oh no, not here! Women take forever in here!" he jokingly exclaimed in horrified pretence. Her perfectly set curly hair bounced as she jogged into the store, receiving both annoyed and confused looks from passers-by. But he laughed away with her: he couldn't help it. Her laugh was infectious.

After Pennies, they whizzed through a variety of shops, some Mark recognised, some he had never heard of, but he didn't complain. In fact, he found it fascinating to watch a girl be so interested in the clothes hanging up on some metal poles. She took him to the men's section of many of the shops too and began fishing out clothes which she thought would look good on him. He liked some of them, but he didn't want to buy anything. He didn't have a debit card yet – Mammy said he didn't need one – and all he had was thirty euros, twenty of which Sylvia had insisted upon giving him. Three hours of shopping had raced by before Alice asked if he wanted a break and something to drink. He gladly accepted. They went into the nearest Starbucks and found an empty table at the back of the store. Mark told Alice to mind the table whilst he bought them both a drink. She said he didn't have to, but Mark said that he wouldn't take no for an answer. She smiled at him sweetly and thanked him. Mark watched her from the back as she sat at the table. He realised that he *really* liked her. He could feel himself being hooked in like any other teenage boy and was surprised to feel glad about that. And he relaxed a little knowing that he had adhered to the golden rule of a first date: pay for her too.

"She's way too good for you, mate"

He turned around and saw a tall, broad ginger-haired man smirking down at him. He looked about thirty. Mark could tell that the man meant no malice and he scoffed, saying "I know," and turned around again.

"First date?"

Mark turned around again. He didn't really want to have small talk – or talk in general to anyone else other than Alice, right now – but he didn't want to be rude either. Maybe Alice would look over and see him talking to a stranger, thus giving off some sort of confidence girls always seemed to look for in a guy. Most likely not, but maybe she'd think that.

"Yeah... well, I think it is a date. Can't really tell."

The man found this hilarious, chuckling for a good ten seconds. Mark realised that the man was younger than he first thought, perhaps mid-twenties rather than mid-thirties as he had originally assumed. "Girls are like that buddy. Making us lads stay on our toes until we make a move."

"Tell me about it," Mark agreed. And he meant it. He really wanted to know what she was thinking.

"My woman still has me on my toes. They all know what they're doing."

It was Mark's turn to laugh. He would usually have killed any conversation by now, polite or otherwise, but talking seemed like a natural thing to do now that Alice had started him off.

"How long have you been with her for?" Mark asked.

"Ahh nearly – what – seven years? Christ, yeah nearly seven years! I think… Jesus, I hope she didn't hear that!"

A troubled look came across the man's face. He turned around looking over towards a table in the corner where a dark-haired woman sat. She was looking down at her phone.

The fact that the man didn't know for sure made Mark laugh.

Alice looked over then. Smiling at him sweetly.

It seemed like hours had gone by before he was carrying the tray over with two lattes, panicking for a moment about whether or not he had ordered what she wanted even though he knew it was right. The man wished him good luck.

"That's a nana's drink for a man," Alice taunted.

Again, a jokingly snide comment but Alice had never been nasty to him; it was refreshing to enjoy humour with someone other than Mammy.

"Yours is the same drink too, grandma!"

"Yeah well mine has a shot of caramel, yours is just boring."

Shit. He had forgotten to ask for a caramel latte!

Alice noticed the change in his face and asked what was wrong. He grudgingly confessed and she snorted, bemused.

"It's all right. Men never listen, anyway. Although, you're a nan with that drink so I'll let you off," she said, winking.

Inwardly, he cursed himself. But he smiled appreciatively.

It was the only mistake he made all day.

Chapter Sixteen

Alice's phone ran out of battery whilst at the cinema, so she used Mark's phone to call her mum to pick them up. As it was Alice's mother drove them back, Mark felt more comfortable ending the 'maybe-date'. There was no worry of lingering awkwardly at his driveway, just a simple goodbye and thank you. He was definitely not going to go in for a kiss, not in front of a parent, and not even if he wanted to. Which he did want to, he thought. An instinctive desire. It was the most normal he had ever felt.

Alice waved at him through the glass, looking happy but sad that they couldn't spend more time together. Seamus had trained her well over the years to put on a mask to suit any situation; Alice luring Mark in with her charm and manipulation for his fragile character was an easy task for her. It was easy to draw boys in, generally speaking, if you said the right things and looked good. Boys were sitting ducks.

However, what she felt was another matter entirely. Alice had had boyfriends in the past but the situation with Mark was different. She was not given much of a choice but to use Mark—she never had to do that before to a boy since she was always allowed to have ordinary relationships with boys, although Seamus was always present if they were in the house. He didn't want any stranger to discover his factory, even if his drug lab was well hidden and disguised. Eliminating all risk was essential for Seamus. Alice also thought that there was a degree of paternal protectiveness coming into play too but since using Mark was Seamus' idea, he didn't really feel the same way about Alice 'dating' Mark.

Alice never would have made much of an effort with Mark had it not been for John Formby being his father. Secretly, she knew that she wouldn't have liked the overly nervous boy that had sat in front of her on that field in ordinary life. She would have been put off. Girls liked confidence in guys, Alice included. And all her ex-boyfriends were confident, some far too much, which she discovered was just as bad as having no confidence. Alice had to persist with Mark though, and she got to discover who Mark was, what he liked, what he found funny, what sore points he didn't like talking about. He was a more interesting a person than she had originally thought and just as genuine and kind as she had first thought too. It made him likable. She had wondered how a whole day in Dublin might be like with someone as shy as Mark—she was even dreading it to a degree. Yet she still felt excitement – real excitement – in spending time with him. That was exactly what she felt when she was with him today too. Was today a date in the ordinary sense as well as being staged? She wasn't overly sure, but it felt like it could have been. If *she* wasn't sure, she could only wonder what Mark thought. It wasn't a chore to be around him at all, like when you're forced to make conversation with someone who has nothing in common with you at all. No, he was pleasant, easy going. Nervous, but not annoyingly so, at least not now he was more comfortable around her. And he was a gent the whole day, offering to pay for coffees, lunch, holding her bags, and she found that very cute. She realised that a truly nice guy can make you feel good, not like the fake happiness you felt on occasions with cocky guys. It made Alice happier in her task set by her father since it was definitely helpful that she liked Mark but when she thought of why she was spending time with him, a twinge hit her stomach every time.

Guilt.

The more time Alice spent with Mark, the more she liked him and the more she felt guilty about using him, but Alice did feel a wave of satisfaction to know that she was helping his self-confidence. It made her feel like a better person too, which really surprised her. She would even go as far as saying that she would still want to see Mark at this point even if her dad said not to bother. She didn't have to put on an act around Mark even though she was in role as Chief Manipulator. She could be herself. She liked who she was around him and could definitely tell herself that he was a genuine friend to her. After a whole day in Dublin, she realised that was definitely the case.

Vanessa looked at her sharply as soon as Mark got out of the car and she had drove ahead. She could tell something was off in the way she looked back at Mark. She watched Alice closely in the rear-view mirror, but Alice quickly caught on that she was watching her, so she turned on her iPod, scrolling to Rihanna without saying a word to Vanessa, trying to look casual. Alice despised her. Not a single word was uttered for the duration of the detoured journey back to the Brensons' house.

Chapter Seventeen

"How did it go, princess?"

"Really well, I think he likes me."

'Good, very good', Seamus mused, "did you get to talk about his father?"

"No. He doesn't like talking about him when I try talking about his parents."

Seamus looked at his daughter curiously. "Why do you think that is?"

She considered. "I don't know… but I don't think they get on at all."

He hummed to himself in thought. This was interesting but quite irritating if he was going to find out anything.

"It explains why they argued when we saw them on our CCTV," he said more to himself than Alice, but she nodded in agreement.

Seamus got up from his desk and strode to the window of his office. Rain was pouring quite heavily and the drumming sound it made on the glass was therapeutic. Watching the rain trickle downwards helped him to focus his thoughts.

"I don't like this feeling in my gut," he announced still facing the window, his right hand clasping his stomach to emphasise the point. "The way John Formby was looking at the house… it isn't normal. He suspects something. I can't get that out of my head. I can't ignore it."

Again, he seemed to be talking to himself, Alice thought, but this wasn't out of character. Seamus was a man who tended to zone out when he was working something out but once a decision was made, his concentration was razor sharp. It was the mind of a great businessman.

And a criminal.

"What are you going to do?" Alice asked.

He turned around slowly, looking her in the eye as if searching for any ideas Alice had a slight sensation of satisfaction pass through her.

"I don't know," he answered.

"Have you had anyone follow him in his car?"

"Not yet, actually," he replied, "I should have as soon as I had my suspicions, but I wanted to wait and see if you found out anything first. More subtle that way. If John is as shrewd as I think he is, I don't want to risk him being conscious of being followed, which he might if he has the guard's third eye in the back of his head."

A thought then occurred to Alice. It would be hard to do but if it could be done it would be subtle, just as Seamus wanted.

"You could bug his car," she offered.

For a moment, Seamus said nothing, as if he had not heard her. He became rigid where he stood, though. There was that glimmer in his eyes that appeared every time a good idea was presented to him. Personally, she was surprised he hadn't thought of it before, but he wasn't really a technologically minded as much as he was a businessman. It was probably his weakness. He could handle calls, texts, turning on WiFi to access the internet but when it came to technology, he didn't really have any clue. Seamus was old-fashioned that way and was distrusting of anything digital. Why type when you can write it down? Why online bank when you can do it in person? He was also paranoid over the idea of being hacked by cyber-criminals. The digital age was dangerous for someone hiding a secret. Sometimes though, technology was needed. For example, he had got someone to come in to put up all his security cameras around his house and his factory which connected to monitor screens in the security room. Despite his lack of know-how, he knew what 'bugging' meant.

"Like sticking a thing on the car?" he asked. Malcolm did something similar to the boy's mother's car but that became a dead end and he had forgotten about it. He didn't expect any leads there anyway.

"Well, yeah. They did it on *'Breaking Bad'*, where Walter White put a GPS Tracker on Gus' car. It might be worth putting a tracker on John's car."

"How interesting," he mused thoughtfully, "but John's car would be risky."

"Yeah, he put it on the inside of the car body just above the wheel. Gus wasn't fooled but still, it was clever," Alice continued.

"But how would we be able to get a device on John Formby's car without him noticing? Or someone else noticing?"

Alice had considered that already. Admittedly, it would be hard.

"Well, there is definitely a high risk during the day of being spotted… but during the night, maybe not. If they had a motion sensor light around the house then that may raise an alarm but I'm sure you could run away again before anyone came out of the house if that happened," Alice reasoned.

"Hmm it is very risky though," Seamus said, frowning.

Alice laughed. Seamus looked at her.

"Because you never do anything risky, do you?" she asked sarcastically.

He smirked briefly. "This is a bit different though," he replied seriously, "I like low profiles. If whoever went to bug the car was caught, that would be bad news for us."

There was silence for a minute or two before Alice spoke, offering to be the one to try applying the GPS.

"Definitely not!" he said firmly, "I'm not having my own daughter being put at risk. You can spy on the boy, but John is another matter. If he knows who we are, he'll know who you are."

"Dad, Walter put a GPS on a car in a car park outside Gus' diner. Broad daylight, too."

"Alice, real life is not *'Breaking Bad'*. Besides, didn't you just say this Gus fella wasn't fooled?"

Alice went quiet again. She was afraid what her father would do if he didn't at least try the GPS Tracker idea. Life wasn't like *'Breaking Bad'*, but there were still similar things that happened where Seamus Brenson was concerned. A drug business is a drug business, and it is a dirty business. People got killed if someone wanted a person 'taken care of' and Seamus wanted John Formby taken care of. Maybe not to the extent of murder – yet – but taken care of in some way. She didn't want that to happen, she was afraid that her father had the ability to kill someone. She had often thought it but because it disturbed her, Alice convinced herself that he didn't go that far. But her father was very precious about his business and he took threats very seriously. And when he felt threatened, he didn't always think with an ethical mind. He was ruthless and cunning when he wanted to be.

Alice cared for Mark too much now to wish harm on John Formby, not that she would ever want to harm anyone. She would have felt awful... even half-responsible if Mark's dad was killed. Mark didn't get on with his father, but he was still Mark's father at the end of the day. How would she be able to look Mark in the eye again if something did happen? She wouldn't be able to. But what could she do? It was obvious that Seamus wasn't going to let her apply the GPS. She just wanted to do something that didn't endanger lives, and the GPS idea was the best she could come up with.

She had to do something.

She realised she could always do it herself anyway, without Seamus knowing. She excused herself from her dad and went to her room. She switched on her laptop and went on Amazon to search for GPS trackers. There were some that were over two-hundred euros but there were some too for about thirty. She wondered how good the cheaper ones would be, but she found one that looked decent enough for eighty euros plus postage. She looked at it for a fair few minutes before deciding that she had to try it. It would mean going behind her dad's back, but it would be worth it if it meant Seamus didn't do something... drastic. So, she read the description, checking it was high quality then read the reviews. Mostly positive. It could be tracked pretty much anywhere on earth, and it could be connected to an app on your phone to let you know where the GPS was in the world...which would naturally mean knowing where John was going if she managed to put a tracker on his car.

Alice proceeded to the checkout and used Seamus' credit card details that were saved on her Amazon account, which was really useful. But in the delivery address section, she put down the address they used for the fake or rather, second house, so that it would be delivered without Seamus knowing. Hopefully. She thought she may use the excuse of meeting Mark at 'her house' in order to pick up the package. There was a small post-box just outside the fake house, American in style, which had survived for decades. She figured a small GPS would fit in it, even with its packaging. It could be a great plan. It could also be a terrible one though and she might get found out, but she had to try. She could handle the risk of a lecture from her dad, but she wasn't sure she could handle feeling responsible and guilty for the possibility of John's death if things really got out of hand.

She wanted to see Mark again, anyway; if she did all she could, she would feel slightly less deceitful. The thought of not seeing Mark again strangely niggled at her insides.

The thought troubled her. If she was beginning to feel something for Mark, that was dangerous.

Five days later, Alice got an email from Amazon asking what she thought of her recent purchase. Seamus had not yet realised that she had used his card to buy the GPS and Alice doubted he would until he got his next card statement, but she wasn't sure how much he studied them even then, particularly his personal bank account. He was much more studious over his business accounts. Maybe he wouldn't notice. The email suggested that the package had arrived, so she went into Seamus' office to say that she was due to meet Mark at the fake house in an hour. Instead of being suspicious, he was delighted. He was convinced that she was only doing more of the work that he had assigned for her. It just so happened that Seamus was having a quiet day, so they went into his Audi and he dropped her off. He left as soon as she got out of the car, especially since in his head, Mark was coming. As soon as Seamus was out of sight, she checked the post box which stood about fifty metres away from the house. The lid was a bit stubborn because rust had crusted around the edges, but it opened after force was applied, creaking as it did so. Stuffed inside was a small parcel with an Amazon logo imprinted into the cardboard, but it was dented in places – it was clear the postman had jammed it inside – probably out of irritation that he couldn't put it in easily and got no answer from the house. She hoped the postman hadn't damaged what was inside, but the main thing was that it had arrived. She checked it was the tracker when she got into the house. It was. She smiled triumphantly.

She then found a plastic bag in one of the kitchen drawers and stuffed the packaging of the GPS tracker in the bag instead of the bin. It would be a basic error to make by leaving packaging in the bin. She had the charger for the magnetic tracker and made sure it was fully charged. She read in the instructions that there were twenty days of usage in between charging so she was aware that she may have to somehow return to John's car if she needed to recharge the tracker, but she felt it was worth the risk if she was careful.

The good thing about the fake house was that it was made to look lived in—all part of the disguise. The kitchen was fully fitted with all appliances and even the fridge-freezers were full of food. In the garden shed were several old mountain bikes Seamus had kept for years from when he was younger. Whilst the tracker was charging, Alice got the shed key from the key rack that hung in the kitchen. Three bikes were inside. She would have preferred the purple woman's bike, but the tyres needed pumping and she couldn't really be bothered finding a pump. She grabbed the bike with the firmest tyres – her father's old bike – and took it out of the shed. She was planning to cycle to Mark's house and see if John's car was on the driveway and most importantly, if she had an opportunity to fit the tracker. She would have the tracker in her pocket ready just in case, but she was very aware that it was a big risk to try and apply the tracker to John's car, if it was there, in broad daylight. There were plenty of bushes around the Formby house so she was sure that if need be, she could nestle herself in them out of sight and wait for an opportunity.

Alice had told her dad she may be cooking dinner at the fake house for herself and Mark, maybe watch a film afterwards too so he would not expect her to be back early. She had told him that if Mark asked where her parents were that she'd say they were having dinner at their friend's house, which Seamus agreed was perfectly plausible. Normally, he would probably be uncomfortable with a boy alone with Alice in a house, but this was not a normal situation. Seamus clearly didn't see that she was actually interested in Mark he was too focussed on getting information for that. Although, one thing Alice inherited from her father was the ability to wear a poker face.

She took her time on the bike as she headed towards the Formby house. Staying in third gear and pedalling slowly, Alice was carefully watching out for possible hide-outs and going over in her head how she may approach the car without being seen. If the car was there at all. About —two hundred metres from the driveway of the house was a wide, big tree on the side of the road. Alice applied the brakes and came to a stop. An idea had come to her to prop her bike behind the tree—if she needed to hide and wait for John's car to arrive, having a bike with her would be problematic. It would be difficult to hide a bike as well as herself.

Alice's thoughts were distracted by the sound of a car engine in the distance ahead of her. There was a bit of a slope before a decline so she couldn't see the approaching vehicle yet, but it was definitely getting closer. Intuition told her to hide behind the tree too, poking her head out slightly to see the car. Finally, the car came into view. It was a black Mercedes; Alice retreated further back slightly but was still able to see the car. Then the car's right indicator flashed as it turned into what was the Formby's driveway. It was John Formby's car then, that's for sure. She knew he owned that kind of car. She was pretty sure she wasn't seen, but now the enormity of the task she had set herself stared her in the face and her heart raced. Alice was relieved she had stopped when she had, it was probable that she would have been seen if she hadn't. She stayed where she was for a few minutes, allowing time to compose herself and refocus. When she judged that sufficient time had passed for John to have left his car and enter the house, Alice emerged from her hiding place and walked slowly up the road until she reached the first stretch of bushes which stitched around the perimeter of the Formby property. Checking that the road on both sides was clear, she pulled aside some foliage and looked through into the garden. She couldn't see any cars within yet, so she repeated the action further down. She could see two cars now, parked near the front door of the house. She recognised both cars; one was the one that had just entered, the other was Mark's mother's car. Everything appeared still, and the house seemed quiet with no lights shining from what she could see.

From behind her this time, she could hear another car approaching. As the car approached and passed by, Alice bent over her shoe, pretending to tie her shoelace. She realised she couldn't hide on this side of the road, she would be seen by drivers, and look conspicuous. Her peering through bushes into someone's garden wouldn't look normal so she knew she would have to hide elsewhere. She quickly scooted past the entrance to the Formby drive and climbed over the small stone wall which gave onto the field and looked for a hideout spot in the bushes. Alice found a small bit of hollow ground, a small dip which she could nestle herself into. Slowly, she moved aside the seams of nature's knitted hedge, stitched together by twigs and leaves. She could see all the windows of the front of the house from where anyone would see her if they happened to look.

Now that she was there the plan took a more complicated turn; in her head, she thought it would be relatively easy to slip in and out given the perfect time to do so. She wasn't so sure now, but she was determined to go ahead with it now that she had gone this far. She fiddled with the GPS and made sure it was on and connected to her phone. On the screen of her Samsung, she pressed on the app which she had installed, and a small blue dot flashed at her on a map, showing that the GPS was working. She smiled slightly. This was the easy part. The hard part was still to come.

Everything was still, save for the slight breeze passing through the leaves. It reminded Alice of one of those times in horror films when everything was too quiet. As if the silence was luring you into a trap, watching and waiting for you to make the fatal move.

The longer Alice sat amongst the leaves, the harder she found it to do what she had set out to do. Giving yourself more time to think was the hardest part, it allowed you to consider all the consequences, all the things that could go wrong. Doing it straight away was probably the best thing to do in most cases, like jumping into an ice-cold pool. The more you think of the cold, the more you torture yourself. Alice wished she applied that logic now.

The air had cooled since she took her position. Alice could feel the chill through her tie-up hoodie and her hands were freezing. Another thing to distract her. She knew she had to do it soon otherwise she would be too distracted to find the right moment. She needed to concentrate and ignore the cold.

One of the lights had been on since she had hidden in the bushes. This was another reason why she didn't dare yet go forwards. It was likely that someone was in that room—she couldn't tell what room it was other than that it didn't seem to be the kitchen. Another light then flicked on the other side of the house and this time she could see someone through the window. She didn't need to look twice to know it was Mark and in a weird way, seeing him relaxed her slightly. It didn't look like he was doing anything in particular, just pottering about and then he was gone, and so too the light. The light that had been on the whole time didn't look like it would go off any time soon. Had she been waiting to make a move when all the lights were off? Really? Even when all the Formby's were in? She laughed at herself. What had she expected? It was dark enough now to need lights on. Besides, she needed some daylight to see what she was doing and enough that any motion sensors didn't come on—in darkness, you're looking for trouble doing that. Not that it was any less risky in daylight. Alice looked at where the black Mercedes was and judged that if she crouched at a half-run, she'd get there in no more than ten seconds. If she didn't do it now after psyching herself up, she feared she never would.

She pulled the hood of her hoodie over her head. If she were seen, at least she could run away with most of her head covered, obscuring her features. She could then spring on her bike and get away as quickly as possible. It would be better for one of the Formby's to see a hooded girl, obscured from view, running away instead of seeing that it was *her* running away. As her dad mentioned, it was possible that John knew who Alice was and if he did, Alice would be creating trouble for him.

Alice carefully made herself a clearing, trying not to make too much noise. On a quiet late afternoon like today, any sound made seemed amplified, at least in her paranoid head. After making a large enough space to pass through she crouched silently forwards into the garden, not wanting to rush any move but now that she was completely exposed, she needed to act quickly. Very quickly. She looked at the hole in the bush where she passed through briefly and tugged on some of the branches to cover the space, just so it wasn't apparent that someone had been there. Seamus had always drilled that into her head: cover your tracks. Then, with her head tucked down and giving herself a view of the way in front of her and the nearest windows, she crept forward and reached the car in what felt like a lifetime. John had forward parked meaning she was ducked behind the rear of the car. The night stayed still with no one shouting, "Oi, who is there? Get off my property!" After another few seconds of silence, Alice slipped the GPS out from her pocket and reached her hand under the inside of the car and soon felt the pull of the magnet within the GPS wanting to attach itself to the car body. Once attached, she made sure it didn't move too much. It didn't, the magnet seemed to be a strong one. Then she rushed forward again, this time towards the driver entrance. She was back on the main road and still no one followed or shouted after her. She allowed herself a grin and ran towards her bike. "That wasn't so bad, was it?" she mumbled quietly.

Back at the tree where she dumped her bike, Alice glanced at her phone, checking the app which connects to the GPS. A small blue dot flashed on the map.

"Just like '*Breaking Bad*'," she said aloud, smirking to herself.

Chapter Eighteen

John Formby stood near the window of his blackened-out bedroom, looking down at front garden, although, it would be more accurate to say he was watching. Did the stupid girl not realise he knew that she was lingering outside? A lot of people may not have spotted someone one hundred metres away hiding behind a tree whilst driving but his well-trained eye would never miss that, especially when that person was doing a bad job of staying hidden. The girl had been practically swinging her body away from the tree as if she was pole dancing before he turned into the drive.

What an amateur.

Still, it had made him curious. He knew the girl was Alice Brenson and he was pretty sure she knew who he was, aside from just being Mark's father. It didn't take a genius to work that one out, he was in the local news for quite a while as the guard who had survived five bullets. For a while, he couldn't see anything or anyone after he parked and got inside the house. He didn't see the girl run past the entrance and jump over the wall, but he did know she was there somewhere. It was an occupational hunch; all he had to do was wait. Why else would she be hiding behind the tree if not to get up close to the house? He just didn't know exactly what she was up to yet, but he knew she was there for a reason. Had her father sent her? Probably.

He was waiting patiently in the dark and just as he thought the girl had done a runner – he was about to turn away – he saw her in his peripheral vision, emerging from a bush nearly out of view from where he stood but the movement caught his eye. Another schoolboy error (or schoolgirl error in this instance) was to assume no one was in a room that was dark. In his experience, people assumed a room was empty if no lights were on, as if it was somehow impossible for someone to be in a room that wasn't lit. In the girl's defence, she probably thought she hadn't been seen standing so obviously by that tree and thus have no reason to assume someone was then going to be watching her from a dark room. Still, he enjoyed watching the girl incriminate herself so effortlessly. She underestimated him and she certainly wasn't the first person to do so.

He peered down at the driveway as he watched Alice crouch by the car, as if squinting would allow him to see more of what was going on. A light clicked on in his head though as he watched her insert an arm under the body of the car and he smirked. He knew exactly what she was up to and it confirmed his suspicions that he was being watched in a broader sense. Seamus Brenson was investigating him, he must be, and using his own daughter to do the dirty work for him.

"Feckin' coward!" he snorted.

Then the girl got up quickly, stooped over and ran off towards the main road.

"Run rabbit, run," he murmured happily, "while you can."

He made no haste to get into his car, taking each stair leisurely as he made his way downstairs and shaking his head at the sound of Mark shouting, "Get the bloody ball!" That Xbox of Mark's drove John mad—Mark seemed to be obsessed with FIFA at the moment. He acted so casually on the way to his car that he even allowed himself time to swig a few gulps of orange juice from the fridge and grab an apple, which he used his teeth to carry. Sylvia gave him a blank look as he made his way to the door. He thought she looked tired.

"I'll be back in a bit," he said indifferently, but frowning as if having to remove the apple from his mouth was an inconvenience.

"Where are you going? It's getting dark," Sylvia enquired.

"Out," he replied.

Before she could say anything else, he walked across the threshold and closed the door behind him.

Instead of getting into the car straight away, he reached under the body of the car padding his hand within the interior for the GPS which must be there. He found it after a few seconds, noticing a small red light flashing when he retrieved it. He was going to follow the girl to see where she went but he couldn't faff about if he was to catch up with her. Understandably, she wouldn't think that the car had moved since she had attached the tracker to his car only mere moments before. However, if she did check, he wanted to make sure the flashing on her phone indicated that his car had not moved even though he was hot on the girl's trail. He left the GPS device on the gravel and got into his car.

John drove one-handed, holding his apple in the other as he munched away. He glanced lazily behind the tree the girl had been hiding but in the dying light he couldn't see anything, not that he expected to. He drove slowly, staying in second gear, not wanting to be seen or heard if possible, for safe measure and thankfully the night appeared to be still. The road was a long, straight one, meaning he could see up ahead for a mile or two during the day but even as the sun was shedding its last rays for the day, he reckoned he could still see ahead for a couple of hundred metres. After a minute or two he could just about see the outline of a cyclist, helped by the frequent flash of the red bike light. He dimmed his front lights and continued forwards, slowly catching up with the cyclist but maintaining a good distance still. He knew Mark went to Alice's house for dinner—that much he had overheard, so he planned to follow her at a distance and see if she went to another house. From the direction she was travelling it seemed that she wasn't traveling to the actual Brenson household across the field, although there was no easy place to get a bike onto a field, he supposed. The cyclist must be her; he watched the direction she went. Maybe she was taking a detour home, but he doubted it. Another of his hunches. He thought it strange for Mark to be invited into the Brenson home across the field considering the array of secrets that could be exposed in the house if Mark went prying. For all Seamus knew, Mark was also being used in some sort of investigation against himself. Would Mark betray his own father? How could he, anyway? He knew nothing. The whole scenario felt like a Cold War, where both sides sat in their trenches in the form of a house with a field in between; armed but not doing anything. Just waiting. Each side thinking there could be enemy spies. This field between them was not a battleground, he thought to himself, it was a field of spies.

He soon caught up with the girl, now only just under a hundred metres away, he estimated. He was aware that the girl might sense a car behind her now, maybe even hear the sound of tyres against the stony surface. He turned off all the car's lights now and cut the speed, now rolling at a snail's pace. He knew the road well enough and the red backlight of the bike ahead was all the guidance he needed; the distance between bike and vehicle now remained constant. They were approaching the end of the road at last and the bike took a turn left, which he followed.

Eventually, the bike swerved right into a driveway of a medium-sized detached house. "Definitely not the Brensons' then," he muttered. He had not actually been able to see the cyclist's face in all the time he followed but who else could it be other than the girl? It was getting dark and it was a quiet country road… if you could call it a road, it was more like an undeveloped lane. Not many cyclists would choose to cycle it in the dark on such a lane, but it was a quick enough getaway if you were up to something like the girl was. After a few minutes, he found the nearest possible spot to park the car, hidden from view should someone in the house look out from a window. There was a cluster of thick, tall trees before you reached the house on his right and on the same side was a small patch of ground that cut inwardly into the bushes. It looked like it had been used by vehicles to park in the past and nature had obliged to allow enough space to accommodate a car. John got out and played the girl at her own game, skulking around the premises of a house whilst she was inside. Assumedly inside anyway. He crouched low and approached the side door of the house which would lead to the back garden. He silently pulled on the handle, expecting it to be locked but it was not. He tiptoed through, silently closing the gate behind him. He kept to the shadows and crab-walked around the perimeter of the garden, which was long but not so much that he couldn't see inside the house with relative ease. He was conscious of the risk of being horribly exposed if there were sensor lights in the garden though. So far, he was in luck but if he kept going in the direction he had been travelling he would reach the shed, which would most likely have a motion sensor light if there was one somewhere in the garden. He had risked not activating one already. John crouched and retreated into the bush, pulling his black hoody over his head. He could see the inside of the kitchen and a few seconds later a teenage girl entered. It confirmed that the cyclist was Alice Brenson. He allowed himself a small smirk. The Brensons had a second home after all. Before any more thoughts could circulate his mind, Alice looked up sharply, staring directly towards him. For a panicked second or two, he thought she could see him and was about to sprint away, but she then looked away slowly just as he was braced to run. Had she sensed someone in the garden? Like the indescribable sensation of being watched. Quite possibly. He had to be very careful here.

The night was on his side as he watched, the increasing darkness helped him to be unobserved. He inched even further back into the bush for good measure. He didn't want to risk any move now until she left the kitchen. As he crouched still, he wondered why Seamus Brenson needed a second house other than an unlikely safehouse; usually safehouses were far away from the place you were known to be. Was it a house he had bought and rented out in the past? It wasn't something he had come across in Brenson's records. When was the house bought? He didn't know the address of this house either—he may have to resort to Google to help him out. There was no doubt that this was the house where Mark had had dinner. From what he had heard, Mark had met the girl's parents. Would Seamus and his wife make appearances with his son though? It didn't make sense if they were spying on him through Mark. They would be risking Mark talking to his own parents once he was back home. What was going on here?

A few minutes later, Alice left the room. He waited another minute or so before striding back to the side gate. It was a very still, silent evening. Every small sound he made seemed amplified in his mind, but nothing stirred even as he made his way back to the car. It was pitch-black now and he looked back at the house as he stood facing it from the side. Then he heard a door open. It sounded like it came from the house. Another few seconds, he caught a glimpse of a beam of light. It looked like the beam was pointing in his direction with only the closed wooden gate blocking it; the light travelled in between the cracks. Someone was in the garden with a torch. Alice must have sensed something and decided to check; he would almost certainly have been caught out if he had lingered any longer. He counted his blessings and headed to the car, wary now that the girl could come through the gate and spot his car leaving. He manoeuvred a three-point turn on the narrow road in near silence and drove away.

No beams of light followed him.

Chapter Nineteen

Alice left the kitchen to find a torch. She could have sworn she had seen one somewhere. She had an unsettling sensation of being watched when she was in the kitchen; admittedly, it was probably paranoia but until she checked she would remain uneasy. It felt like someone or something was watching her from the back garden but the night closing in prevented her from seeing anything. She soon found a big torch in the pantry that worked and headed out of the side door of the house. The gate, she noticed, was unlocked and she remembered that she had forgot to lock it after putting her bike back into the shed. She cursed herself for being mindless and attached the lock. She turned the beam towards the back garden, now aware that if anyone was in the garden, she had locked them in. Her heartbeat began to beat faster, and her breath seemed deafening as she exhaled into the cool air, producing puffs of visible breath. She then turned back towards the gate sharply as she heard the sound of stones being crushed against the weight of moving tyres. Someone was driving away from the alcove that fashioned a car parking space for postmen. Having locked the gate with what was a fiddly and awkward lock she now raced back into the house, fumbling for her keys to unlock the main front door in order to get out quicker but she was too slow; by the time she got outside, any sign of a car was gone. She wandered out to where a car must have been parked, shining the torch's light at the ground. Were the faint tyre marks she saw fresh ones or old? She couldn't be sure—but she was sure that she had heard a car leave. She saw no car parked nearby as she cycled back from the Formby house, so why would a car be parked there for a few minutes before leaving again? She had only been in the house for ten, maybe fifteen minutes max. It seemed strange. Maybe someone just made a wrong turn so used the alcove as a means to manoeuvre back to where they came? That seemed like the most rational explanation, but she couldn't shake off the feeling of it linking to her sensation of being watched. She knew it sounded ludicrous, but she wasn't a girl from a typical family. Her father held secrets and she was being used as a pawn in his game, to spy on someone for whom she now had increasingly strong feelings, friendship or otherwise. And Seamus probably had many enemies too. Was it so crazy to think that one of his adversaries had found out about his safehouse and begun investigating? She shuddered. All these thoughts were freaking her out and she forced herself to shake it off. She glanced down at her phone to check the GPS tracker hadn't moved. It showed nothing had changed. She relaxed slightly and made herself a cup of tea.

Meanwhile, John had a lot to think about. He tried to work it out in his head but couldn't, nothing made sense about the second house, and he felt it never would unless he asked Seamus Brenson himself. It was pointless asking Mark; for one, Mark would most likely think the 'second home' was the only home; secondly, he had already tried questioning him about Alice – any further attempts and it might raise Mark's suspicions and result in him being watched even more closely by Mark, which would be enormously irritating. Mark had already been watching him closely recently, making mental notes of his movements and when he was absent too no doubt. If he had been close to Mark, he would have tried to adopt a normal father's stance of having a man-to-man talk about girls, asking how it was going with Alice and all that kind of thing. Eventually, he would be able to determine what Mark and Alice talked about; crucially, if they ever talked about him. He would soon have worked out what Seamus Brenson was thinking. However, there was no point dwelling on what could have been. He would still find out what Brenson was up to, it would just take more time. A small part of John wished that he could bond with Mark, but he could not look at Mark without feeling anger towards Sylvia.

A new idea hit him then. Could he use Sylvia in some way? Perhaps he could coax her into investigating the Brenson house if he spent time being nice to her, apologising for his behaviour towards her. Maybe. It surprised him that she had not already tried to leave him; he wouldn't stop her, but he knew that she was too weak to throw away the security he provided; a home, money, food on the table. It was all from him, she didn't earn anything now after his incident and even when she was working it wasn't a lot she got as an admin assistant. Her mind wasn't in the right place. No, she wouldn't leave, particularly now that he had received a sizeable amount from the Force as a token of loyalty. Maybe she would do his bidding if he played his cards right but what would he even ask her to do other than monitor the house from across the field? He thought if he got her to ring Alice's mother – should Mark get the number – that he would be able to then ask Sylvia to find out Alice's mother's name. He knew her name was Kristen so anything other than that name would mean Seamus was paying someone to pretend to be Alice's mother. It was as good a plan as he could think of for the time being—it may result in a few of his questions being answered. Equally, it could also result in further questions and he would also have to come up with a reason why Sylvia should ring without rousing suspicion, but it was worth a shot, he felt.

So, it was time to be nice to his wife. Every now and again he felt a stab of guilt at the way he treated her as he did love her once, but he had just as quickly dismissed that feeling when it was outweighed by the undying anger he felt towards her. He knew that if he was reasonable, he would have just left himself after learning about the affair. Most men, he knew, would have done that. Perhaps, he too, was weak. He had convinced himself that the power he had enforced upon Sylvia by making her life difficult was his motive to stay with her but at the back of it he was a man who didn't like change or admitting defeat. Leaving would have been like a defeat to him, allowing Sylvia's actions to drive him out of his own home that he had worked hard for. Hidden in the huge wall of hate were some solitary bricks of love, though. He still loved her, in his own warped way, or was it just the memory of being in love? He remembered the happier times with Sylvia, before the affair and the arrival of Mark, and those memories clung to him during those moments when he had an urge to walk away. As he parked in the drive, he remembered being sat on Kilmurvey beach in Galway with Sylvia. The year was 1996. They had taken a long weekend there and it had been an unusually hot summer's day on their second day. There wasn't a cloud in the sky and Sylvia had stripped down to her bikini to make the most of the sun, whilst he wore shorts and a long-sleeved top. He didn't particularly like the sun and he burned easily but he remembered that day how happy he was, sitting on the beach for hours, laughing away with each other. Back then, Sylvia had an amazing body; all he could do was stare at her and think how lucky he was. That had been a good day, possibly the best day they had ever had together. How far away they were from that now.

Perhaps that memory would be helpful. It might make it easier to be nice, even if it was to fulfil an agenda. If he was honest with himself, he hadn't tried to be nice to his wife, not for a very long time. He had been too focussed on his bitterness to consider it. He knew that his anger continued to simmer away on a heat because he held onto that bitterness. If he let go, tried to move on, he knew there was a chance that his anger would begin to ebb, be pushed back into his subconscious. The fury inside him though was inbuilt, it was routine. It was like trying to give up smoking. The longer you smoke, the harder it becomes to give it up. Anger was his nicotine and he chain-smoked it. To force himself to change would be impossibly hard. He knew how to be nice though, he had beguiled many people before.

Sylvia was in the living room watching TV. *MasterChef* was on but she gazed vacantly towards the screen. She had been unhappy ever since moving here, distancing herself from everything that was normal, and he had often spotted her expressionless. It was irritating to watch so he normally walked away from her. He forced himself to think of the day on the beach and he smiled at her.

"What?" she asked blankly, without looking away from the screen.

"Would you like a cup of tea?" He had to force down his disgust at playing nice.

A flicker of surprise flashed in her eyes and it was enough to break the daydream-like trance. She glanced at him.

He asked her again and for a few seconds she said nothing.

"Okay," she replied cautiously, as if wary of a trap. Which in a sense it was, he thought.

John nodded and made two teas. He couldn't remember the last time he had made her a cup of tea. She used to joke how bad he was at making them, he remembered. She told him once that she had never seen anyone mess up a cup of tea like he did, and he had laughed along at the time. It was another small memory that proved to him that they had once been happy. It is often said that a portfolio of small memories means more than the occasional bigger memory. He could see the logic in that.

The TV was on low volume but the silence between husband and wife made it seem much louder. He hugged his hot mug whilst balancing it on his knee. He tried to act like he was comfortable, pretending that he was enjoying Gregg Wallace's over-expressive face as he tucked into a cheesecake. Sylvia's expression was much more strained, taking in this sudden normalness from John, her eyes now trained on his calm face instead of the TV. It was almost as if the past sixteen years of hostility had never happened, as if they were a normal couple enjoying each other's company. She couldn't remember the last time there was a comfortable silence between the two of them, nor the last time she had seen John's facial expression be anything other than stony in her presence. She had dreamed of such a moment for years, even though it was not a memorable moment just both being in the same room, watching the TV together. She had all but given up on that ever happening but here they were, out of the blue, sitting together in the same room as if the past sixteen years had actually just been a five-minute spat about something mundane and had now reached wordless, mutual reconciliation. The whole thing felt so bizarre that she almost allowed it to overwhelm her. She so wanted this to be a gesture to tell her that he had decided to let go of all the anger and resentment, but she knew better. She wasn't always the most pragmatic, but she was not fooled, hence her wariness. Even though John had not expressed any strong feelings of happiness, just casual contentment, it was still a huge change in his manner. It was a little bit like when Scrooge became a better person overnight and startled everyone with his sudden good nature. She was pretty sure John was not a transformed Scrooge though so, she stared at him, not wanting this moment to end, not wanting an icy stare to break the wonderous ordinariness of the situation by her speaking. What could she say? What could anyone say right now?

John was not oblivious to being stared at, but he pretended not to notice for a few minutes. Act comfortable and after a while she may start to feel comfortable too. That was the first step, doing enough so that she was no longer alarmed or surprised by his being nice. He didn't know how long it would take—weeks, months, years… never, but he had to try. He had to start somewhere plausible, not just be nice without explanation, which wouldn't wash. He had to apologise first and foremost. He might even want to believe it too, maybe if he said the words out loud the cracks in his wall of anger would start to form.

"I'm sorry," he stammered. "For everything." The words felt alien to him, like reading aloud an unfamiliar language from a piece of paper. He felt like the words didn't sound right and he could tell he wasn't the only one. Sylvia continued to stare, now her whole body frozen in shock. Her lips were moving as if she wanted to say something but couldn't. He didn't know what to do now, the whole thing felt like an over-emotional scene from a film. He kept talking. "If I don't say anything now, I feel like I never will, so here it is. I'm sorry. I know I've been a complete arsehole and you probably don't believe a word I'm saying, and I wouldn't blame you if you didn't, but I've been wanting to say something for ages but… well, I don't know, I have no buts or excuses. I've acted…" he tried to find the right word, any appropriate word, "terribly, I've acted really fecking terribly towards you and to Mark. I still love you, Sylvia."

Despite the pretence and the weirdness of saying the 'L' word, he surprised himself at how much of what he said was true; he did feel bad sometimes about his actions and he did still love her. He thought he might a bit anyway. He felt guilt even that he did not love Mark, he had not allowed himself to love him. As he sat there, he felt the 1996 version of himself trickle to the surface, but he was still fuelled on his new tactics. His anger was a misguided response to love being challenged by an affair; love and hate were not a million miles apart after all. During his bitterness, Mark was born. He knew about the affair before his birth and so his feelings had always been barriered by resentment. He never allowed himself to love Mark in the same way and for that he did feel guilt, but his pride had prevented him from ever acknowledging it to himself, let alone Mark himself. The DNA test confirmed Mark was his, but it so easily could have been different. The pregnancy was the worst time of his life and so when Mark was born, everything felt like it had been tainted by betrayal and Mark was the final product.

Tears cascaded down Sylvia's face. Maybe she had seen truth in his eyes. He couldn't deny it either, not completely. His outburst was not totally untrue despite it being deceptively rehearsed.

He felt like his head was spinning.

Chapter Twenty

That night, Sylvia awoke. It was just after two a.m. and she had only managed an hour's sleep. John breathed heavily beside her. The past few hours were surreal, completely out of the blue and after the initial wave of relief and happiness that John's behaviour had changed so dramatically, she now only felt bafflement. It didn't make any sense. Before tonight, they had not spoken more than a few things to one another in anyone sitting and now they had spent several hours chatting away as if they were star-crossed lovers again from the mid-nineties. She hadn't questioned any of it, his sudden desire to want her company. She basked in what she thought was too good to be true in case he suddenly decided to return to his aloof ways. Okay, at first, all she could do was stare at him but under the circumstances she didn't think this was unusual. But why now? Why wait sixteen years to make an effort, on a random night when nothing in particular was happening?

As she brooded, she also started to recollect her other recent broodings. Why had he taken such an interest in Mark and his new friend Alice? The whole incident with his demanding-to-know-answers attitude and the alarming conclusion of Mark hitting John over the head with a vase was not exactly the picture of domestic bliss. Even amongst dysfunctional families such a scene was bizarre. If she thought that he was being protective of his son she would, perhaps, understand but she knew that John couldn't bear to spend too much time in the same room as Mark. Why now care who Mark talked to or spent time with? In the many days of her sullenness, she knew that there was a reason for the interrogation, she just didn't know what. Due to the nature of his job, she considered the possibility that Alice was part of some investigation he was conducting, but what investigation? He hadn't returned to work and as far as she was aware the force had not asked him to start digging into any cases in the meantime. He wouldn't have told her if they had asked him, but she would have sussed something out by now; she would have been able to tell if

John was working from home. In fact, John rarely was home. If not in his small office, he was out. She never knew where and she didn't bother asking. Maybe he was working on something without her knowledge. For all she knew he was having an affair, which would not have surprised her. That sounded like the kind of thing John would do – maybe not the old John that she had fallen in love with – but definitely this not-so-new-now John she had become accustomed to. No doubt he would have enjoyed sleeping around knowing that all the power was in his hands, knowing that she had nowhere to go and nowhere near enough money to leave and start afresh. It would no doubt feel like the perfect revenge, but she was not sure how much she would have cared if he had; maybe if he knew that, the taste of victory would taste less sweet.

Occasionally, Sylvia allowed John to have sex with her. She never enjoyed it because she knew that he was doing it for his own sense of selfish release, she had always felt like an object these past sixteen years. He always made her face downwards, away from him so that they couldn't look at each other and he always made sure the light was off. There was no sign of intimacy, it was more of a need must kind of thing. She knew she deserved better than that, she knew she had a right to enjoy sex and to enjoy it with a man that wanted her for her. One night had ruined everything. One stupid one-night stand and now she was too weak and too helpless to get out of the rut. Inwardly, she felt weak but had also felt this all started because of her so she always lay there limply and let it happen. And even away from the bedroom, Sylvia blamed herself most for all that had happened; she had made her bed as the saying went and now she would have to lie in it, no matter how uncomfortable that bed was. She had stepped through a trapdoor that she later realised she could never return, all because she slept with one of John's old friends from school. She was out on a work do and when most of her colleagues had gone home, she had stayed with two of her closer friends, Clara and Katie. It was a Friday night, so they figured, why not? Both of her friends were single, and both had found men at the bar before too long, leaving her alone with a vodka and coke that she had barely drank. That's when he came over. Brian Nicholls his name was, a charming man who had got her too drunk for her to say no. It wasn't rape, though. Even though she was happy in her relationship with John she had been successfully seduced—

she didn't know why she felt the need to sleep with this man other than the fact that he was very attractive, but she could recall wanting to sleep with him. At first there was nothing untoward or risqué about what they talked about. As more drinks were consumed, she became more susceptible and when he said something about a girlfriend dumping him out of the blue – she couldn't quite remember the exact details – she reached out and took his hand comfortingly. She didn't remember an awful lot more after that—shots, some more mixers and before she knew it, she was in his bedroom. That was the worst part for John when she admitted everything the day after. She felt awful and knew it couldn't stay hidden forever. She would never forget the moment John's face clouded over, a cloud that had never allowed sunshine to pass again.

These recollected memories left her sat up in bed, wide awake. She just wanted to cry about all that had happened. How different everything would have been if she had just got a taxi home once her friends had started dancing with men in the bar. It was not so long after admitting her one-night stand to John that she cut ties with Clara and Katie. She felt as though they were reminders of her mistake. It was easy to try and put some blame on someone else even though she knew it was wrong to. She didn't just cut them out of her life either, eventually she cut everyone out because all that seemed to matter was making John forgive her. When she found out she had become pregnant, she tried even more to get her happy John back. Everything surely couldn't have been lost, they had been such a rock-solid couple! If she could have foreseen the future that awaited her back then, she would have tried to get away or at least try to accept that the relationship with John was now irretrievably damaged and somehow move on. She would have been able to move back in with her parents or a friend. Now she had no one.

But could things be turning? A newfound hope spread through her and she felt like if she made an effort again in the wake of this happier John, maybe they could pick up at least some of the pieces of their woefully withered marriage and learn to love each other again. Maybe if they could get things back on track, she would be able to look at herself in the mirror each morning and justify to herself that she had made the right decision all along. Perhaps, too, she would be able to say that she wasn't a failure of a mother for not raising Mark in a home that was full of love instead of one where there was only resentment and tension. Of course, she had always shown Mark love and she got it in return but even despite that, she could not help but wonder what Mark was really thinking. Did he think her a weak person? A bad mother? A good mother but unable to provide for him properly? She was too afraid to ask him and it had never come up. She knew that Mark loved her and that was enough. Without Mark, she would have been a completely empty soul. Mark had given her a reason to live or at least enough of a reason to keep going. Maybe they could start to become a normal family, perhaps family harmony could still come out of all this.

Calm down, you naïve woman, you know what John is like. A few hours of him appearing nicer to you doesn't mean all is forgotten. It doesn't mean he's forgiven you.

She knew she was naïve and that she always had been. An idealist, a dreamer. The horrible thing with anger was that it could resurface at any time, like a black cloud that would inevitably return after a few days of blue sky.

She got out of bed then, restless, knowing that she wasn't going to go back to sleep any time soon. She walked along the hallway towards the stairs, making sure to dodge the creaky floorboard so as not to wake Mark. She tried the knob of Mark's door, but it was locked so she went downstairs to make herself a glass of hot milk. The evening was mild, so she decided to sit on the edge of the stone patio floor, enjoying the light breeze through her hair. She felt better for the fresh air, she needed it. She wanted to escape her mind, compartmentalise her current worries and think of nothing but the still night, the stars and half-moon which gently illuminated the night's sky.

When Sylvia was a girl, she kept a diary. Writing every evening after dinner, she would pour out her thoughts about anything and everything; sometimes, she talked about local boys and her desire to meet her celebrity crushes, particularly Patrick Swayze and Harrison Ford; other times, she wrote about school life and growing up as a teenage girl. She tried to cover different subjects in her writing so that when she came to reading it later in life, it would not be dominated by predictable things girls her age would write about boys, period pain, trying to fit in with the popular girls, make-up, etc. Despite this intention, the same topics tended to crop up often.

Being an idealist, she wanted to sound interesting and, in many ways, she had failed. She did not think she was anything special, but she did have fun when she was younger. She laughed with her friends, flirted with boys her age and sometimes played the rebel. She had a personality, something that made her interesting. Where had that girl gone? As she sat looking at the stars, she pondered the thought of returning to diary-writing. The liberation of it had once made her feel better when she was sad and even though she was the only reader, the act of writing helped to get a weight off her chest. She could see why it was something encouraged by mental health specialists for their patients; thoughts locked in one's mind was corrosive and lately she felt like there was so many emotions within her that she needed to vent. Occasionally, she talked to Mark about her feelings, but she knew in herself it was not nearly enough, and it would be unfair to burden her son with the catalogue of feelings going on in her head. Mark was a teenage boy himself; he had his own worries. She didn't need him worrying about her too. Soon, Mark would need to return to education and that would be something good for him to focus his thoughts on. It was currently late July and school had finished since May. Of course, Mark had not been to school really since John's accident; initially, she had arranged for homework to be sent electronically to Mark but Mark was distracted by everything that was happening. She expressed her concerns to Mark's school in Dublin and they advised he take time away from education, concentrate on his wellbeing and recover psychologically from the trauma of John's accident as well as the stress of media attention. Mark was ready to go back to school, she felt. Especially now that he had newfound confidence after being with Alice. Sylvia had already arranged for him to repeat the year – his final Junior Cert year – with a local high school in Dunshaughlin. She had not really talked about his return to school with him, she knew that she would need to soon. Mark going to a new school and mixing in with other kids was another thing she worried about; he was not naturally sociable; he was a social introvert, which was why she was so surprised by his making friends with Alice, particularly because she was a girl. He didn't have friends and the few times she had tried to encourage it, nothing happened. He was happier by himself, which wasn't unusual, but she wished he at least had one friend he could talk to. Sylvia hoped that Alice went to the same school, just so that Mark would know someone, but she wasn't relying on it. She'd casually bring it up with Mark at some point. Sylvia had every reason to like Alice. She had seen a change in Mark since he had met her. Mark didn't trust people as a general rule so the fact that Alice had managed to get close to him spoke volumes about the type of person she must be. Her only worry was that Alice was a girl, a very pretty girl too, and that Mark would become infatuated by her. Despite a rise in confidence in him, which was fantastic to see, what would happen if Alice crushed any hopes of romance? She would just have to see how things panned out, she supposed. Besides, what if Alice did like Mark as more than a friend? Being his mother, she'd always say her son was handsome, but she knew too that he actually was a decent-looking boy. She was unsure how well Mark would be able to deal with strong feelings.

From preoccupying her thoughts about John to worrying about Mark's transition into school life, she suddenly felt a wave of tiredness as she hugged her now cold and empty mug. With a sigh, she stood up and returned to the house.

Chapter Twenty-One

Mark probably would have left the house even if the heavens cascaded never-ending rain. At that moment, he didn't know what to think, what was going on between John and his mum and how did he feel about that? He had seen them talking yesterday. Pleasantly talking. Like a normal couple. Like normal parents.

Like everything was normal.

He did not feel normal though, far from it. How are you supposed to respond in his situation? Nothing made sense and he felt exhausted trying to keep up with what his mum was thinking and feeling. One the one hand, it was strangely nice to see her relaxed in John's company. That was not something he was used to seeing but he did see a sort of desperation in her eyes that suggested she was clinging on to that moment as much as possible, expecting a swift return to the old ways at any second. On the other hand, he felt anger and frustration towards her. Did she not realise John was clearly up to something? His agenda is not going to be to set aside differences and start again afresh. Surely, she must know that. From what he could see from her face, he didn't think so. Hope and desperation does crazy things to the mind, he thought. She was too engrossed in her desires for happiness that she couldn't see what was literally staring her in the face. His real, manipulative father, not the father she had always wanted him to be to Mark. He didn't think either of them had seen him watching them, they hadn't acted like they had anyway.

Mark didn't like to admit it but whatever game his father was playing he was playing it well. He was convincing, charming and his sickeningly sweet but unmistakably nice smile was winning her over. For a woman like his mother, a tactic was always going to work. It annoyed him how easily fooled his mother could be and time after time he put those feelings to one side. If he let frustration rule his mind, he'd have no parents to confide in, which was important when you had no friends. It was worth it because he loved her more than any other person and it was easy to forgive her.

That didn't mean he felt like tearing his hair out sometimes.

Right now, he didn't want to see either of them, let alone talk to them. The sight of this fake happiness was driving him crazy and he couldn't wait to forget about them for a few hours.

He was walking towards the kitchen to make some sandwiches for his picnic when he stopped instinctively at hearing his parents talking at the dinner table. Mark inched towards the closest blind spot and listened in. He wasn't sure why as he was sick of them talking nicely to each other, but he couldn't resist.

"It's nice to see Mark making a friend," John said.

"Yes, it is. She seems like such a lovely girl," she cooed.

Despite himself, he couldn't suppress a small smile knowing that she approved of Alice. It was weird to hear John act like he cared about him though, which turned the smile into a scowl.

"Have you spoken to her parents?" John asked.

"Not yet, there's not really been a need to."

A slight pause. "Huh, fair enough. Has Mark talked about either of them? He had dinner with the family, didn't he?"

Here's the start of the interrogation again, Mark thought.

"Erm, not a lot. I think Mark mentioned that Alice's mum insisted he call her Valerie… or was it Vanessa? I can't quite remember. Not a lot more than that though apart from that they both seem nice. Why?"

There was another pause, longer this time. John seemed to be surprised.

"Oh, no reason, just curious. Valerie, you say?"

"Yes, I think it was that," she replied carefully, now starting to notice that it was a bit odd for him to keep asking.

Finally, she was starting to question why he was so interested, Mark thought. It was about time…

John must have sensed this change in tone from Sylvia as he then changed the topic, casual as you like.

"Well, it would be grand if the girl and Mark stayed friends. I fancy a coffee; you want a tea?"

He had heard enough. Mark strode into the kitchen without looking at either of them, heading directly for the breadbin.

"Morning, son!" John cried happily.

Mark glanced at the two of them and grunted in response. Sylvia's content smile made him want to scream.

"Wanna drink? I'm making."

"No," Mark replied curtly.

He quickly spread butter on a few slices of bread which he knew wasn't evenly spread and slapped some ham from the fridge onto two of the slices and made a rushed job at wrapping them in clingfilm. He then grabbed crisps, an apple and a Pepsi and headed away from the kitchen as quickly as he could.

"Mark...where are you going?" Sylvia called after him.

"Leave him," John replied with a sigh.

Mark practically ran past the threshold of the front door, throwing a backhander from behind to shut the door. He couldn't take it, this happy family business facade, John acting like a calm dad who was trying to understand his son and was now no doubt telling Sylvia that he was just being a teenage boy, nothing to worry about. How dare he! Who did he think he was acting like he knew him, like he knew his son inside and out. That man – if you can call him one – had no idea. And what was she thinking, too, buying into all this nice behaviour shite? He could feel a stinging at the edge of his eyes that had nothing to do with the breeze hitting his face.

It didn't take long for him to feel better. The further behind the house he got, was the better he felt. The countryside had done wonders for him, he was grateful that they had moved here for that reason... as well as meeting Alice. He got butterflies in his stomach just thinking about her. Her smile made him feel weak in the knees as if he was literally going to fall for her. It terrified him. He had let in this girl that, let's face it, he still didn't really know well. He fought with his emotions to build a barrier that was normally so natural to him. He had no friends, but he had the opportunity to have at least one or two. In primary school, there was a guy called George who was also quiet and at first there was potential for friendship. They spoke and even shared a joke or two—it felt great, he felt accepted and he found that it was nice to have someone to talk to his own age. Then, one day, as he was walking down a corridor one of the main bullies in his year shoved him to the ground from behind. The smaller kids referred to him as Scary Simon.

"Stand and deliver, quiet boy," he demanded with a sneer.

"W-what?" he whimpered.

Scary Simon mimicked in delight. "Ohh look at that, you can talk! Now shut up again and give me your tuck shop money."

"I don't have any money," he pleaded truthfully. He had forgotten it was tuck-shop day. Sometimes he bought a Twix or some Haribo.

Scary Simon stopped smirking and ducked down to grab his shirt.

"Don't lie to me, I've seen you getting sweets before."

"I forgot it was tod—"

Before he could say anything more, Scary Simon punched him in the stomach. An explosion of pain hit, the shock making him breathless.

"Fight, fight, fight!" chanted some amused onlookers who had quickly circled them, laughing and shaking fists in the air for encouragement.

Luckily for Mark, the commotion attracted a schoolteacher, and everyone scarpered, including Scary Simon, who got away before the teacher could see it was him punching Mark.

He turned his head in time to see George laughing with the others.

Mark had never again let anyone else in. He thought friendship was a thing that would never happen for him and he accepted that quite quickly. He didn't mind being alone, he was better off, or at least he had always assumed that to be the case. Kids his age tended to be much less mature than himself anyway, so was he really missing out?

However, he wanted to talk to someone now. Mammy would be his go-to normally, his only go-to, but she was part of the problem this time. He was nearing the field where he was planning to sit under the big tree, but he stopped at the wall where he was going to climb over onto the field.

After a quick deliberation, he turned around and started walking in the direction he came. Striding past his own house without a glance he headed onwards in the direction of Alice's house. It took him longer than he thought but the walk had done him good. His heart began to pound when he saw her house, even more so when he reached the front door. There were no cars in the drive, so perhaps she had gone with her parents somewhere. There was no answer after several attempted knocks on the door. There was no doorbell and for a split second he wondered why there wasn't. Not all houses had one, he guessed. He was downbeat that she wasn't there but also slightly relieved at not having to awkwardly ask if Alice was in if one of her parents had answered. He would have felt like a lapdog chasing after her but his need to see her outweighed his embarrassment. That alone was a surreal feeling.

Instead of walking back towards his house he headed right. He knew the way roughly and if he did get lost, he could just retrace his steps. In truth, he wanted to avoid walking past his own house again. Maybe if he took a diversion this way he'd end up on the other side of the field, parallel to the road he had walked up to get to Alice's house. The roads were all straight around here, there was a strong likelihood that he'd end up walking past that same house his dad has been so interested in.

He took out a Pepsi and a sandwich from his bag. The excursion was hungry work, and he wasn't really in the mood for sitting down now. He needed the walk so that he could walk away from his own issues. Considering it was a nice day he saw little sign of human activity other than hearing the occasional car in the distance. The birds filled the silence, though. He didn't know what birds they were, he had little interest in them as far as identifying them, but they were pleasant to hear on a country walk. The chirping was happy and conversational.

After ten or fifteen minutes of slow walking, he was in no rush to get back home, he reached the end of the road. Right would take him to… God knows, but he was pretty sure turning left would eventually take him to the other side of the field, where Alice had appeared from when they first met. This must have been the walk she took on the day they met. Even taking what must have been the same route made him feel a bit closer to her. Was that sad? Weird? Pathetic? He didn't know. At least no one else would know. He took the left and sauntered down, taking in the blossomed bushes and large overhanging trees. Would he ever get bored of seeing so much greenery? It was surely better to be bored seeing fields and trees than ugly factories, the same worn-out looking corner shops and estates that he once had grown to loathe the sight of. What suited him out here more than anything was the sense of isolation and tranquillity. He didn't have to look at people walking by or have to block out the sound of urbanisation with his iPod. He liked being left to his own devices, where no one could see him or give him any grief. Yes, the county life definitely suited him. He would definitely live somewhere like here when he was older, he promised himself. He'd get a dog and they'd play fetch for hours in a field without a care in the world. He'd always wanted a dog, but Mammy said no. John couldn't care less so long as he didn't have to look after it. If he could choose, he would get a golden retriever or maybe a springer spaniel. If he was alone that would be fine, a dog was all he needed for company he felt. But would that be enough? Would he want or need a woman in his life when he was older? If he asked himself that a few months ago he'd probably have said no but now he did doubt it. Right now, all he wanted to do was see Alice and just talk to her all day. If he felt like that now, wouldn't he get similar urges when he was older and more mature? Wouldn't his own company bore and depress him eventually? Maybe any future wife he had wouldn't be Alice... it was weird even considering that at all but someone like Alice? Maybe he did need a woman in his life. He'd been fine up to now because he'd always had Mammy to talk to, but she wouldn't be around forever. Not to sound Freudian, but he may need a woman in his life that was a little bit like his mother. At least in terms of her kindness, the love she showed unconditionally, etc.

Don't overthink it, you're sixteen.

His inner voice always was more pragmatic than his conscious one. His own guidance was sometimes the best advice he could get. Mammy had never been a teenage boy so sometimes his own intuition was best.

Don't let yourself get too attached, either. You don't want to get hurt, do you? She's an attractive girl and you know it. You also know that she can crush your heart too, right?

Sometimes, his inner voice also felt like a paternal voice telling him to be careful. Here he was walking to Alice's house without invitation after all. A boy who, until recently, shied away from anyone he didn't know. Maybe Alice saw him coming up to her house and got creeped out. He didn't think she would, but the thought lingered all the same. How uncharacteristic it was of him to put himself out there like that! Was he setting himself up for disappointment? There was no denying that deep down it wasn't just a friendship in his mind. He wanted more. He wanted to kiss her. It scared him but he it felt good to feel something strong for someone.

His thoughts had caused him to zone out and wander aimlessly down the long, stony road he had pursued. In the distance he saw the house John investigated. A surge of curiosity ran through him again. Perhaps, he could take a look himself now. What could have been so interesting about this seemingly deserted house? His pace seemed to increase now his thoughts had momentarily forgotten about Alice and attached itself instead to the house ahead. With perhaps another fifty metres to go he shrank away to his left, pressed the backs of his legs against the cool stone wall and leant forwards behind a small tree that just about hid his narrow frame from anyone walking in that direction. Someone was walking his way. If he had been any bigger, the person that was heading in this direction from a distance probably would have seen him, despite staring fixedly down at their phone.

It was pure instinct. He had become used to hiding recently, what with listening in to his parents talk as well as following John. He became suddenly conscious of how exposed he was. He chanced a quick glance and realised it was a female figure. How was it going to look if the girl noticed him hiding behind a tree? He felt a little ridiculous now. He took another peek. What reason had he to hid—

It was Alice.

He kept quiet, resisting an urge to run out to her. How was he going to explain hiding behind a tree next to a big, deserted house? Hurriedly, he tried to think of a plausible reason, but it wouldn't come. 'I just wanted to check out this spooky looking house because I saw my dad look around it recently'? How could the truth sound like such an atrocious lie? He knew how it looked: odd. There was no way of explaining himself here without him looking like a weirdo. Why was he still bothering to try and hide when she'd notice him as she walked by? Would it be easier to slide out of his hiding place now whilst she was on her phone and act like he'd just been strolling down the road?

Probably. It made him happy too to know Alice hadn't avoided answering the door to him. She had been out.

He was just about to step out when Alice made a sharp turn towards the house. She then knocked softly at the door. She knew the house then,

Mark observed, confused. The way she had walked to the door was not that of someone who was unfamiliar of the house. She was too relaxed and still looking down at her phone, and the short two-tap knock at the door suggested she was expecting a quick entrance from someone who was expecting her. Back in his old house, the back door locked you out if it closed behind you. On the odd occasion he put out the rubbish bin out for collection, the door closed behind him and he'd do the same two-tap knock on the door on his return and Mammy, who would usually be in the kitchen by the back door, would let him back in.

If she had come unannounced or if she was a visitor most people would knock at the door strongly at least four times and readying themselves to explain their presence, right? Just like he had done a little earlier at Alice's house. You wouldn't do a soft knock twice without looking or paying attention, would you? Or was that just him?

Sure enough, within a few seconds of her knock the door opened. So, someone did live there then. He poked his head forward an inch and just about made out the outline of a blonde-haired woman who looked a lot more like Alice than her mother.

It was possible, he supposed. Alice hadn't mentioned any nearby relatives but why would she if it didn't come up? What struck him as odd about the few seconds between the door opening and Alice entering was that no words were spoken. No greeting whatsoever, not even a smile or facial acknowledgement of each other.

His foot snagged on a twig, making a sharp snap under his weight. He held his breath and silently jerked backwards as well as suck in his stomach. He could feel the gaze of the blonde-haired woman pierce through the small tree trunk but after a painful ten seconds of being stabbed by foliage and jagged rocks of the wall, he heard the door softly shut. He let out a relieved breath and realised he was gripping his Pepsi in his left hand. He downed the rest of it, which now felt uncomfortably warm against his throat after being warmed up by the sun.

What was going on? Was it just a coincidence that his dad had shown interest in Alice and of this house? Now Mark realised both were connected in some way and it raised more questions than it answered.

Chapter Twenty-Two

Kirsten Brenson stared long and hard towards the small tree that stood against the rocky wall. She got the sense that someone was hiding behind it but decided not to go and check. It was probably just a bird.

"Mammy, what's up?"

Kirsten shut the door and turned to Alice.

"Nothing… just thought I heard something behind the tree outside. Must have just been me. Would you like a tea?"

"No thanks," she replied, distracted.

She walked into the living room and took a quick look out of the window. Her eyes widened in surprise. Mark was stood there, shaking off bits of twig and leaves from himself and walking away. Her mam had been right to sense someone there. She wondered how on earth he managed to conceal himself from view even if he was rather skinny; that tree was not very thick. More importantly, why was he there? Did he know that this was really where she lived? If so, that was very bad news. How was she going to explain that one when he next talked to him? Was he stalking her? Should she let her dad know? She panicked a little bit then; the cover had potentially been blown. If only she hadn't been on her phone, she probably would have spotted him and could have acted like she had no intention of going into this house. What was she going to do?

She asked herself if Mark had actually seen her but almost instantly dismissed any doubt. Of course, he must have seen her. The fact that he was hiding suggested he might have been waiting for her; by extension, that also meant he must have known she was going to appear. How else could it be explained? Perhaps she could say that Kirsten was an auntie? Mark had never met or seen her real mother until just moments ago so that could work. However, John Formby would know Seamus' wife was Kirsten; how much could she rely on Mark not to mention what he had seen to his dad? It helped that they didn't get on, so she was pretty sure he'd say nothing. It would have been even more awkward to explain if it was Valerie answering the door—who, of course, Mark thought was her mother. Still, it wasn't ideal, and she felt like she needed to talk to Mark soon. She took out her phone again and texted Mark, asking if he fancied coming round for a coffee soon. Next, she opened up the GPS tracker app she had installed and checked the location of John Formby's Mercedes. She had been checking her app out of habit for two days now and according to the system, the flashing icon told her the car hadn't moved since she attached it. Even on the night she had attached the tracker and later sensed someone in the garden before hearing a car leave, she instinctively checked her tracker then to see if she had been followed. She had panicked slightly at the time because if she had been followed it was no doubt because she had been seen attaching the GPS to the car since she had only just fitted it to the car. She had breathed a sigh of relief when the flashing icon hadn't moved from the Formbys' drive. Little did she know that she had in fact been seen and John had followed her after detaching the GPS and dumping it on the gravel to make it appear like the car hadn't moved.

About half an hour later, Alice's phone buzzed. Mark texted back saying he liked the sound of a coffee, asking what time. She suggested in an hour. The following answer was immediate. He accepted. Alice collected her handbag and left for the second house, all the while thinking of what she was going to say.

Chapter Twenty-Three

It was a horrible feeling for Alice as she conceded that control was slipping away from her. She thought she could steady the ship and keep tabs on John Formby with the GPS and on Mark by befriending him and being his confidante. She was careful not to give anything away about where she actually lived and who her family really were; now, she had potentially risked everything because of a few moments of complacency. Also, she still couldn't forget when Mark told her that John had been asking questions about her. He seemed to know who she was, and he seemed to be interested. Surely that meant he was at the very least keeping an eye on her family, particularly her dad. It was he who had a lot to hide; running a drug business on the side was never an easy operation. It was a twenty-four-seven job and always risky, constantly looking over your shoulder. John Formby was in the Gardaí, was he now working from home? Mark didn't seem sure if he was still on active duty or not, even if it was office work whilst he fully recovered from his injuries. The media had treated him like a hero for getting involved in a drug raid and it was a miracle he had survived. Fishing out the dealers and the makers was in his repertoire, catching them was in his blood. Alice's dad was swimming in a lake of an experienced fisherman; before long, John was surely going to catch him. That made him very dangerous for the Brensons but sometimes Alice wondered whether her dad was too stubborn and proud to see that if John Formby was investigating him, it was only a matter of time before he got a warrant or found proof of his business. Yes, it might take time, especially seeing as the clever disguise of being a medical lab made it ingenious to run an illegal drug business too. But Alice felt it was inevitable that he would be found out eventually. If things got out of hand, she was worried that Seamus would simply 'get rid of the problem'. She knew it to be true deep down, she feared considering the fact that her own father was capable of murder or at least arranging it. She didn't want anyone to die and she was afraid that Mark could end up getting hurt if he found out too much.

If Mark did start to wise up or began to ask the awkward questions, what the hell was she going to do? She could only pray that Mark accepted the story of her visiting her aunt's house without question. It was a plausible story, after all.

If Mark didn't just accept it though, if he could look into her eyes and see that she was lying or hiding something, she was struggling to think of plan B. Usually she was a good liar, which was partly why she had been able to form a bond with Mark without suspicion but there was a first for everything. A lie is always a well-kept secret until one thing, no matter how small, doesn't fit the story. One small detail and suddenly there's no going back, like a domino crashing into the next until you reach the inevitable end of the lies being exposed. The pressure of the following few hours piled onto Alice's shoulders. She could feel that a lot rested on it. She would have felt less uneasy had Mark not been hiding behind the tree. That didn't strike Alice as someone who was oblivious. Either way, Mark now knew she had associations with her real house which, no matter way you looked at it, was not a good thing.

She had now reached the safe house. No sign of Mark so far. She was half-expecting him to be there waiting for her and she was glad he wasn't. She needed time to think things through and she also needed to look her best. It was the oldest trick in the book but if she looked good, Mark was more likely to be a typical guy and be distracted. The less thinking she could make Mark do, the better. Although she would have to be cautious so as to not go overboard, he was only coming around for coffee after all. She smirked to herself as she thought of 'coming around for coffee' meant something else entirely but with a boy like Mark she could be almost certain that it would, in fact, be just coffee. He was too innocent, and, in a way, that made things more complicated for Alice. They were both sixteen and sex would normally be at least a possibility. She needed to keep Mark comfortable in her company; the prospect and allure of sex would usually do the trick, but she felt it would do the opposite for Mark. She smiled affectionately at his honest nature; she couldn't help but think he was the sweetest boy she'd ever known. However, she did get a certain vibe from Mark, the vibe a girl always gets when a guy likes them. She could tell she had won his trust and that was a huge advantage for her now, she just had to be careful not to lose that trust.

She slipped on a figure-hugging jumpsuit and then applied a light coverage of make-up and lip gloss. Just the right amount of effort as she had set out to do. She also couldn't deny that it felt good to make an effort for him; she wanted to. He deserved to find a girl that made an effort and so she felt like she was doing a nice thing by doing so. Whenever she'd met Mark, he had always made an effort too. Sometimes she questioned his fashion sense, but it was clear that he had tried which was more than enough for her, she thought it very cute. He was also a real gentleman, insisting on paying for her coffee, holding the door open and letting her go first and he even pulled out a chair for her when they sat in Starbucks. The last bit was slightly embarrassing, it made one or two people look over as if to say 'he's keen to impress' but she didn't care much. It was an endearing characteristic. She had never been made to feel special like that before.

Her thoughts were interrupted by the sound of the doorbell. She went downstairs slowly, fully aware of her racing heart and the small bead of sweat journeying its way down her hairline. She swiped it away, took a deep breath and opened the door with a winning, genuine smile.

Mark did all he could not to gawp but not completely, Alice noticed, pleased.

"Hey, thanks for coming! I've been so bored today that I needed a friend to distract me."

Mark beamed.

Keep it together, warned his inner voice.

"No problem, it's good to see you."

Alice cooed. "Aw ain't you a cutie, you too. Anyway, come in. What would you like to drink? We've got a load of soft drinks and a coffee machine if you fancy a hot drink, so you don't have to stick with kettle coffee."

He laughed, informing her that he'd never heard of the expression 'kettle coffee'.

"Mark, the more you get to know me, the more you'll discover I've got loads of weird expressions."

He shook his head, bemused.

"So, what's your poison? See, there's another one."

"At least I've heard of that one. Could I have a latte then please if it's not too much trouble? I'll trust you not to actually make it poisonous…" he jeered mockingly. He was surprised at himself at how at ease he was when around her.

It was Alice's turn to laugh. It was genuine too.

The latte exceeded expectation and he told her it was good.

She gave him an evil look. "The best poisons always taste the nicest," she said menacingly.

After more laughter, there was a moment of silence. Both looked into their mugs. Mark was determined to break the silence.

"I saw you earlier."

"Oh yeah? Was I walking towards a big house?"

He nodded.

"Yeah, that's my auntie's house. She's nice and all but I don't see her that often. She can be a bit irritating."

She watched his face with bated breath, but he wasn't giving much away. Alice hoped she wasn't either.

Mark didn't want to say that he'd watched her approach the house, he didn't want to sound creepy.

"She seemed to be expecting you anyway," he said with a jolty laugh. It was too late. He sounded creepy, like he was intentionally stalking her. "I only saw the back of you go in at the last moment, I wasn't hanging around waiting for you."

He was drowning in a sea of embarrassment now. He could not look more like a freaky stalker if he tried. He still thought it was odd though how they didn't seem to acknowledge each other. She must have read his thoughts.

"She's not exactly a woman for pleasantries if that's what you're wondering about. She hates small talk and she always expects punctuality. I knew she'd be waiting at the door ready to answer but then she'll do this thing where she won't say anything. It's just the way she is. Like I say, she can be a bit irritating. She's a little strange, really. I think she might be on the spectrum to be honest."

It still didn't explain why Alice walked to the house from the complete opposite direction. Why was she walking from that way? Was he overthinking again? Maybe she had just gone on a long walk. It was a nice day after all. He decided to let it go. No doubt he was exploring something that didn't need examination. Plus, he didn't want to irritate her with questions that didn't concern him.

Alice saw the confusion and for a second, she thought she was going to not be believed but he quickly changed the subject, thankfully.

They spent the next hour chatting away about nothing in particular and Mark was amazed how well he was doing to keep a conversation from going stale. He had a fluttery sensation in his stomach the whole time, unable to get over how beautiful Alice looked. He wasn't one hundred percent certain, but he could tell she was wearing make-up. She always looked great in his mind but somehow it was clear that she had made more of an effort this time. Did that mean she liked him in that way? Or was she merely being polite, looking nice for a guest? He wished he knew but it was also a weird feeling for him to work out what was going on in a girl's mind. It was good to feel like an ordinary teenage boy for once.

Alice couldn't help but be impressed by the difference in Mark. He was a nervous wreck when she first met him, hardly able to string a sentence together. Now, he was chatting away with her so comfortably and not forced either. She worried that it might get boring talking to him, he was such an introverted guy after all, but she couldn't help but feel a bit of pride towards him. She didn't want to torture the poor fella any more though. He had only just gotten used to talking to someone his own age and now he must also be wondering if she actually liked him or not. It was impossible not to think that. She knew how much of a big step this must be for him. He also looked more attractive now that his body language was so much more relaxed.

She got up from the armchair she was sitting on and sat next to Mark on the sofa. Her heart raced slightly as she sat down again, she knew that she liked him in that way now, no other guy she had met had been so genuinely lovely without any agenda. She could see in Mark's eyes that he was nervous as she sat next to him. This time, she completely understood. If he wasn't, it would have crushed her slightly as he couldn't have fancied her. Nerves were good sometimes and this was one of those moments.

She didn't want to say anything corny; she hated that smushy stuff. She leaned in and kissed him full on the lips. What was only a few seconds seemed like a long time for the two of them, it felt as though time had slowed right down. They looked at each other, both smiling.

"Well, that was nice," she said softly.

Chapter Twenty-Four

Relief flooded through Alice as Mark left the house. Despite seeing flashes doubt in eyes, Alice was pleased that he did not heavily question her. Thankfully he was too distracted by his feelings to give it more thought. Typical boy, she thought, but she was pleased. Never before had she thought Mark a typical boy until now.

She continued to dwell on that look in his eye though, maybe later he would revisit it when he has had chance to recover from the daze he was in after kissing her. On that note, she was pleasantly surprised, she had enjoyed the kiss more than she thought she would. There was a small flutter in her stomach. She knew what that meant. That was not good given the circumstance. She pushed it away though very quickly, it was important to err on the side of caution.

Still, she was comfortable enough by how she had handled the situation to not have to admit the ordeal to her dad. After a few hours of enjoying a bit of alone time in the house she had left the safe house again and headed back in the direction of her real home, this time more alert of her surroundings. It was unlikely Mark would take the exact same route as he had earlier but after the close call today – and one that would continue to occupy her thoughts – it was best to stay alert.

Kirsten could sense something was on Alice's mind when she returned home but when she asked Alice had said she was fine.

Instead, she approached the topic all children hate talking to their parents about. "I can't help but notice, Alice, that you're spending an awful lot of time with Mark, which is great by the way, but your homework better take priority when your home-schooling resumes. The summer break is over soon, and this year is really important."

Alice rolled her eyes at her, telling her she didn't have to worry. She had given no thought whatsoever about her education, she knew she was bright enough to pass with decent marks. She didn't particularly like Mrs Finlay who taught her one-to-one, but she tolerated her lessons. Having her back was not something she was looking forward to that much, although she had been rather preoccupied this summer anyway. She wondered how much harder it was going to be to keep tabs on Mark once she was back doing schoolwork though, maybe by then Seamus would stop asking her to see Mark if John Formby had made no other action to make him suspect that he was being investigated. She would have to find an excuse to see Mark and if her dad no longer suspected John he'd actively discourage her from any future contact with Mark in case John himself started to suspect something. Alice didn't want that to happen either, she had genuinely found a friend – at least – and something special in Mark. She decided then that she needed to work quicker and chase the GPS lead. Why had the car not moved since she had attached it? Was it faulty? Was her app that was connected to the GPS not functioning? Apps could be a liability at the best of times. She found it hard to accept that in nearly three days the car had not moved at all. She took out her phone again as it was fresh on her mind and checked the app again, expecting to find that nothing had changed. However, the GPS had moved! It was located down the road from where the Formby's lived. She scowled when she saw the flashing red light was not moving. Where it was flashing suggested it was in the middle of the country road with very few places to be stationary for drivers… yet there it was flashing in the same spot. What was that all about? She quickly grabbed her coat and a torch but acted as normally as possible when passing her parents in the kitchen, who happened to be both at the table. "Just fancy a walk, do you two want to come?"

She hoped that by inviting them along it would avert suspicion away her; she didn't want them thinking she was hiding something from them. They would easily find out if they did suspect, her dad would make sure of that. Her approach worked though, she had been casual enough to avoid a funny look or awkward questions. All they said was to not be out for too long.

As soon as she was out of the house and had climbed over the brick wall and into the field she began to run.

Little did she know that John Formby was waiting for her.

Chapter Twenty-Five

Nothing could pollute Mark's exceptional mood after leaving Alice's house. He felt euphoric, finally able to understand why other boys his age felt when they got with a girl they liked. It was an odd thought, to be on the same emotional page as other kids. No way did he think Alice could possibly like him and as hard as he fought to build an emotional wall for protection that was normally so natural to him, he could not help but let feelings slip through. He almost skipped home, with a spring in his step that was not characteristic of him.

Sylvia was stunned to see this burst of happiness. She knew it must have been because of Alice. A maternal pride flowed through her veins and this new confidence in Mark made her feel just as happy as Mark. It was some sort of miracle—and to think where he was only a few months ago!

"Why are you so happy?" she asked beaming. She already had a hunch.

"I kissed Alice!" Mark proclaimed joyously.

She knew he was elated but that took her by complete surprise. She hoped that there might be something between them, Alice seemed like a sweet girl, but she knew how Mark was; she knew how protective he was over his own feelings. How on earth had Alice managed to work her way in? But a kiss? It was unbelievable! But believable or not, she didn't mind.

"So, are you two going out now, then?"

A look of confusion washed over Mark's face that dented his joy ever so slightly. He had no idea how to answer.

"I don't know. I mean, I don't think so. What do you think?" he said, now worried.

Sylvia's laugh to that irritated him, it made him feel stupid and naïve. Was he, though? Should this be something he should know? Or is this why boy-girl relationships were notoriously complicated?

"Well, not everything is said out loud. It can be hard to know for sure," she replied with a shrug.

Didn't he know it…

The two of them had wandered into the living room and he slumped into the chair, still dazed from the kiss but now preoccupied about what it all meant. Sylvia followed suit and sat down beside him, pleased when Mark leaned his head against her shoulder.

"Mammy, what do I do?"

The helplessness in the question was bittersweet for her; on one hand, it was a good problem to have and most people faced the same dilemma in their lives. It was just something both he and Alice would have to figure out. No one could provide the answers but them. However, the question reminded her just how innocent he was. It was like a rite of passage he had to take before he could grow up—it was strange to see her boy growing up and facing ordinary adolescent feelings. It left him vulnerable though, and it wasn't as though Mark were an ordinary teenager. He was sensitive, unused to having friends, unused to talking to anyone who was not her, if she was honest with herself. Was that her fault? Did it make her a poor mother to not throw him in the deep end so to speak and make him tread water in social spaces? She had always taken him under her wing when he didn't want to do something that was out of his comfort zone and she now realised she had been too protective. She needed to let him figure things out for himself, so this very situation now wasn't quite so overwhelming.

"I'm sorry Mark but the truth is that I can't really help you reach the answer. Only you and Alice can answer what happens now, I'm afraid." She stroked his face affectionately, genuinely sympathetic to his feelings although secretly glad that he was facing them.

For a while Mark said nothing. He knew she was right.

"I really like her," he eventually replied. "I just don't know if I'm ready and I'm nowhere near as confident as she is."

Sylvia ruffled his hair. "That's all in here," she said shaking his head, making him smirk. "If you keep telling yourself you're not ready then you never will be. You just need to ask yourself two things. Do you want it? And does it feel right?" She turned around to him, looking him in the eye. "If the answer is yes to both then you've got your answer. Some of the hardest questions have the simplest yes or no answers sometimes, you know."

He let it sink in, glad of her counsel. It did make sense.

"Thanks, that's actually useful," he said giving her a one-arm hug. A thought came to him then. "What about you though, Mammy? Can you answer yes to both questions with John?"

She gave him a sharp look. "You'd do well to call him 'dad', Mark. And your dad does make me happy… in fact, he has been really nice to me recently."

It was Mark's turn to give her a stern look. "And don't you find that a bit odd?"

She couldn't deny that it was a bit, but she was just happy that things had changed. How had the conversation turned to her anyway?

They were both good at pointing things out and often giving each other advice. They were similar in that respect. What they lacked was following their own advice.

"Things between me and your dad are… complicated," she finally replied. "But it's fine. Now let's get back to you. You have to talk to Alice." she continued curtly. She stood up and made to leave.

"Ma," Mark called after her. "I'm sorry, I didn't mean to annoy you. I just want you to be happy."

Sylvia's face softened, turning back to him. "I know, my love, I know." She bent to kiss his head. "I'm sorry too. Now, what do you want for dinner, Romeo?" she teased.

"Oi!" he exclaimed, grabbing a cushion and playfully hitting her arm with it. They both laughed.

Mark's eyes darted right towards the door as John walked into the room in what was a purposefully casual manner. He even smiled. Mark did not.

"What do you want?" he asked bluntly.

"Mark…," Sylvia pleaded weakly.

John's reaction was, again, uncharacteristic. Instead of rising to being spoken to in a rude, abrupt manner he acted like he was unfazed. Instead, he just shrugged as if he was confused by Mark's chipped response. "What are you two laughing about?" he asked in a friendly tone. Mark could still see the icy expression in his eyes even if he was doing his best to mask it. He was no fool.

Sylvia tried to ease the tension. "We were just chatting about teenager stuff, right Mark?" she asked invitingly. Mark grunted, feeling bad slightly that he was now being rude to her but not letting up.

"Ahh girls and stuff?" John enquired, taking the invitation instead.

"Yeah, and Mark actually has had a good day in that regard, haven't you Mark?"

Mark could see the desperation for her to start a happy family. Nor did he appreciate her sharing the news.

"Oh yeah? That Alice girl, is it?" John asked, looking directly at Mark.

"Like you care," Mark mumbled back. Then anger took hold. "Actually, you really *do* care, don't you *Dad*?" he said as if he were saying an ugly word. "You care way too much. Not about me, oh no, but you really bloody cared when you realised, I was talking to Alice in the first place didn't you! Are we all just going to forget when the last delightful discussion about this ended up with me whacking a vase at your head? Because we seem to just be acting like that weird feckin' mess didn't happen!" He had built up anger as he spoke to the point where he was fully yelling towards the end, like a hurricane that built in strength and speed.

He stormed out but still had time to register John's blank expression. He was playing a game, but Mark was too angry to care why. His happy mood had been torn to shreds. He just wanted to be alone and try and block out his parents on one side and his dilemma about what to do with Alice on the other.

Chapter Twenty-Six

John had heard the whole conversation between Mark and Sylvia. He was now convinced that he had to act and the first part of that was to not act at all when Mark's emotions exploded from him. He could understand Mark's reaction perfectly, the same short fuse was in-built within him too. He then locked himself in his office to go over the plans and his notes. The business with Alice and Mark was evidence that eyes were on him. He was being spied upon, albeit unsubtly.

One: there was a safe house that Alice Brenson went to whenever she met up with Mark. Why do that in normal circumstances? He knew it was Alice Brenson, he had seen her full profile and he knew where the Brensons really lived. He suspected Seamus Brenson knew that John thought something was off too, he must have been aware of him sneaking around their real house. That was the first thing that was odd.

Two: apparently Alice's mother was called either Valerie or Vanessa. He knew that wasn't true. It was too dissimilar to Kirsten, which meant Alice's real mum was not the same person that Mark was introduced to. Odds are the same with the dad. John found this strange—why take the risk of Mark talking about the parents when he got back home and mentioning their names? Why do that if you were spying on someone? Surely keeping to the same parents would be less suspicious to him, a guard. It would draw less attention. Seamus Brenson must be relying on the fact that he and Mark didn't communicate or see eye to eye. That was true, in fairness. Still, it was a risky move.

The third clue was of the persistent contact between Alice and Mark. Of course, this could be purely coincidental, the typical boy-meets-girl scenario, but Mark was not a typical kid. He knew little about his son, but he knew that much. It was probably initially Seamus' main grand plan to infiltrate the Formby household and find out his business through Mark. He and Mark not getting on must have thrown a spanner in the works, so it was a surprise in a way that Alice still met up with Mark if that was a dead end. Not to be cruel to his own son but he had little confidence in his ability to talk to a girl. Why not quit with that field of attack if Mark was unable to pass on information about him? Was there something real happening between them? Or had he completely underestimated Mark who actually *did* know what John was up to. Mark would never reveal to him if he did know anything about his… enquiries.

All those clues held no water to the big giveaway. The GPS. He had seen the stupid girl apply the magnetic device to his car plain as day. There was no other possible reason other than the fact that Seamus was using his own daughter as a means to track him. Getting his daughter to do his dirty work seemed low but given what the Brensons had to hide, it was not all that surprising.

Collectively, the evidence was damning. It was incontrovertible. He needed to act. Something about Mark and Alice kissing was the final straw for some reason, like they were taunting him in some elaborate game in which had been losing up to now. The girl was using Mark to get to him in some way—if she became Mark's girlfriend this gave her an excuse to be in the parameters of his home, this must be why she persisted in meeting Mark. He even felt a bit sorry for the kid; Mark's confidence would shatter all over again when this all reached its inevitable, sticky end. A pang went through him at the thought. Anger, he thought to himself, surprised.

If the whole thing hadn't been so clumsily done, he'd almost say their plan was a clever one. Or was it done clumsily on purpose so as to set a trap? They had left too many gaps in their plan for John not to peek through them and see exactly what they were up to. He needed to act now despite the risk of a trap.

Questioning Alice Brenson directly without alarming Seamus Brenson was the step he needed to take. He knew how he would do that. But first, he needed to confirm everything with his boss.

But not his boss at the Gardaí headquarters…

Chapter Twenty-Seven

It took John ten minutes to sum up the evidence to his other boss. They were in mutual agreement that something had to be done now; John's plan was given the green light. It was all in his hands, things were about to hit the fan. Standing up from his desk, he went to his bedroom to fit on his bulletproof vest. He then unlocked the drawer near his bed and took out his Glock and loaded it with the bullets that were also stashed away before locking it again. No one knew about this gun; it was one he had been given by his second boss. Of course, it was illegal for him to have a gun outside of official constabulary business, all firearms were handed back into a controlled unit after use. All records of weapons going in and out of the Gardaí station got recorded. He tucked the Glock in the back of his jeans, with the safety on. He put on a baggy old sports hoodie that hid it from view and went downstairs.

His face was fixed in concentration, as it often was. He had always been stoic, no one could tell what he was really thinking. The truth was that he was starting to get nervous now. His heart had noticeably started beating harder but there was also a surge of excited energy running through him. This was his chance to make amends for his injuries; his second boss was not at all happy about the publicity it received when he went in on a drugs raid. It was reckless going in alone, he knew that. No one in the Gardaí knew why he did either; he got reprimanded for it once he got better but no further action was taken.

John got into his Mercedes, firing the engine into life. This was it. He knew he had to be careful, this had to go to plan. The GPS was safely in his hoodie pocket, ready to be used as bait. What if the girl didn't check it on her phone? What if she weren't alone? There was still so much that could go wrong. That was probably why he was nervous. Swerving left from out of the drive, he drove slowly down the country road. About two hundred metres ahead, he came to a gentle stop, cutting into the left slightly off the road so that there was still room for cars to pass; luckily, the road wasn't quite as narrow as others nearby. He hoped he wouldn't be parked there for too long anyway. Aware that the darkness was closing in fast now he crouched low, nestling into the shrubbery in an attempt to hide himself as much as possible. The car was far enough away to make it unclear. He then took out the GPS and put it down on the gravel next to the road, making sure that passing cars wouldn't accidentally roll over it. The difficulty he faced was hoping that no cars would come past; all the lights of his car were turned off and it was crucial that they remained so. This also meant that any cars coming by would not see his parked car until their own lights brought it into focus. He really wanted to avoid arguments or angry beeps from drivers; any sign that he was there presented a risk. It made him uncomfortable how many risks he was taking tonight. John then shuffled ahead for another ten metres or so before stopping.

He was crouched there for fifteen minutes when a flash of light grabbed his attention to his right straight ahead of him, coming from the field. It was only for a fleeting moment, but he saw it; he thought for a second, he might have imagined it, but it came again, this time closer. He sat upright, now fully alert and making no movement whatsoever. Everything seemed to freeze around him, time holding its breath as if distracted from its continuous duty to watch what would happen next. This was it. He was relieved, the girl must have checked her phone and noticed the GPS had moved. If she was clever, she must suspect she was walking into a trap but if she was obsessed by curiosity and perhaps a determination to get a result for her father maybe she wasn't thinking clearly. The light flickered in all directions for a few seconds, probably looking for a car, then the figure climbed over the wall of the field and losing a steady control of the torch he or she was carrying. Luckily, he had parked the car slightly away from the GPS otherwise the beams of light would no doubt have shone onto his car, which of course is what Alice would be looking for. That was where she fitted the GPS after all. No doubt the girl was now questioning herself, perhaps panicking at not seeing the car and realising too late she had run into a trap.

The figure jumped down from the wall, torch pointing down towards the ground whilst another soft light appeared in their shadowy hands. *It must be the girl*, he thought triumphantly, she must now be checking her phone to where the GPS tracker would no doubt be flashing. The torchlight kissed the stony road ahead, reaching the other side of the road now. Very soon she would spot the GPS, which was emitting its own red light.

Slowly, John braced into position like a lion does before chasing its prey. He waited until she spotted the GPS.

Which she did.

John positioned his own torch with his left hand on top of his Glock which he was holding in his right after slipping it out of his jeans. He then clicked the safety off and clicked the on switch of his own torch.

"Hello Alice," he whispered shining the light at the figure who gasped in surprise, rooted to the spot.

Chapter Twenty-Eight

She looked like a rabbit in the headlights, he thought. She seemed to be in a state of bewilderment that she had been caught, perhaps even thinking herself clever. In fairness, it was a fairly sound plan in principle. It was the application of the plan that let her down.

Alice straightened, looking at him impassively now. If she was afraid, she didn't show it, she had shaken off her shock quickly and decisively. A flicker of admiration etched through him.

"I just want to talk to you, if that's okay?" said John. He tried to sound relaxed, he needed to control this situation.

"Yes," she replied, outwardly unmoved.

He smiled. "Great, do you mind getting in the car? It's just there." He motioned to the car with his torch's beam. "Don't worry," he added, "I would rather be off the road too."

This time she hesitated. She sensed danger but she also didn't really see that she had a choice. Odds were no one would hear her even if she screamed.

"Where?" she asked.

"Oh, not far."

She glanced back towards the GPS and then headed towards John's car. She looked back to see John scooping up the GPS and pocketing it.

Neither of them spoke in the car. John drove without so much as a glance towards Alice whilst she in turn stared sullenly out of the window. The darkness closed in more rapidly, entombing Alice in impending danger but she was trying to not show any sign of weakness.

She should have known it was a trap, it seemed odd for John to have suddenly stopped in the middle of a country road. As soon as she reached the other side of the field, she felt something was off. She couldn't see a car, but she was too curious to turn back around. What was her dad going to say? He would kill her for his... if she wasn't killed before then. She suppressed a shudder but was feeling progressively more panicked. She tried to think quickly, find any plausible reason that could explain the GPS John obviously knew about, but no cover story came to her. A slight sound came from behind her then, or so she thought, but she dismissed it.

Alice couldn't tell where they were or where they were driving to. John turned down a narrow lane, just about wide enough to fit his car. Eventually, they took a right turn into what looked like a deserted farmyard. A single light shone outside one of the doors, one that was more foreboding than welcoming. Everything felt very surreptitious, as if John and whoever else was in this place were up to no good, trying to keep as low a profile as possible.

"Follow me," instructed John simply, once the car came to a stop. She obeyed. They headed towards the door with the single light hanging above it, the gentle breeze making it squeak as it swayed.

Scared though she was, Alice tried to take in her surroundings, looking for any possible escape routes or even good hiding places should she manage to run. She also noted that John did not lock his car. If she managed to get in the driving seat could she get away? The roads were very narrow, but it gave her a sliver of hope. This was no social call; she was prepared for a hard time. Maybe that was why he didn't bother locking it, he'd probably reach the car before she managed to drive off down the tiny lanes, she said to herself.

The room they walked into was large enough and in the far corner was another dingy light, one of those bulbs where you could see the swirly lines of light within the bulb. There was no way this whole set-up was improvised. John had every intention of bringing her here, assumedly to interrogate her.

"Sit," he said softly, gesturing to the cheap fold-up chair.

She did so. The room had a damp smell. It was abandoned all right. Either that or it was not looked after. She looked up at John Formby properly for the first time. There was a look of Mark in his features, also handsome, with the same strong jaw only lighter haired. Physically he was built more like a rugby player, whereas Mark was much slimmer, although Mark had the same broadness in his shoulders that suggested he could have a strong physique if he so chose. She hoped Mark was okay and hadn't gotten into trouble. Surely John wouldn't hurt his own son? But looking into the eyes of John Formby made her less sure of this. There was a ruthlessness, or rather, a lack of feeling that made him impossible to read. He'd been in this position before, she realised. Of course, as a guard he would have questioned people before... but not like this, not in a set-up like this. Perhaps the very chair she sat on had hosted other people who were in need of interrogation. John wandered over to a table with a drawer, retrieving a piece of rope.

That confirmed it. This was definitely not going to be a standard chat. He seemed to see pleasure in the momentary look of concern that came over Alice's face. He was enjoying making her squirm.

"Don't be alarmed, it's just a precaution." He then tied up her hands behind her, tight enough to be secure but not to the point where it hurt.

"Sorry, do you want a drink? You'll have to use a straw if so."

She shook her head, saying, "How hospitable of you."

He looked at her, surprised but bemused. "You have a sense of humour then. Good. Humour is important. Although, my son is very sensitive, so I imagine you're not sarcastic with him?"

She remained silent.

"You know, it was very amusing watching you fit that tracker onto my car. I did wonder why you buried yourself out of sight for so long, but I had a good little chuckle..." he paused at seeing her flicker of disbelief. "I was watching from my bedroom window the whole time. Crazy how people can still be in a room with no lights turned on, isn't it? You made it too easy, Alice. Daddy hasn't trained you very well has he?"

"So, you saw me, so what?" she asked bluntly, ignoring the jibe.

"Was it your idea or your dads to spy on me?"

Silence.

"It's easier if you just answer my questions. You don't think I've done this before with people more stubborn than you?"

Therein lay the trouble, Alice thought. He had done this before and didn't mind doing so. But this felt more than a guard trying to bust her dad for criminal activity, there was more to it. Bringing her to this dingy, damp place was obviously illegal. Why not just question her at the station? Who was John Formby, really? Her dad didn't go into his business too much, so she knew very little but the impression she got was that he had no idea who John Formby was apart from being the guard who'd got injured in a drug bust. Was he afraid to bust her dad in case he was killed? Alice didn't think so, the man in front of her looked like a man who cared little about things, a man who even enjoyed danger.

"Why have you brought me here?" she asked, trying to sound more confident than she felt. He merely smiled as if she had just told a joke.

"It's convenient," he replied, shrugging. "So, are you going to be a good girl and answer my questions?"

She glared at him in response.

"I'll ask you again," he continued, now serious. "Whose idea was it to spy on me?"

She knew it was pointless to refuse to reply. "Mine," she said quietly.

"And did you tell your dad about your idea? Did he convince you to do it?"

"He has no idea."

John studied her for a few seconds, and she looked down at the ground. She didn't want to look at him. The floor felt a bit sticky under her feet though neither wet, nor dry. The room had a peculiar smell. In the distance she could hear a sporadic dripping of water.

John stepped away for a second, the scraping of a chair he went to get echoed around the damp walls. He sat close to her, probably in an attempt to intimidate her. A thought seemed to come to him.

"Where's your phone?"

"Why?" she asked carefully.

"I don't want anyone tracking us," he replied, "so, where is it?"

Alice looked at him in the eye, not really wanting to give up her phone straight away. Who knows? Maybe those extra few seconds would make a difference. Enough time had elapsed for either of her parents to wonder where she was, particularly now night had fallen. Once she couldn't hold his gaze any longer, she gestured to her jeans pocket with her eyes. John leaned over and retrieved her phone, a little too slowly for her liking. *Pervert*, she thought inwardly. Annoyingly, he could turn off her phone without unlocking it—she could have spent another few valuable seconds messing him around otherwise. She couldn't help but feel this was important somehow, that tracking her phone was the only way anyone would know where she was. John turned it off and carelessly threw it on the grimy looking table behind him.

"So, why is your dad so interested in me?" he asked, almost pompously. She did not like John Formby; it was incredible to think he fathered a boy like Mark.

"Don't flatter yourself," she jeered.

John snorted. "I like you, Alice. You've got attitude, which is good. But," he paused, probably for effect, "don't mess with me, I don't have time to play stupid games. Of course, he's interested in me, he's got you to do his bidding by talking to Mark. I assume to try and get information about me?"

Again, she just looked at him, trying not to give anything away.

"But, you see, me and Mark don't see eye to eye. He's a stupid kid, annoying too. I wouldn't tell him anything."

A spark of anger triggered inside Alice. "Mark's not stupid and you'd be stupid to think so!"

His eyes narrowed. "What do you mean?" he urged.

She shrugged, in control of her emotions again. "He's intelligent is all and probably sees right through your bullshit, doesn't he?" A newfound cockiness rose within her. "Is your head still sore from where he smashed a vase against it?"

For a moment, the only thing that could be heard was the distant drip of a leaking pipe in the background. "A bit of a gossip then, is he?" he eventually replied. "He's a coward, I'm surprised he is even talking to anyone that's not his mother."

Again, a flash of anger hit her, harder this time. "He's not!" she exclaimed.

"You actually like him, don't you?" he mused.

She looked down at the floor again.

"Why is your dad interested in me?" he asked again.

She looked up at John again, searching for something in his face. Surely because he was a guard was enough of a reason? If you're doing something illegal and a guard starts looking around your house wouldn't it be obvious that you were suspected of being a criminal?

"Because you're a guard", she replied carefully, still watching his face.

"That's it?" he asked, still looking at her intently.

What is that supposed to mean? she thought. *What other reason could there be?* And then an idea hit her that shocked her, what if he *wasn't* just a guard? What if he was something more?

Chapter Twenty-Nine

After Mark had stormed off, he managed to calm down. He stayed in his room though because he couldn't bear facing John. He began to read to try and distract himself; escapism was normally an area of expertise for him, but he couldn't quite concentrate this time and then he heard it, a faint but unmistakable sound of a hushed voice, he thought from outside. On instinct he got up and went to the back door, intending to head towards the voice he knew to be John's. He was curious to know what was going on, why John was being so secretive? No one was in the kitchen as he headed towards the back door, Mammy must have gone upstairs. He tried the door, and it was unlocked. He pushed slowly, let himself out and softly shut the door again so no one could hear. He tiptoed around the side of the house and stopped at the corner of the house, straining to hear anything. He chanced a quick peek around the edge he was hiding and saw John on the phone. He was about ten metres away from where John stood. He had to really strain to hear but he could just about make out a few words.

"It's her... Brenson's daughter... yes... they know... investigated... question her... yes... can now..."

Mark had to put two and two together, but he thought John planned to question Alice. He might be off the mark completely, but he had a hunch that whatever John was going to do now involved Alice. What *was* he going to do? Mark had his phone in his pocket, and it was getting dark. He was confused about how John intended to get hold of Alice seeing as it was getting dark. And then he heard John's words more clearly as he walked towards the front door. "One second, I'm just going to grab a torch."

Mark peered around the corner again and watched John head towards the main front door, raising an arm over his shoulder as he pressed a button.

The Mercedes flashed as the doors were unlocked. Mark had seconds to decide what he was going to do. For a split second he wondered why

John unlocked the car before going back into the house, but he didn't dwell on it. He probably assumed no one was around. Impulse took over and Mark ran to the Merc as silently as he could, opening the back door quietly and nestled himself behind the driver's seat. Luckily, he was slim enough to fit and he curled up so he could be as compact as possible. The looming darkness did its part in concealing him too. He briefly thought about texting Mammy with an excuse as to where he was, but it was difficult to reach his phone and then the thought was forgotten as he heard the front door shut and he replicated a foetus as he tucked himself away.

John drove for what must have been less than five minutes. He felt sure they must still be on the country road leading up to their drive when he stopped. He heard the soft click of the headlights being turned off. For a horrible moment, he thought that John knew he was there and was going to grab him at any moment but instead John seemed to sit, seemingly unaware that he was there. Finally, the door opened, and he felt the weight of the chair relax as John got out. What an earth was going on? Faintly he could hear his footsteps ebb away and he risked another glance, this time out of his passenger-side window. Even though it was dark, and he looked for no more than a second or two he could tell he was right about where they must be. John must be meeting someone here, he thought, although it's an odd spot to meet. And so close to their house that he may as well have walked instead.

Mark then heard someone getting into the car with John. It was strange because neither of them spoke, and as he crouched, he wondered what this had to do with Alice. Curiosity got the better of him and so relaxed his position so that he very slowly straightened up. He was glad that he happened to be wearing dark clothes to help hide him. Just as slowly, he then chanced a look into the passenger seat. At first it was too dark to see anything other than a silhouette but then a streetlight momentarily flashed a light into the passenger seat to reveal more of the figure. He nearly gasped in shock as he realised Alice was sat in the passenger seat. He saw her head move towards him as though she sensed someone was there and he quickly ducked away again, praying he hadn't made any noise. Nothing else stirred so he thought he must have gotten away with it. *Where the hell were they going? And how did John know*

Alice would be there? Everything felt weird and confusing, all he could do was sit and wait.

Eventually the car stopped again, and John then asked Alice to follow him. Where were they going? After he thought it was safe, he got up and peered out the opposite passenger door's window. It was dark but he could tell it was some sort of barn. Just one light could be seen here... Mark didn't like where this was going, and he feared for Alice. Would John actually hurt her? He didn't know what his father was capable of, but he never actually thought he'd be the type to hurt someone. Maybe he was making up a subconscious fairy tale in his head that his dad couldn't possibly as monstrous as he acted sometimes. He had to do something. He felt helpless just sat here. Should he call the police? What would he even say? "Hello, guards please, my dad's gone crazy and kidnapped Alice Brenson in some abandoned barn? I'm in the back of the car and hid away as it happened..." Hardly believable. Perhaps it wasn't like that though... maybe if he found out what was going on there would be a simple explanation. It seemed far-fetched but he'd see how things panned out first. He had Alice's mother's number on his texts when Alice used her mum's phone to text him once when she has lost her phone... he was pretty sure he hadn't deleted it.

He tried the door and it opened. Again, his father was being clumsy locking his car... did he know Mark was there? Was he luring him into a trap in some way? Probably not, he was pretty sure he hadn't been seen. He looked around, nothing but high unkempt hedges enclosing the abandoned farm. It didn't look like anyone else was coming. Perhaps there was no need to lock your car here. He inched forward, heading towards the entrance with a single light hanging loosely over it and peered inside. He could see another glow at the other end of the room, this light being far dimmer, but he could tell Alice was sat there. He waited to see if he had been heard and stayed, again straining to hear. Fortunately, the simple bricked room was damp so any conversation would be amplified by the echoes. He could make out what was being said, it was an interrogation all right. He was then horrified to see that Alice was being tied up. This was no normal questioning; the very act of tying Alice up was threatening despite his father's attempts at convincing her it was just precautionary. Other than the dim light at the far end of the room it was very dark. He figured he might be able to slip inside and back against the dark wall undetected. But that's not to say there wasn't a main light, in fact there probably was. However, it seemed unlikely that John would turn this on if there was one. Surely, he would have already, and the darkness would suit him if he was trying to interrogate in secret, which he no doubt was. Still, Mark felt he couldn't sneak in unarmed. He sensed danger on Alice's part, and it was everything he could do to think rationally. He had to relax. If he didn't, he would only endanger Alice even more.

He listened for another minute or so, trying to figure out what was going on, to see if he could make any sense of this whole thing. He was shocked to hear questions of being spied upon. What was that about? Then his own name was mentioned, John insinuating that Alice's father was in some way using Alice to gain inside information about John by her befriending him. It sounded like paranoid drivel. Mark had met both Alice's parents and thought they were lovely. Not what John was suggesting. These thoughts were distracted when Alice defended him in front of John, then John jeering her that he actually liked him. He was touched, despite the danger she was in – and he too by extension – it was nice to hear her defending him. But that soon turned to the need to do something. Again, he was just stood there doing nothing but eavesdropping. What could he do anyway?

He slipped away from the threshold and looked around desperately for something he could use as a weapon, anything that might at least be used to defend himself or even help untie Alice so she could maybe get away. He was wasting time. He quickly got his phone out, ignored the missed calls notifications from Mammy, and turned on the torch. He hoped that the beams wouldn't attract John's attention, but he felt he had to risk exposing himself at this point. Maybe by doing so it would give Alice time to untangle herself and run for it. What was John going to do with his own son? Kill him? He seriously doubted it, despite his hatred for Mark.

Then a beam caught hold of a fairly large, irregular shaped rock and he quickly ran towards it. It wasn't too heavy, but it was awkward to hold. It was better than nothing.

He again considered calling the guards, but the sirens would give the game away and John might do something rash. He wanted to keep Alice as safe as possible. He also didn't want to make it obvious to John that he was calling anyone. Even if he hushed his voice on the phone it might be enough for John to hear and do something stupid. It was the first time in a while that he felt fear towards his father, though it was not for himself this time. He had a hunch that if pushed John would do something on the spur of the moment whether he intended to hurt someone or not.

Then he remembered the car. He quickly went towards it and carefully got back in the passenger seat, easing the door so that it wasn't fully closed and make a sound. He quickly rooted out the number Alice had called from his phone whilst they were in Dublin and was thankful now that she did. He pressed the call icon. The car should hopefully contain his voice. He prayed that there was an answer. After three rings there was a worried sounding woman.

"Hello?" asked the woman tentatively. It didn't sound like Alice's mum.

"Hi, it's Mark," he said quietly but firmly, "listen, Alice is in trouble. I can't explain it all, but my dad has just kidnapped her and is questioning her… she's… she's tied up. We're in an abandoned farm somewhere, I'm not sure where but it's not far away. I didn't wanna call the guards in case the sirens made my dad do something stup—"

Another voice came from the receiver.

"Mark Formby?" he asked authoritatively.

"Yes," Mark answered, "Alice's dad?"

"Yes," came the reply. Again, it didn't sound like Alice's dad.

"Look, my dad has just grabbed Alice and we're in som—"

"I know, I heard the first time," he barked gruffly. "You have no idea where you are? Is Alice okay?"

"I think so… I just know she's tied up, he can't hear me, and he doesn't know I'm here. I don't know where we are for sure."

"Okay," Alice's dad said firmly, "stay on the line, keep talking. We'll track you. How come your dad doesn't know where you are?" he asked with a hint of suspicion, as if he was wondering if Mark was setting him up.

Mark quickly explained how he'd overheard John saying he needed to act and heard Alice's name mentioned. He acted out of impulse and snuck into the backseat without him realising. Alice's dad listened carefully, taking in everything he said.

"Okay, keep calm Mark," he said, clearly now satisfied he wasn't involved in Alice's abduction in any way. "Have you got anything you could maybe use as a weapon?" His voice was oddly cool for someone whose daughter was in danger but perhaps that's how he acted under pressure. Perhaps he was just a logical man.

"Erm," blundered Mark, "I found a rock. I've got it now, but I feel so helpless, I don't know what to do, I—"

The whole situation consumed him, and he was suddenly speechless, and tears were falling down his face, his breath erratic. Still Alice's dad remained calm. "It's okay Mark, you did well to get away to call us. I can see why you didn't want to call the Gardaí. I agree, the sirens could give the game away. Good thinking."

Through his own despair he could now detect alarm in the man's voice, as if he was reassuring himself rather than Mark. There was a strange silence for the next minute whilst Mark held the line. It felt like the longest minute of his life. He didn't question how Alice's dad was tracking his call. Right now, he wasn't in a state to care. After what felt like a half a lifetime, Alice's dad returned to the phone. "Okay Mark, we've managed to track where you are. We tried to track Alice's phone before you rang but we couldn't detect anything. We're coming. Don't worry, no sirens. Just us. We won't make it obvious we're there. We'll be there very soon. And Mark?"

"Y-y-yes?" he stammered, still unable to re-control himself.

"Thank you so much for ringing us. It couldn't have been easy given that it's your father."

Then the line went dead. Silence resumed save for his own snivelling. He forced himself to get it together, wiping his face with his sleeves, now slightly relieved that he wasn't feeling quite as helpless knowing that help was on its way. He quietly returned to the doorway of the dilapidated barn and when he saw his father facing Alice instead of in his direction he slipped into the room and crouched by the wall and shrouded himself in the darkness once again.

But this time he wasn't in a foetal position. He was ready to launch at his dad if he tried anything life-threatening towards Alice, crouched as if he was preparing to hear the gunshot for a hundred-metre sprint. He was also holding the jagged rock.

Chapter Thirty

Mark's legs began to hurt in this position, but he ignored it, willing himself not to get pins and needles now of all times. He listened as much as he could to what was being said, happy that there was a discussion now. Conversation killed time and it gave John less time to do anything drastic.

"Why bother with the other house?" John asked. "Why have a pretend family... I know your mother is actually called Kristen, not Vanessa or Valerie or whatever Mark said her name was."

Mark was confounded. What the hell was going on? What was John talking about?

"How do you know?" Alice asked, genuinely surprised.

"Oh, come on, girl, I'm a feckin' detective!" John Formby spat. "And you don't need to be Sherlock Holmes to do a bit of digging like finding out people's names."

Mark sat in absolute shock. *Was it true? Was Alice deceiving him all this time? Could she really have different parents and live in a different house? It didn't make sense, John was right on that score, why go to all the bother?* His hand began to shake, he wasn't sure he was able to cope with much more of this. He held the rock with both hands now in a desperate attempt not to accidentally drop it.

He wasn't sure how much time had gone by but figured Alice's parents must nearly be here, whoever they were. He had thought their voices different, were they really different people? Was Alice's mum really called Kristen like John had said? If so, the whole thing went right over his head. He couldn't fathom a reason why if so. It was crazy. But he couldn't think about that now. He thought he could hear a slight scuffle outside but wasn't sure if he was just wanting to hear something instead of actually hearing it. He had stopped listening to what was going on, all he cared about was that Alice was okay. He didn't care if he had been deceived, he was genuinely just wanting her to not get hurt, plus he was sure what they had, whatever they had, was genuine. He wished so much that he was older and strong enough to tackle his dad on his own. He felt like a schoolboy who called for a bigger brother to help beat up some bullies for him, but he didn't regret sending for help.

And then he heard something. Suddenly two men stormed through the door holding torches and, to Mark's horror, guns. John whipped around, genuinely shocked by the new arrivals. No one had seen Mark yet.

"Let her go, Formby!" ordered one of the men.

"Well, well, well, Mr Seamus Brenson. We meet at last!" John returned, sardonically and laughing manically.

Mark looked blank. Who was Seamus Brenson?

"Who's your friend?" John asked.

The man called Seamus Brenson ignored him. Instead, he turned to his accomplice and told him to get out a knife and untie Alice. John made a swift movement. In a split second two shots were fired. John's Gardaí training came into practice as his two quickfire shots hit Seamus Brenson's shoulder, the second hitting the accomplice squarely in the chest. The second man crumpled into a heap on the damp floor, his gun clattering to the ground with a dull thump. He was dead. Alice screamed. Mark sat stunned at what just happened.

"Let's calm things down a bit, shall we?" John said conversationally, like nothing had happened.

Seamus crouched to his accomplice, checking for a pulse but it was useless. Seamus looked up angrily at John, holding his shoulder. Just a flesh wound, Mark thought vacantly. But he couldn't help but stare at the man in a heap, dead because of his father. Mark had sent for help and now a man was dead. The shock took hold, but he knew that he would feel guilt for the rest of his life. He would never be able to shake off this moment.

"Who are you?" Seamus asked helplessly. "I know you're a guard, everyone knows that. But what's all this about?" he asked, gesturing with his head to indicate the room they were all in.

John laughed. A hearty, chilling laugh

"I have been investigating you." John cocked his gun at Seamus "You're correct there… but not as a guard."

Again, he laughed, this time barely able to contain himself. He gathered himself. "Sorry, sorry, but it is quite funny."

Seamus said nothing. He continued to glare at John, waiting for him to say more. Both men still had their guns directed at one another in what was the most bizarre, unexpected standoff anyone would ever see.

"You may have heard about my little incident?" asked John.

Seamus grunted in reply.

"Yeah, well everyone thinks I went in as a guard... which technically I did, I guess, but that isn't really why. Oh no, the 'drugs raid', as people have labelled it, wasn't an official raid." He paused again, this time wiping his smile from his face. Seamus' face dropped as he realised what he was going to say next. "It has been a hassle convincing everyone why I went in alone, but I was under orders from my own... gang. I'm a rival of yours, Seamus. I'm your enemy. I work for your rivals. That raid was just business... but admittedly I handled it poorly. But here I am. I'm alive!" he exclaimed, like an overly dramatic amateur actor.

Seamus was shocked though he tried not to show it. It took a few moments for him to say anything.

"Why go in alone?" he asked.

"Like I said, it was stupid on my part and I was under orders anyway," John replied, shrugging. It didn't seem to matter to him that he was disposable, sent in with the expectation that he would die.

"Did you move to your new house to investigate me?" Seamus asked neutrally. Mark couldn't tell what the man was really thinking or feeling. He was a closed book with a soft, crooning voice. He was the type of man, Mark thought, who could be full of rage whilst sipping nonchalantly on a coffee-to-go at a bus stop. He supposed someone involved in the drugs trade had to be like that.

John looked at Seamus for a few seconds, perhaps trying to intimidate him, perhaps trying to work him out. "No, not entirely," John replied quietly. "I did fancy being out of the way, away from the city. But I was aware that you lived nearby, my boss clarified it but never really gave me an address. He knew I was getting media attention and wanted me to keep a low profile. Although I didn't realise just how close you lived. When I realised that, it's hard not to sniff about. It's what I do after all." Mark couldn't help but shiver at how relaxed both men seemed whilst talking. You would think they were old acquaintances who had bumped into one another. The tone in which they spoke was disturbingly relaxed, thus accentuating the extreme situation he now found himself in.

"Who is your boss?" Seamus chanced.

John merely smiled pleasantly back as if to say, 'nice try'.

You could sense something was building in the air, this was just a lull before the storm. Mark willed himself to stay focussed, at least for now before another wave of shock hit. He knew deep down how much this was going to affect him, but he had time to deal with that after this… assuming he wasn't killed in the process. As far as he was aware, no one was aware he was there. That was something, it gave him some sort of advantage, but Seamus must be wondering where he was. Perhaps he thought Mark had entrapped him. More likely, he was too concerned to save Alice to worry to wonder where Mark was. To Mark, whatever advantage he had didn't really feel like one. All he had was a rock and a certain element of surprise but what was that worth? To use the rock would give away all element of surprise and he still had to run across towards John to even use it. Every step forward would make him more visible to John, he was just glad the darkness shrouded him enough to conceal him from view. Also, the very thought and realisation that the rock was intended for use against his own father sunk in. The vase was different, it was self-defence in the heat of the moment to protect Mammy. He knew it was a temporary distraction. A jagged rock could easily kill a man, however. As much as he hated his father, he didn't wish him dead. He wouldn't wish anyone dead. All he could really do right now was be a spectator.

Alice could only watch too. She wasn't sure but a slight movement earlier made her wonder if someone else was here. Maybe she was just hoping for a saviour and only wished to see something, but the feeling remained. She was desperately relieved when her father appeared but was shaken when the shots were fired. She feared the worst at first, wondering if her dad was going to topple over too. Thankfully, she wasn't searched before she was tied to the chair. John probably assumed he had caught her by surprise and didn't need to. In fairness, this was true. She was taken by surprise, now berating herself that she was stupid enough to blindly follow the blinking light on her phone to what seemed an odd place for the GPS to be. To make up for her foolishness she subtly searched the compartment of the passenger side's door of John's car with her hand. Expecting to find nothing she clasped her fingers around a dusty penknife amongst a deposit of scrunched up bits of paper and copper coins. Clearly it lay there forgotten, a strange place to leave it though, Alice thought, instead of the driver's door compartment. She took no more time wondering about that though and slowly pushed the penknife up the sleeve of her jumper. There it remained and when she was tied to the chair, she managed to let the penknife slide down the inside of her hoody. Had John not been efficient tying her to the chair he may have discovered it, but she was quick enough that the knife didn't fall out of her sleeve. The rope was tied around her bare skin, so she had to angle her hand in a way that didn't result in the knife spilling out. Talking to John at the same time made things more difficult and she had an intake of breath when the knife did finally slip out but managed to close her fingers around it. Luckily John didn't notice and as he spoke she clicked open the knife, angled it towards her other arm and began to softly grate the blade against the rope, the strain of trying not to let go of the knife making her wrist ache but she ignored it.

Just before her dad came storming in, she had managed to cut enough fibres to wriggle her hand out of the rope's grip. The gunshots had distracted her momentarily but once she realised her dad was okay, she used the opportunity of John being side-tracked to slip the knife into her free hand and cut away to free her other hand. She had to be careful to not do it too quickly in case it attracted John's attention. She was grateful her dad kept him talking but he had no idea she was halfway to freeing herself. She could feel the fibres being cut through quicker this time though and finally she was free. She had to wait for her moment though, she now had the element of surprise. She needed to use that advantage at the right time.

The conversation between John Formby and Seamus Brenson continued. Seamus was starting to understand what had been taunting his mind since John moved to the house across from him but still some questions remained.

"Why investigate me then?" he asked. "To bring me down and give your boss the opportunity to take my business?"

John gave Seamus a wicked smile. "And see you behind bars. That would be a perfect result, don't you think?"

"But how did you expect to do that?" Seamus continued, "all you've done is sneak up on the house."

John sighed, shaking his head. "It's true," he agreed, "that's all I've been doing." He gave him another smile.

"What do you mean?" Seamus asked sharply. "Someone else is giving you information. Who?" He knew he should have trusted his gut. Someone had set him up. They probably knew what was going to happen.

"How's your wife doing, Seamus?" John responded, as though he hadn't been following the conversation.

Seamus was silent, waiting for him to continue. But the meaning behind John's words hit him. Surely not Kirsten, he pleaded inwardly, she wouldn't betray him!

"Well," said John, taking the cue, "she's a very lovely woman… but I hear you've not been a loving husband to her." John tutted patronisingly. "I have a very charming friend who is a very old friend of hers. And let's just say they continued to be friends."

Seamus had gone white, the gun pointed at John now feeling like a deadweight and Alice was staring open-mouthed at John.

"Yes, yes, it is shocking," John continued, waving it off like it was nothing, "but my friend is great at extracting information. Annoyingly your good wife didn't know anything herself, but she was persuaded to sneak into your office and clone all your files from your computers and a wallop of damning other files. At first, she was adamant that she wasn't going to root around your private business, but my good old buddy foresaw this so gave her something that made her very angry. Oh my, she was angry when she was given the photos of you hooking up with your bit on the side… Valerie was it?" he asked, feigning curiosity.

Everything then happened at once.

Alice screamed, this time in unequivocal rage, flew out of her chair and yelled "YOU LIAR!" as she raised the knife and cast it downwards towards John's looming figure. A flicker of surprise registered in both John and Seamus' eyes. John then moved to aim a shot at Alice out of protective instinct, but Seamus was quicker this time, shooting twice at the first instant of John's movement. Seamus' first shot hit John below his ribs which would no doubt bruise despite his vest, the second thundering into the very top of his chest, just above the bulletproof vest. He stumbled back slightly, rupturing through the scar of one of his previous wounds. Alice managed to stab John in the arm and went in for another stab when she was knocked aside by the flinging left arm John threw at her. She fell backwards, the knife falling out of her hand. With his other hand John managed to get a shot at Seamus, hitting him in the throat as John went backwards, eventually falling to the floor. Within the three seconds that this happened, all three were now on the floor injured.

Mark stared in horror as the three of them scrabbled around on the damp floor, although he thought Alice was just winded. In the process of falling backwards and lashing out at Alice, John's grip on his gun had slipped. With one hand clutching his throat, Seamus slid forward kicking the gun out of John's hand. Alice attempted to retrieve the penknife she had dropped but John managed to grab it before her and thrust the knife towards her; the knife would have hit her on the side of her head and surely would have killed her but her quick movement meant his thrust missed. The second time he tried, the knife penetrated her jeans and went into her thigh. She whimpered in pain, retreating away but John kept going. Seamus let go of his throat and put all his energy into fending John off Alice by tackling him backwards. John grunted as he was forced backwards and the pain in his chest flooded through him. However, he still had the penknife in his arm and managed to draw it around Seamus and stabbed him in the back, forcing the blade to go in as far as it would allow. Seamus groaned but kept all his weight downwards to pin John down. He then gave Alice a pleading look and managed a croaky "Run"" He stretched out a hand and Alice dragged herself to meet his outstretched hand, crying uncontrollably. With his last ounce of energy, he mouthed another "Run" at Alice whilst his eyes looked at her with a deep love that was agonising for her to see.

"Dad!" she wailed, paralysed.

But he was gone. Seamus was dead.

John then tore the knife from Seamus' back and slowly hauled the limp body off him and wriggled away. He swiped again and slipped but the knife went through Alice's right foot. There was no way she would be able to run on that foot now. John gathered himself and tried to hoist himself up but the pain in his chest was too much, blood pouring through in an increasingly bigger volume. Instead he eyed the nearest gun and with all his might began to crawl towards it. He still had Alice to take care of and he was starting to lose consciousness. Alice saw what he was doing and looked towards the doorway across the room to see if she could escape quickly enough but with her injured foot and thigh, she knew it was hopeless. Instead, she tried to deter John, grabbing his leg. She had to try and get to that gun first. He threw another wild swipe with the knife, this time nowhere near Alice but he managed to kick her off and aimed another kick at her which sent her backwards. He was slowly gaining on the gun.

Mark, who was petrified, saw what was happening and something in him clicked back into the world. John was going to shoot Alice when he got to the gun. He was nearly within reaching distance now. Alice was too weak to catch him. With hands still trembling he stood shakily and clumsily ran forward, almost tripping but continued his approach. He had no time to think, he just had to act. John managed to get a hand to the gun, it was his good arm that hadn't been stabbed and as he realised someone was running at him, he managed to take aim at the figure coming towards him who John could now see was holding something solid despite his increasingly blurred vision. He steadied the gun as much as possible.

And took a shot.

Out of pure reflex and survival instinct Mark held the rock in front of his face when seeing the gun being raised and he crouched low as he ran. The shot was fired, and it hit the edge of the rock—it would have hit him in the forehead were it not for the rock. The bullet went ricocheting away. Without another thought he kept running, risking getting shot at again and brought the heavy rock downwards with as much force as he could muster once reaching John. There was no time to think, thinking would have made him hesitate and get both Alice and him killed.

John's face took the full impact despite him trying to raise an arm in defence.

There was a dull thud, a splatter of blood and the terrible sound of bones crunching.

Then there was silence.

Chapter Thirty-One

The hospital was eerily quiet. Sylvia Formby had been to see him, and she was distraught. The story of what happened was retold to her by Mark who had witnessed the whole thing. This time there was no media, it was all hushed up on account of the authorities. She had no idea about the drug gang John had been involved in and had no idea about the Brensons. She wailed in despair after being told John shot at Mark, but Mark tried to reason with her saying that he was sure John didn't know who it was when he shot at Mark. He really did believe that. He tried to be as strong as he could in front of her, tried to ignore his own turmoil. He was thankful when the nurses came in and advised her that he needed rest and that she needed a break too. Mammy was unwilling but eventually left, promising to be back soon. In the moments he had to himself he wept inconsolably into his pillows, allowing the trauma to consume him. Therapy had already been promised to him and he felt that he needed it. He witnessed three people die in front of his eyes, one by his hand. He had killed his father, but he knew he needed to do it too. That didn't make him feel an awful lot better but saving Alice's life would help him to cope in time, he thought.

Alice was in another ward for physical injuries to her leg and foot—nothing too serious but like Mark would need extensive therapy. Mark still hadn't seen Alice, but she had not wanted to see anyone. It had only been two days; everything was still quite raw. He did ask how she was getting on to nurses tending to him and they were sympathetic, but they told him she was doing fine. He completely understood her need to be alone, he took comfort in being alone too. But he did want to talk to her at some point, both of them were in the same position who had been through the same thing. It would be good to talk. Perhaps after everything Mark meant nothing to her at all, he was ready for that possibility. In the grand scheme of things, it was not the most important thing. He really liked her and still did but he would also find a way to cope without Alice in his life if he had to. He felt he had grown as a person and he'd still be thankful to her for giving him more self-confidence. And anyway, he needed to focus on himself too.

On the third day, he was told that he was ready to be discharged but support would be given to him for posttraumatic stress and anything else he needed. Someone must have told Alice that he was ready to leave because a nurse wheeled her into Mark's ward and when their eyes met, she gave him a small smile. The nurse parked her chair next to Mark's bed and walked away, giving them both a knowing smile and telling them someone would be back in half an hour or so. For a few moments there was a stiff silence, neither one knowing what to say to the other. But Mark managed to break the silence.

"I can tell the nurses have been treating you well."

"Thanks... I think." They both smiled weakly.

"How is your foot and your leg?" he asked.

"Sore...but it could have been worse." Alice rested a hand on his. "Thank you for what you did. I – I know it can't have been easy." She looked at him seriously. "How were you even there?"

He fed her the story of how he came to be in the back of the car when Alice was taken and everything that happened after. She listened in amazement, telling him she thought she saw movement in the car and in the dank farmyard but had dismissed it.

"Will you be okay?" Mark asked.

She nodded seriously. "I think so. In time." She paused, tears now streaming down her face. "I can't help but feel my dad is dead because of me, I fell into a trap too easily. It's all my fault...and I'm not sure I can forgive my mum for what she did either" She couldn't control it now, she dropped her head and wept.

Mark squeezed her hand in reassurance. "You weren't to know what would happen."

She squeezed his hand back.

Eventually Alice stopped crying. She looked up at him again. "And I'm sorry for what I've done to you, Mark," she spluttered, "I can't believe you'd want to speak to me at all after everything"

"You're the only friend I've ever really had," he replied honestly. "How can I be mad about that?"

"Because I was *told* to be your friend!" she burst out. "But please believe me, I didn't enjoy deceiving you and I really did feel comfortable around you. You're such a lovely guy."

He could tell her guilt was genuine and he knew that he probably would have done the same for Mammy. He knew he was right to question why Alice paid such an interest in him and in a way, that helped. "Will you be okay? I can't even imagine how you must feel about… you know…"

Mark nodded. He knew what she meant. It had not really sunk in that he had killed his own father, but he knew deep down it was necessary under the circumstances. He knew he'd feel remorse and even have nightmares over what happened, but he was happy knowing he also saved a life.

Mark smiled sadly. "I'll be okay in time, I'm sure. With some help. What's the plan now?" he asked, changing the subject.

"I don't know. I haven't spoken to my mum and I don't really want to," she responded, this time with an edge to her tone. He could understand why but he hoped that anger would pass.

"When we're both out, you will come and see me won't you?" she pleaded.

"Yes, of course, if you want me to," Mark returned. "Which house?" he added, smirking. The whole story of the two houses was unfurled over the past few days he was in hospital, apparently Alice had wanted to get the whole truth out because Mark deserved to know. Alice's face remained serious. "I'm sorry about that too –And in a way I suppose it was useful…" she looked away absently. The mention of her father clearly got her lost in her own thoughts. Mark still thought it was a strange place for a safehouse but it hardly mattered.

"Have the Gardaí spoken to you yet?" Mark asked. "They came in for my statement and I guess they did with you too?"

"Yes" she replied simply.

They both looked down. A natural silence ensued. After a while, the nurse came back and said they both needed to rest, and that Mark needed to get ready to be discharged. Alice squeezed Mark's hand again before she was wheeled away. "Take care of yourself," she said, "I'll text you when I'm back home and everything's settled down again, okay?"

He smiled back. "Sure."

Alice returned the smile, more like the one he was used to seeing and she kissed his hand before she left. It seemed everything would turn out okay in the end with Alice after all, even if it was just as friends. Time would tell but there was something good to look forward to after all this washed over.

After all that had happened, he felt things could have been very much worse.

Acknowledgements

I would like to thank my partner, Natalie, first and foremost for her continuous support and for dealing with me, particularly the long days and nights where I've done nothing but write, delete and write again. Thanks for your patience and listening to my ramblings about where I am going next with the story. Completing a novel for the first time was a real challenge and experiencing that head on can't always have been easy.

Secondly, thank you to my parents for their belief in my writing and giving me some interesting insights. Away from my writing, they have been great supporters in everything I do. I really do appreciate everything you do. Extended thanks to my sister too.

Pete and Lydia, thanks for your thoughts on the book design, it was good to have a second and third pair of eyes.

Lastly, thanks to my editor Coleen for taking the time to scrutinise my novel and to my very good friend Brad Linney for his time and effort in his spare time to make some fantastic points and for making tweaks that made this story as strong as it can be. His services will definitely be required for my next novel.

I wouldn't have gotten very far without all these people; I really am grateful.

Seán Cassidy